AN INTRODUCTION TO MATHEMATICAL ANALYSIS

H.S. Bear

Department of Mathematics
University of Hawaii

THE BLACKBURN PRESS

Reprint of First Edition, Copyright 1997

Copyright © 1997 by H.S. Bear, with corrections

An Introduction to Mathematical Analysis

ISBN-10: 1-930665-88-1
ISBN-13: 978-1-930665-88-0

Library of Congress Control Number: 2003112697

THE BLACKBURN PRESS
P. O. Box 287
Caldwell, New Jersey   07006 U.S.A.
973-228-7077
www.BlackburnPress.com

AN INTRODUCTION TO
MATHEMATICAL ANALYSIS

This book is dedicated to
A.M. Gleason, a veritable
Fountain of Truth.

Acknowledgment

The author is indebted to G.N. Hile who found a distressing number of inconsistencies in an early version of the text. I would also like to express my profound thanks to Sherene Hayase, Pat Goldstein and Susan Hasegawa, whose wizardry transformed my scruffy hand-written manuscript into an object of stunning legibility.

CONTENTS

ANNOTATED TABLE OF CONTENTS

natural numbers are proved. Any nonempty subset of $\mathbb{N}$ has a smallest element. The school algorithms for the arithmetic of decimally represented natural numbers are explained. The strong theorem of induction is a problem, as are the laws for positive integer exponents.

5. Finite and Infinite Sets

Functions are defined (set theoretic notation is assumed as part of the language), and then finite and countable sets. The elements of a finite set can be arranged in ascending order, and finite sets have largest and smallest elements. There are uncountable sets. Every infinite set has a countable subset.

6. Long Division and Prime Factorization

The division theorem for natural numbers is proved, and the fact that every natural number has a unique decimal representation. The natural numbers have a unique factorization into prime powers. (The uniqueness part of this is a genuinely hard theorem.) The material of this chapter may be skipped, skimmed, or skimped by those anxious to get on with the analysis.

7. The Completeness Axiom and Sequences

The least upper bound form of the completeness axiom is given, and we finally begin the study of analysis. An example shows that $\mathbb{N}$ is bounded (and $1/n \nrightarrow 0$) in some fields without completeness. Convergence of sequences is defined, and the elementary $\varepsilon - N$ theorems are proved slowly and carefully. Intervals are characterized, using the Completeness Axiom, as those sets that contain with any two elements x and z, all numbers y such that $x < y < z$.

8. Three Heavy Theorems on Sequences

The three theorems are:

    (1) Every bounded increasing sequence converges.
    (2) The Weierstrass–Bolzano Theorem.
    (3) Every Cauchy sequence converges.

There are numerous problems to encourage some serious reflection on these basic results.

9. Alternative Completeness Axioms

This chapter is devoted to showing that each of the three theorems of Chapter 8 is equivalent to the Completeness Axiom. This is not a necessary part of the development, but it will certainly help the student gain a deeper understanding of the completeness concept.

10. Continuous Functions

Continuity is defined with no preconditions on the domain. If f is a function of one variable, then f is continuous at $x_0 \in \mathcal{D}(f)$ provided $f(x_n) \to f(x_0)$ whenever $\{x_n\}$ is a sequence in $\mathcal{D}(f)$ such that $x_n \to x_0$. Similar definitions are given for functions of two or more variables. The limit-of-a-sum theorem for sequences (Chapter 7) is therefore the same as the statement that addition is a continuous function of two variables. The Intermediate Value Theorem has the following form, using the characterization of intervals of Chapter 7: $f[I]$ is an interval if f is continuous on the interval I. Maximum/minimum theorems are proved, along with the frequent continuity of inverse functions.

11. Uniform Continuity

We start by proving the $\varepsilon - \delta$ criterion for continuity; this complements our definition of continuity in terms of sequential convergence. A function continuous on a compact interval is uniformly continuous. Uniformly continuous functions map Cauchy sequences onto Cauchy sequences. If f maps all Cauchy sequences onto Cauchy sequences, then f is continuous — and uniformly continuous if $\mathcal{D}(f)$ is bounded. A uniformly continuous function on a set S extends to a continuous function on the closure of S. The definition of x^r is extended from rational exponents r to all real r.

12. Closed Sets; Compact Sets; Open Sets

Closed sets are defined first as those sets that contain all limits of sequences in the set. A compact set is a bounded closed set. Closed sets are closed under finite unions and arbitrary intersections. The continuous image of a compact set is compact. A set is open if its complement is closed. The Weierstrass–Bolzano and Heine–Borel characterizations of compactness are proved.

13. Derivatives

The theorems of first semester calculus are proved — correctly this time. Convex functions are defined and shown to be continuous over intervals. The fact that derivatives have the intermediate value property is a problem — with hints.

14. The Darboux Integral

We give the calculus-text definition of the integral of a bounded function as the common limit of upper and lower sums. There is nothing innovative in this chapter, but the details are handled carefully, since the same proofs are used later (Chapters 32–34) for the Lebesgue integral.

15. The Riemann Definition

We define directed set, and net (a function on a directed set), and show that f is integrable (i.e., Darboux integrable in the sense of Chapter 14) if and only if the net of Riemann sums converges. General theorems on convergent nets are proved, and these now have some legitimacy since there is a real example. Later, we use nets extensively in the treatment of unordered sums. The Lebesgue integral will also be shown to be the limit of Riemann sums with the same proofs. The linearity of the integral is immediate from the Riemann-sum characterization.

16. $\log x$ and e^x

We define $\log x$ as the integral of $1/x$, and prove that $\log xy = \log x + \log y$. $E(x)$ is defined to be the inverse of $\log x$, and we show that if e is defined by $\log e = 1$, then $E(x) = e^x$ for all rationals x. The continuity of E shows that $E(x)$ is the only extension of e^x from rationals to reals.

17. Unordered Sums and Infinite Series

The unordered sum is presented as the legitimate generalization of summation to infinite sets of numbers. The unordered sum is the limit of the net of finite sums, where the finite sets are ordered by inclusion. The unordered sum $\sum x_\alpha$ converges if and only if $\sum |x_\alpha|$ converges. It is shown that an infinite series — the limit of finite sums of numbers taken in specific order — can be anything, depending on the order of the terms, unless the series converges absolutely. We show that rearranging or grouping are legitimate for unordered sums, and hence for absolutely convergent series. In particular, the order can be reversed in an iterated series, and the grouping necessary to show (in Chapter 27) that $e^z e^w = e^{z+w}$ is legitimate.

18. The Calculus of Series

Here we prove the standard calculus-text theorems for convergent series. The root test is used as motivation to introduce the concepts of lim sup and lim inf, and these are characterized as the largest and smallest limits of subsequences. Cesàro summation is introduced for the sake of variety.

19. Sequences and Series of Functions

Pointwise and uniform convergence are defined, and we prove that the uniform limit of continuous functions is continuous, and the uniform limit of integrable functions is integrable. If f_n' converges uniformly and f_n converges at one point, then f_n converges uniformly. We give a novel and more intuitive treatment of Taylor approximations. The estimates are motivated one n at a time starting with the observation that if $|f''|$ is small on $[a, x]$, then f' does not change much between a and x, and hence the difference between f and the tangent line (the first Taylor polynomial) is small. This "small" is made quantitative, and the same argument is made if $|f'''|$ is small on $[a, x]$, and so forth. The Taylor *series* theorem is a problem that is accessible following the careful motivation of the estimates. The example $\exp(-1/x^2)$ is a problem with hints.

20. Topology in $\mathbb{R}^2$

$\mathbb{R}^2$ is provided with a norm and inner product. Convergence in $\mathbb{R}^2$ is defined in terms of the norm-distance. Open sets and closed sets are defined and shown to have the usual properties. Compact sets are defined as closed bounded sets. The Weierstrass–Bolzano and Heine–Borel theorems are proved. A continuous function on a compact set is uniformly continuous. A function f on an open set in $\mathbb{R}^2$ is continuous if and only if $f^{-1}[U]$ is open for every open set U. Dini's Theorem is a problem with hints.

21. Calculus of Two Variables

Partial derivatives, differentiability, directional derivatives, and the gradient are defined. Sufficient conditions for a relative maximum or minimum are given. The Implicit Function Theorem is proved.

22. Complex Numbers

We introduce a multiplication in $\mathbb{R}^2$, change $\|(x, y)\|$ to $|(x, y)|$, and call the result the complex plane, $\mathbb{C}$. It is emphasized that $\mathbb{C}$ is a construct from the reals with their axioms, and that the algebra and topology in $\mathbb{C}$ follow from our theorems about $\mathbb{R}$ and $\mathbb{R}^2$. Continuity of the algebraic operations in $\mathbb{C}$ is proved. The mapping properties of $(z - a)/(1 - \bar{a}z)$ are given. Fractional linear transformations $(az + b)/(cz + d)$ are shown to map lines and circles onto lines and circles, and to form a group of transformations.

23. Curves in the Plane

As a prelude to line integrals, we have to make sense out of plane curves. Rectifiable curves are defined, and the length of the smooth curve is shown to be the appropriate integral. The unit circle is parameterized, and π is defined to be four times the integral that gives the length of the arc from $(1, 0)$ to $(1/\sqrt{2}, 1/\sqrt{2})$. The length of the circle is thus 2π. Connected sets and components are defined, and connected open sets are shown to be arcwise connected.

24. Trigonometric Functions

The functions $\cos\theta$ and $\sin\theta$ are defined to be the coordinates of the point P on the unit circle such that the positive arc from $(1,0)$ to P has length θ. All this is made tediously precise. There are slicker ways to define $\cos\theta$ and $\sin\theta$ (e.g., as series), but then all connection with angles and geometry is lost. With our clumsy approach we can define angles in terms of translation and rotation maps of any three points onto the points $(x,0)$, $(0,0)$, $(\cos\theta,\sin\theta)$, and make the connection between $\cos\theta$ and "adjacent over hypotenuse." The identities for $\cos(x+y)$, $\sin(x+y)$ are proved using Taylor series.

25. Line Integrals

Line integrals of complex functions are introduced in preparation for complex analysis. Differentiation formulas for a complex function of a real variable are proved. The line integral of a complex function is defined as the limit of Riemann sums and then reduced to real integrals using the parameterization of the curve. The integral of z^n is shown to be independent of path if $n \neq -1$, and the integral of z^{-1} around the unit circle is shown to be $2\pi i$.

26. Power Series

Power series are (of course) treated for a complex variable. The radius and circle of convergence are defined, with the $\limsup \sqrt[n]{|a_n|}$ formula for the reciprocal of the radius. Series converge uniformly on closed subdiscs by the Weierstrass M-test, and the identity theorem holds for series.

27. The Transcendental Functions

We showed earlier that the Taylor series for e^x, $\cos x$, and $\sin x$ converge for all real x. These series are used to define e^z, $\cos z$, and $\sin z$ for all complex z. Euler's formula appears. To prove that $e^z e^w = e^{z+w}$ we first prove that if $\sum x_\alpha = x_0$ and $\sum y_\beta = y_0$ (unordered sums), then $\sum x_\alpha y_\beta = x_0 y_0$. This allows the grouping necessary to show the exponential identity. The trigonometric identities then follow easily from Euler's formula. The hyperbolic functions are defined, and identities like $\cosh iz = \cos z$ are left as a problem. The function $\log z$ is defined as the inverse of e^z on $-\pi < \operatorname{Im} z \leq \pi$. No mention of multiple-valued functions is ever made.

28. Analytic Functions

The complex derivative is defined and the usual differentiation formulas given, along with the Cauchy–Riemann equations. The integral of a polynomial is independent of path, and more generally, the integral of a derivative is independent of path. The sum of a power series is analytic, and its derivative is the derived series. The integral of $1/(z - z_0)$ around the circle $|z - z_0| = r$ is $2\pi i$. The inverse of an analytic function is analytic.

29. Cauchy's Integral Theorems

The integral of an analytic function around the boundary of a triangle is zero. If f is analytic in a convex domain, then $\int f = 0$ for every closed curve in the domain. These theorems are then extended to continuous functions in a convex domain that are analytic except possibly at one point, and in particular to the difference quotient $\big(f(z) - f(a)\big)/(z - a)$. Cauchy's integral formula follows, and then the power series representation. Conformality of analytic maps is explained. We tippy-toe around the complexity of plane topology and give proofs of slightly restricted theorems.

30. Lebesgue Measure in (0, 1)

The (outer) measure μ is defined as usual to be the inf of the total lengths of coverings by intervals. We stick to subsets of $(0, 1)$ initially for simplicity. The intervals can be any kind, and so for compact sets finite coverings suffice. The measure of an interval is its length. This is shown first for compact intervals and finite coverings, and then for other intervals by monotonicity. The measure μ is countably subadditive. We construct a (nonmeasurable) set to show that μ is not countably additive.

31. Measurable Sets

We stick to sets in $(0, 1)$ initially, and since additivity is our goal, we define E to be measurable if and only if $\mu(E) + \mu(E') = 1$. We then show that E is measurable iff E splits every subinterval of $(0, 1)$ additively, and then show this holds if and only if E splits every subset T of $(0, 1)$ additively; that is, $\mu(E \cap T) + \mu(E' \cap T) = \mu(T)$ for all $T \subset (0, 1)$. Using only this no longer magical criterion of Carathéodory along with the outer measure properties, we develop the properties of μ on measurable sets. The measurable sets form a σ-algebra containing the intervals, and μ is countably additive on the measurable sets. Then μ is defined as an outer measure on subsets of $\mathbb{R}$ in the usual way, and the preceding results and proofs hold with little change for measurable subsets of $\mathbb{R}$ as defined by the Carathéodory criterion. Here "little change" amounts to nothing more than a modest delicacy concerning ∞.

32. The Lebesgue Integral

The integral is defined in Darboux fashion as the common limit of upper and lower sums. A partition of a measurable set is now a finite or *countable* disjoint family of finite measure measurable sets, the union of which is the given set. Countable partitions allow us to treat all functions (bounded or not) on all measurable sets (finite measure or not) at the same time. The only restriction is that there be at least one finite upper sum for $|f|$ — that is, the graph of f lies in some finite area $\bigcup E_i \times [-M_i, M_i]$, where $\sum M_i \mu(E_i) < \infty$. Such functions are called admissible. This approach may be unique; in any case it is conceptually very simple, and the definition of "integrable" does not have to proceed in layers. A function f is integrable if and only if f^+ and f^- are, and $\int f = \int f^+ - \int f^-$; this is now a theorem, not a definition, and it allows us in the final two chapters to restrict proofs to nonnegative functions. An admissible function is integrable if and only if the net of Riemann sums converges. Riemann integrable functions are Lebesgue integrable, and nonnegative improperly Riemann integrable functions are Lebesgue integrable.

33. Measurable Functions

The proof that a continuous function is integrable on an interval is recalled to motivate the definition that f is measurable if and only if $\{x : a \leq f < b\}$ is a measurable set for all a and b. The same proof then shows immediately that a bounded measurable function is integrable over a set of finite measure. The equivalent conditions for measurability are given. Sups and limits of sequences of measurable functions are measurable. The "almost everywhere" idea is explained, and the limit theorems are extended to a.e. convergence. If $h \geq 0$ and $\int h = 0$, then $h = 0$ a.e. This is used to show that integrable functions are limits of simple functions, and therefore measurable.

34. Convergence Theorems

Egoroff's Theorem is motivated by the sequence x^n on $[0, 1]$, and then proved. The bounded convergence theorem then follows immediately. Examples are given to show that pointwise convergence can yield $\lim \int f_n > \int f$ if the f_n are not confined to a finite area. A function is primary if $g \geq 0$, g is bounded, and $g = 0$ off a set of finite measure. A nonnegative function is integrable if and only if $\sup\{\int g : 0 \leq g \leq f, g \text{ primary}\} < \infty$. This is now a theorem, not a definition. Fatou's Lemma is proved, and then the dominated convergence theorem.

PREFACE

I n this text we develop the basic elementary results of analysis, using only the axioms for the real numbers. The course is logically self-contained; appeal is made only to results that have been proved, or that obviously could have been proved with the methods at hand.

Although every effort has been made to ensure that the definitions and proofs are complete and correct, we have strenuously avoided a formal Satz–Beweis approach. Mathematics is not discovered or understood in a purely deductive way, and so we try always to present the "music" of the subject in the informal and evocative language in which mathematics is actually conceived and understood.

Several different courses may be taught from the text. In those courses primarily designed to prepare the student for graduate work, some of the introductory material of Chapters 5, 6, and 9 can be omitted. The definition of the trigonometric functions in Chapter 24 is tedious, and instructive only insofar as it shows that $\sin x$ can be defined with the same precision — if not the same elegance — as e^x. Of course, series provide an elegant way to define $\sin x$, but then one is hard pressed to show that $\sin x$ occasionally has something to do with triangles.

Courses designed for prospective teachers should go carefully through the first seven chapters, where the mysteries of school mathematics are carefully unraveled.

A rigorous treatment of elementary complex analysis is unusual at the undergraduate level, and the treatment in many graduate texts naturally tends to skimp on the basic theory. We have therefore included a careful treatment of the elementary theory of analytic functions and Cauchy's integral theorems.

The final chapters provide a novel and quite simple approach to the Lebesgue integral. We use upper and lower Darboux sums in much the same way as in our earlier treatment of the Riemann–Darboux integral. The novelty is that now we use the countable additivity of the length function (i.e., measure) to consider countable partitions (of a measurable set) into disjoint families of measurable sets. This allows us to treat all functions — bounded or not — and all domains — of finite or infinite measure — in one fell swoop. We also show that the Lebesgue integral, like the Riemann integral, is the limit of a net of Riemann sums. This is not only a useful fact; it makes the Lebesgue integral appear more familiar and less formidable.

The problems are not segregated at the backs of the chapters, but scattered through the text as an integral part of the mathematical development. Mathematics is not a spectator sport, and so the student is frequently asked to provide part of a proof, an explanation of why a hypothesis is necessary, or a generalization of the stated result. The successful student will have good reason to be pleased with the thoroughness of the development, for he will have been responsible for a large part of it.

NOTE TO THE INSTRUCTOR

This text contains more material than anyone should try to cover in one year. Let's face it – I got carried away. The various sections, however, are more or less separable and provide material for several courses as follows:

Chapters 1 - 6

This material is suitable for a first course in "proofs," using the axioms for the real numbers to develop arithmetic formally. This is a course that every elementary school teacher would take in the ideal world.

Chapters 7 - 21

These chapters cover the material of the usual one-year undergraduate course in real analysis. The standard facts are developed quite rigorously.

Chapters 22 - 29

Here is a concise, one-semester introduction to complex analysis. All the details are spelled out, and the treatment is novel in many respects.

Chapters 30 - 34, perhaps preceded by chapters 14 and 15

This is a clean and simple treatment of Lebesgue measure and integration on the line. The Lebesgue integral is presented as a precise analog of the Riemann integral, with the one difference that the Lebesgue integral uses a better definition of length (i.e. measure), and this allows countable partitions of the domain rather than finite partitions. This is material that every student should see *before* taking the standard abstract graduate measure theory course.

I

EXHORTATION

The purpose of this text, as the title subtly suggests, is to introduce the serious student to the serious study of mathematical analysis. By "serious student" we mean one who takes mathematics seriously and is prepared to spend a great deal of time on this course. This material is elementary, and fundamental, but it is not easy, and there is no way to make it easy.

By "mathematical analysis" we mean the study of the real number system, and the mathematical things that can be constructed from real numbers. These things include various interesting sets of numbers, and sets of sets of numbers, and so on. In particular, we mention open sets, closed sets, finite sets, complex numbers, infinite sets, functions, sequences, sequences of functions, and functions of sequences.

In most undergraduate mathematics courses, there are some definitions, some theorems are proved, and lots of problems are worked. The proofs ideally are based on more elementary facts — simple algebra and the like — that the student is presumed to know. In most cases, however, an important part of the course will depend on some big-gun result the author pulls out of a hat, without proof or apology. The chant that goes with this exercise is "beyond the scope of this course." Our ambition here is to be explicit about all our assumptions; we start with axioms, and we prove all our results from these axioms.

Like most ambitions, this one is a trifle overblown. Although there will be no use of facts that are beyond the scope of this course, we manifestly do not have the time to develop all of school arithmetic, geometry, algebra, and so forth, *ab ovo*. What we will do is start with the axioms for algebra — field axioms — and prove enough of the elementary results so that it becomes clear how all of school algebra can be proved from the axioms. To divide fractions, you invert the denominator and multiply — this is a theorem, and we will prove it. Then we add the axioms about the order relation, less than, so that we are then proving theorems about any ordered field. At that stage we can prove that $0 < 1$, and many equally exciting results.

The next step, once we gain the confidence that we can prove everything we already "know" about addition and multiplication of numbers, and inequalities involving numbers, is to define the natural numbers and the integers. This includes the "theorem of induction," and the concepts of infinite set, finite set, countable set, and uncountable set. This is a big step — the proof that a set is infinite if and only if it can be put in a one-to-one correspondence with some proper subset of itself is considerably more difficult than the proof that $x^2 \geq 0$ for all numbers x.

The final axiom needed to characterize the real numbers is the completeness axiom, and this is where analysis begins. Without the completeness axiom, we do not know, for example, that positive numbers have square roots, or that the natural numbers get arbitrarily large. Another way of describing analysis is to say that it is the study of limits. Without the completeness axiom, we have no way of showing that many of our favorite limits exist. For example, $\lim_{n \to \infty} 1/n = 0$ is not a theorem without the completeness axiom.

It is hoped that our procedure of chasing all our truths back to explicit assumptions will appeal to the mathematically minded student. Indeed, this is one good test of whether or not a student is mathematically minded. Many of you will have noticed that in physics texts terms sometimes disappear from equations because their presence is an embarrassment to the author. If that kind of logic makes you nervous, perhaps you will enjoy this axiomatic approach.

It should be noted that progress in an axiomatic development is initially slow. It may be disappointing to work long and hard only to end up with trivialities like "$1 > 0$," or "there is no integer between 2 and 3." The initial snail's pace does have an additional purpose, however. Most undergraduate students have had little experience in writing proofs. This approach provides an opportunity for the student to get some idea of what a proof really involves, using material that is conceptually simple. Most undergraduate mathematics courses consist mostly of finger exercises designed to create the illusion that the students have learned something. There are no finger exercises here, only proofs, and proofs are hard. After all, if mathematics were easy, it would not be any fun.

II

THE FIELD AXIOMS

T he real number system consists of a set $\mathbb{R}$ on which are defined two binary operations, addition and multiplication, and a binary relation, less than. A binary operation is a function of two variables (i.e., two elements of $\mathbb{R}$) the values of which are again elements of $\mathbb{R}$. Rather than writing something like $A(x, y)$ for the sum of x and y, we use the usual notation $x + y$. Similarly, we write $x \cdot y$ for the product of x and y. A binary relation is simply a set of pairs of numbers (i.e., elements of $\mathbb{R}$). We write $x < y$ or $y > x$ to indicate that the pair (x, y) is in the less than relation. The system $(\mathbb{R}, +, \cdot, <)$ is the system of real numbers provided the nine axioms we list in the sequel are satisfied. We start with the field axioms, which are the assumptions about addition and multiplication.

AXIOM I (Commutative Laws): *For all x, y,*

$$x + y = y + x \qquad and \qquad x \cdot y = y \cdot x.$$

AXIOM II (Associative Laws): *For all x, y, z,*

$$x + (y + z) = (x + y) + z \qquad and \qquad x \cdot (y \cdot z) = (x \cdot y) \cdot z.$$

We call attention to several things before we go on. The phrases "for all x," "for some x," and the assorted equivalent expressions, such as "for each x," and "there exists x," are called quantifiers. In school algebra, the equations

$$x^2 - y^2 = (x + y)(x - y),$$

$$(x + y)^2 = x^2 + 2xy + y^2,$$

3

are understood to mean that these equalities hold for all numbers x and y. On the other hand, an equation such as

$$x^3 - 3x^2 + 2x = 0$$

is interpreted to mean that x is one of the numbers for which the equation is true. In previous courses the context has been allowed to carry a large part of the meaning, in order that students and professors not be burdened with excessive amounts of verbiage. This practice will no longer suffice. We will have to deal eventually with complicated sentences involving three or four variables, and all the variables will have to be quantified. For our present purposes, we will agree that a sentence is any statement that can be proved true or proved false from our assumptions. Thus,

$$\text{``}x + y = y + x\text{''}$$

and

$$\text{``}x^2 + 1 = 5\text{''}$$

are not sentences, but

$$\text{``For all } x, \quad x + y = y + x.\text{''}$$

and

$$\text{``For all } x, \quad x^2 + 1 = 5.\text{''}$$

are sentences. The first sentence is true, and the second is false. Without quantification,

$$\text{``}x^2 + 1 = 5\text{''}$$

is neither true nor false.

We must note that the order in which the variables are quantified is all important. For example, consider these sentences:

(1) For all x and some y, $x + y = 10$.
(2) For some y and all x, $x + y = 10$.

Both these sentences use the same quantifiers, "for all x" and "for some y," but (1) is obviously true and (2) is obviously false.

Now another axiom, and a little mathematics.

AXIOM III (Existence of Identities): *There is a number, 0, such that $x + 0 = 0 + x = x$ for all x, and there is a number $1 \neq 0$ such that $x \cdot 1 = 1 \cdot x = x$ for all x.*

Notice that we now know that there are at least two numbers, since $0 \neq 1$, but we do not yet know that there are more than two. Notice also that addition and multiplication could be the same operation as far as Axioms I and II are concerned, but that Axiom III implies that they are different, as shown in the following problem.

PROBLEM 1: Show that $+$ and $\cdot$ are necessarily different operations. That is, for any system $(\mathbb{F}, +, \cdot)$ satisfying Axioms I, II, and III, it cannot happen that $x + y = x \cdot y$ for all x, y. Hint: You do not know there are any numbers other than 0 and 1, so that your argument should probably involve only these numbers. Did you use Axiom II? If not, state explicitly the stronger result that you actually proved. •

PROBLEM 2: Let $\mathbb{F} = \{0, 1\}$ with $+$ and $\cdot$ defined by the following tables:

$+$	0	1		$\cdot$	0	1
0	0	1		0	0	0
1	1	0		1	0	1

Show that this two-element set with these two operations satisfies Axioms I, II, and III. •

The next two axioms complete the axioms for a **field**.

AXIOM IV (Existence of Inverses): *For every x there is a number y so that $x+y = y+x = 0$, and for every $x \neq 0$ there is a number y so that $x\cdot y = y\cdot x = 1$.*

So far the two operations appear to be independent. The distributive law ties them together.

AXIOM V (Distributive Law): *For all x, y, z,*

$$x \cdot (y + z) = x \cdot y + x \cdot z.$$

From what we have explicitly assumed so far, the formula $x \cdot y + x \cdot z$ is not intelligible. Since there are two multiplications and one addition, and both operations are binary, there are five possible interpretations for $x \cdot y + x \cdot z$, namely,

(i) $\left[x \cdot (y + x)\right] \cdot z,$
(ii) $x \cdot \left[(y + x) \cdot z\right],$
(iii) $\left[(x \cdot y) + x\right] \cdot z,$
(iv) $x \cdot \left[y + (x \cdot z)\right],$
(v) $(x \cdot y) + (x \cdot z).$

If you first add, then perform the left multiplication, then the right multiplication, you get (i). If you first perform the right multiplication, then the left multiplication, then the addition, you get (v). We will now invoke our childhood training and agree that multiplication takes precedence over addition, and interpret $x \cdot y + x \cdot z$ in the usual way. We also agree to use juxtaposition to indicate multiplication, so that we can dispense with dots and write, as usual,

$$x(y + z) = xy + xz.$$

Similar remarks apply to formulas like $a+b+c+d$, which nominally depend on the order of the operations. Since the commutative and associative laws make it clear that the order is not important, we can dispense with parentheses, and we use without comment identities like

$$a + b + c + d = a + d + c + b,$$

$$xyz = yxz,$$

$$xy + yz = yz + yx.$$

PROBLEM 3: Show that the two element system of Problem 2 satisfies all of Axioms I, II, III, IV, and V. It is, therefore, a smallest possible field. Are there other possibilities for $+$ and $\cdot$ in a two-element field? Hint: The only possible change in addition, since 0 is an identity for addition, would be $1 + 1 = 1$, and this would leave 1 without an inverse. There is also only one possible change in the multiplication table, since 1 is an identity for multiplication. Show that the distributive law suffers if $0 \cdot 0 = 1$.　　　　　　　　●

Finally, we come to a theorem.

PROPOSITION 1: *There is only one additive identity and only one multiplicative identity. That is, if z is a number such that $x + z = x$ for all x, then $z = 0$, and if u is a number such that $xu = x$ for all $x \neq 0$, then $u = 1$.*

Proof: Suppose that $x + z = x$ for all x, so in particular $x + z = x$ for some x. Fix one such x, so that we have

$$x + z = x = x + 0.$$

The number x has an additive inverse, y, whence $x + y = 0$. Therefore,

$$x + z + y = x + 0 + y,$$

and, rearranging, we have

$$z + (x + y) = 0 + (x + y),$$

$$z + 0 = 0 + 0,$$

$$z = 0.$$

The second part is left as a problem.　　　　　　　　■

The statement that the additive identity is unique is: "If $x + z = x$ for all x, then $z = 0$." We actually proved the stronger statement: "If $x + z = x$ for some x, then $z = 0$." It is important to check your proofs — and mine — not just for accuracy, but for content. Make sure you get the maximum amount of information out of every argument.

PROBLEM 4: Prove that the multiplicative identity is unique. Can you strengthen the result as we did for the additive identity?　　　　　　　　●

PROPOSITION 2: *Each number has a unique additive inverse, and each non-zero number has a unique multiplicative inverse.*

Proof: We prove the second part. Suppose there is a nonzero number x that has two multiplicative inverses, so that there are numbers y_1 and y_2 such that

$$xy_1 = 1 = xy_2.$$

Multiplying both xy_1 and xy_2 by y_1, we get

$$y_1(xy_1) = (y_1x)y_2.$$

Since $xy_1 = y_1x = 1$, this yields

$$y_1 \cdot 1 = 1 \cdot y_2,$$

$$y_1 = y_2. \qquad \blacksquare$$

In the previous proof we used the following sort of argument: if $r = s$, then for any x, $rx = sx$. This is not mathematics, this is just language. The statement "$r = s$" means that "r" and "s" are different names for the same number. Obviously, then "rx" and "sx" are also just different names for the same number, or $rx = sx$.

PROBLEM 5: Prove that additive inverses are unique.   ●

Since inverses are unique, we are now justified in introducing the usual notation. We use $-x$ to denote the additive inverse of any number x, and we write $y - x$ for $y + (-x)$. We use x^{-1} to denote the multiplicative inverse of any nonzero number x, and we also write $\dfrac{y}{x}$ or y/x for yx^{-1}.

PROPOSITION 3: *For all x, y,*

(i) $-(-x) = x$,
(ii) $-(x + y) = -x - y$.

PROBLEM 6: Prove Proposition 3.   ●

PROPOSITION 4: *For all x, $x \cdot 0 = 0 \cdot x = 0$.*

Proof: Using the distributive law we know that for any x,

$$x \cdot 0 = x \cdot (0 + 0) = x \cdot 0 + x \cdot 0.$$

The last equation says that $x \cdot 0$ acts as an additive identity for the number $x \cdot 0$; that is, adding $z = x \cdot 0$ to the number $x \cdot 0$ does not change it, and so by Proposition 1, z is *the* additive identity: $z = x \cdot 0 = 0$. $\qquad \blacksquare$

PROPOSITION 5: *For all x, y, if $xy = 0$, then $x = 0$ or $y = 0$.*

Proof: We show that if $xy = 0$ and one of the factors is not zero, then the other must be. Assume, therefore, that $xy = 0$ and $y \neq 0$. Since $y \neq 0$, y^{-1} exists, and, by Proposition 4,

$$xyy^{-1} = 0y^{-1} = 0.$$

Therefore, since $yy^{-1} = 1$,

$$xyy^{-1} = x \cdot 1 = x = 0.$$ ∎

PROBLEM 7: For all $x \neq 0$, all $y \neq 0$:

(i) $\left(x^{-1}\right)^{-1} = x$;
(ii) $(xy)^{-1} = x^{-1}y^{-1}$.

Hint: Notice that this is the precise analogue of the theorem on additive inverses (Proposition 3). You will first have to show that $x^{-1} \neq 0$ if $x \neq 0$, for otherwise $\left(x^{-1}\right)^{-1}$ makes no sense. •

PROBLEM 8: Prove the following cancellation laws for addition and multiplication:

(i) For all x, y, z, if $x + y = z + y$, then $x = z$.
(ii) For all x, y, z, if $xy = zy$ and $y \neq 0$, then $x = z$. •

PROBLEM 9: Prove that for all nonzero numbers x, y, z, w:

(i) $\left(\dfrac{x}{y}\right)^{-1} = \dfrac{y}{x}$;

(ii) $\dfrac{x/y}{z/w} = \dfrac{x}{y} \cdot \dfrac{w}{z}$;

(iii) $\dfrac{x}{y} + \dfrac{z}{y} = \dfrac{x + z}{y}$;

(iv) $\dfrac{x}{y} + \dfrac{z}{w} = \dfrac{xw + yz}{yw}$.

In which of the preceding can you relax the assumption that all of x, y, z, w must be nonzero? •

PROBLEM 10: Prove that 0 and 1 are the only numbers x such that $x^2 - x = 0$. Hint: Here you are asked to solve a simple ninth grade equation. The first step is to discard the notion that the solution of an equation consists of a sequence of related equations ending with, in this case, "$x = 0$ or $x = 1$." A solution of an equation is a sequence of *sentences* — and all the variables must be quantified or it is not a sentence — that explain (prove), using previously proved results, that 0 and 1 satisfy the equation, and that no other number does. •

PROBLEM 11: Let $\mathbb{F}$ consist of three distinct elements (numbers); $\mathbb{F} = \{0, 1, 2\}$. Define $+$ and $\cdot$ in $\mathbb{F}$ by the following tables:

+	0	1	2		·	0	1	2
0	0	1	2		0	0	0	0
1	1	2	0		1	0	1	2
2	2	0	1		2	0	2	1

Show that $(\mathbb{F}, +, \cdot)$ is a field. Hint: To verify the associative and distributive laws would require checking 27 possibilities. This is tedious and uninstructive, so just check the six cases where the variables are distinct. Commutativity of either operation is just symmetry about the main diagonal, so that commutativity of both operations is clear, literally by inspection. You do need to check that 0 and 1 are indeed identities for their respective operations, and that inverses exist as required.   •

PROBLEM 12: For all x, y,

$$x(-y) = -(xy),$$

and consequently,

$$(-x)(-y) = xy.$$

Hint: Start with $x(y - y) = 0$.   •

Now we have proved enough of the basic facts of algebra to see how the subject develops from the first five axioms, which involve only the two operations. We will now leap forward and feel free to use all of the standard algebraic manipulations, confident that we could chase back to the axioms any fact that we need. We still must be careful, however. We cannot, for example, assume that there are more than two numbers, no matter how much manipulation we do. (Why?) We cannot, for example, assume that $1 + 1$ always has a square root. This is not surprising, since even in school algebra one must know that a number is positive to have a square root, and we do not yet know what "positive" means.

PROBLEM 13: Show that the equation $x^2 = 2$ has no solution in the three-element field of Problem 11. If we define 2 to be the number $1 + 1$ in any field, does $x^2 = 2$ have a solution in the two-element field of Problem 2?   •

PROBLEM 14: (i) Show that in a finite field with elements $0, 1, 2, \ldots, n$, every row in the addition table must contain every field element, and hence every element must occur exactly once in each row. Hint: Show that for each x (each row) and each k (each field element) there is a field element y (a column) such that $x + y = k$.

(ii) Can you add a fourth element, call it "3," to the three-element field of Problem 11 so that you get a four-element field with the same addition as in Problem 11 for 0, 1, 2? Where does multiplication enter the argument?   •

PROBLEM 15: Show in detail, using the associative and commutative laws, that the 12 possible multiplications of three numbers are equal; that is, show that

$$a(bc) = (ab)c = b(ac) = (ba)c = \cdots.$$ •

PROBLEM 16: Solve the equations (i) $x^2 - 1 = 0$; (ii) $x^2 + 4x + 4 = 0$.   •

PROBLEM 17: The elements a, b, c, d, e form a field with the following addition and multiplication:

+	a	b	c	d	e
a	b	e	a	c	d
b	e	d	b	a	c
c	a	b	c	d	e
d	c	a	d	e	b
e	d	c	e	b	a

$\cdot$	a	b	c	d	e
a	a	b	c	d	e
b	b	d	c	e	a
c	c	c	c	c	c
d	d	e	c	a	b
e	e	a	c	b	d

(i)   Which element is 0 and which is 1?
(ii) What is $-d$?
(iii) What is d^{-1}?
(iv) What is b/e?

•

III

THE ORDER RELATION

As we have seen, fields can be very small and dull — a two-element field is indisputably dull. Once we insist on an order relation, the field becomes a more interesting structure. In particular, an ordered field must have infinitely many elements. Here are the order axioms.

AXIOM VI (Trichotomy Law): *For all $x, y \in \mathbb{R}$, exactly one of the following three relations must hold:*

$$(i) \quad x = y; \qquad (ii) \quad x < y; \qquad (iii) \quad y < x.$$

We will agree that $y > x$ means the same as $x < y$, and write $x \leq y$ for $x < y$ or $x = y$. Similarly, $x \geq y$ means $x > y$ or $x = y$. Numbers greater than zero ($x > 0$) are called **positive**, and numbers less than zero ($x < 0$) are called **negative**. According to the trichotomy axiom, every number other than zero is positive or negative but not both, and zero is neither positive nor negative.

The trichotomy law says that no numbers are neglected by the order relation, and the remaining three axioms relate order to addition and multiplication. Briefly, they say that the order relation is translation invariant, and the set of positive numbers is closed under both addition and multiplication.

AXIOM VII: *For all x, y, z,*

$$x < y \qquad \textit{if and only if} \qquad x + z < y + z.$$

AXIOM VIII: *For all x, y, if $x > 0$ and $y > 0$, then $x + y > 0$ and $xy > 0$.*

11

The transitivity of the order relation ((ii) of Proposition 1) follows immediately from Axioms VII and VIII.

PROPOSITION 1: *For all x, y, z:*

(i) $x < y$ iff $x - y < 0$ iff $y - x > 0$.
(ii) *If* $x < y$ *and* $y < z$, *then* $x < z$.

Proof: From Axiom VII we can add $-y$ to both sides of $x < y$ to get $x - y < 0$, or add y to both sides of $x - y < 0$ to get $x < y$. Similarly, we can add $y - x$ to both sides of $x - y < 0$ to get $0 < y - x$, which is the same as $y - x > 0$. To prove (ii), assume that $x < y$ and $y < z$, which is the same as $y - x > 0$ and $z - y > 0$. By Axiom VIII the sum of these two positive numbers is positive, or

$$(y - x) + (z - y) = z - x > 0,$$

which is the same as $x < z$.   ∎

Now without further ado we will use obvious theorems such as these:
For all x, y, z:

(i) If $x < y$ and $y \leq z$, then $x < z$;
(ii) If $x \leq y$ and $y \leq z$, then $x \leq z$.

PROBLEM 1: For all x, $x > 0$ if and only if $-x < 0$, and $x < 0$ if and only if $-x > 0$. Hint: Since $-(-x) = x$, there is really only one equivalence to prove here.   •

PROBLEM 2: Show that there is at least one positive number and at least one negative number, and that there are at least three numbers. How many ways are there to put an order relation on the two-element field of the last chapter so that the order axioms hold?   •

PROPOSITION 2: $1 > 0$.

Proof: By Axiom III, $1 \neq 0$, so that $1 > 0$ or $1 < 0$, but not both. By Problem 1 we can say that $1 > 0$ or $-1 > 0$ but not both, since $1 < 0$ is the same as $-1 > 0$. If $-1 > 0$, then since the product of positive numbers is positive (Axiom VIII) $(-1)(-1) > 0$. But in the last chapter we showed that $(-x)(-y) = xy$ for all x and y, and so $(-1)(-1) = 1 \cdot 1 = 1$. Thus we have the contradiction that if $1 < 0$, so that $-1 > 0$, then $1 > 0$, contrary to the trichotomy assumption. This leaves only the one possibility, $1 > 0$.   ∎

PROPOSITION 3: *The product of two positive or two negative numbers is positive. The product of two numbers is negative if and only if one number is positive and one is negative.*

Proof: The fact that the product of two positive numbers is positive is an axiom. If $x < 0$ and $y < 0$, then $-x > 0$ and $-y > 0$, and so $xy = (-x)(-y) > 0$. Since a product is zero, and hence neither positive nor negative, if either factor is zero, it follows that a product is negative only if one factor is positive and the other negative. Conversely, if $x > 0$ and $y < 0$, then $-y > 0$, so that $x(-y) = -xy > 0$, and $xy < 0$.   ∎

PROBLEM 3: For all x, y, z, if $x < y$, then

$$xz < yz \qquad \text{if} \quad z > 0,$$

and

$$xz > yz \qquad \text{if} \quad z < 0.$$

In other words, an inequality is preserved if both sides are multiplied by the same positive number, and reversed if both sides are multiplied by the same negative number.   •

PROBLEM 4: For all x, y:

(i) $x > 0$ if and only if $x^{-1} > 0$.
(ii) $x/y > 0$ if and only if both x and y are positive, or both are negative.   •

The **absolute value** of a number x, denoted $|x|$, is defined as follows:

$$|x| = \begin{cases} x & \text{if } x \geq 0, \\ -x & \text{if } x < 0. \end{cases}$$

Clearly $|x| \geq 0$, $|x| = |-x|$ for all x, and $|x| = 0$ if and only if $x = 0$. It follows immediately from the preceding propositions and problems that, for all x, y,

$$|xy| = |x|\,|y|,$$

$$\left|\frac{x}{y}\right| = \frac{|x|}{|y|}.$$

The study of limits, and that is the purpose of this course, amounts to estimating the size (smallness) of the difference between something and its limit. We use the usual geometric imagery and refer to $|x - y|$ as the **distance** between x and y. Our constant companion in these estimates of distance will be **triangle inequality**, which is proved next. The triangle inequality frequently occurs in estimates like this: if the distance from x to z is less than $\varepsilon/2$ and the distance from z to y is less than $\varepsilon/2$, then the distance from x to y is less than ε, since

$$|x - y| = |(x - z) + (z - y)| \leq |x - z| + |z - y|.$$

PROPOSITION 4 (Triangle Inequality): *For all x, y,*

$$|x + y| \leq |x| + |y|.$$

Proof: If x and y are both positive or both negative, or if x or y is zero, then it is easy to see from the preceding results that equality must hold. If none of these conditions hold, then we may assume, to be specific, that $x > 0$ and $y < 0$. There are two cases to consider:

(i) $-y \le x$;

(ii) $-y > x$.

If $-y \le x$, then $x + y \ge 0$, $|x| = x$, and $y < 0 < |y|$, so that

$$|x + y| = x + y < |x| + |y|.$$

If $-y > x$, then $x + y < 0$, $-x < 0$, and $-y = |y|$, so that

$$|x + y| = -(x + y)$$

$$= (-x) + (-y)$$

$$< 0 + |y|$$

$$\le |x| + |y|. \qquad \blacksquare$$

PROBLEM 5: Write $x = (x - y) + y$ and use the triangle inequality to show that $|x - y| \ge |x| - |y|$ for all x, y.   •

PROBLEM 6: (i) For all x and all $\varepsilon > 0$, $|x| < \varepsilon$ if and only if $-\varepsilon < x < \varepsilon$.

(ii) For all x, y and all $\varepsilon > 0$, $|x - y| < \varepsilon$ if and only if $y - \varepsilon < x < y + \varepsilon$ if and only if $x - \varepsilon < y < x + \varepsilon$. Hint: Since $|x - y| = |y - x|$, the second part of (ii) follows immediately from the first part of (ii) by interchanging x and y.   •

PROBLEM 7: Let a and b be two given numbers with $a < b$. Solve the inequality $(x - a)(x - b) < 0$. Hint: You may use any of the preceding propositions or problems in your proof; you may not argue from a splatter of plus and minus signs on a picture of the line.   •

PROBLEM 8: (i) Solve the inequality $x \cdot x > 1$.

(ii) Solve the inequality $x \cdot x < 1$.

Hint: $x \cdot x > 1$ is equivalent to $(x + 1)(x - 1) > 0$.   •

PROBLEM 9: Solve the inequality $x \cdot x - x - 3/4 < 0$. Hint: See Problem 8.   •

An **interval** is any set which is of one of the following forms:

$$[a, b] = \{x : a \le x \le b\},$$

$$(a, b) = \{x : a < x < b\},$$

$$[a, b) = \{x : a \le x < b\},$$

$$(a, b] = \{x : a < x \le b\},$$

$$(a, \infty) = \{x : a < x\},$$

$$[a, \infty) = \{x : a \le x\},$$

$$(-\infty, b) = \{x : x < b\},$$

$$(-\infty, b] = \{x : x \le b\},$$

$$(-\infty, \infty) = \mathbb{R}.$$

The intervals (a, b), (a, ∞), $(-\infty, b)$, $(-\infty, \infty)$ are **open intervals**, and the intervals $[a, b]$, $[a, \infty)$, $(-\infty, b]$, $(-\infty, \infty)$ are **closed intervals**.

PROBLEM 10: Show that if two intervals have nonempty intersection, the intersection is an interval, and their union is an interval.   ●

PROBLEM 11: What can you say about the complement of an open interval? The complement of a closed interval?   ●

PROBLEM 12: Show that if $0 < a < b$, or if $b < a < 0$, then $a^2 < b^2$.   ●

PROBLEM 13: Show that $|x + y + z| \leq |x| + |y| + |z|$ and $|x + y + z| \geq |x| - |y| - |z|$ for all x, y, z.   ●

PROBLEM 14: Show that the square of the average of any two numbers is less than or equal to the average of their squares.   ●

PROBLEM 15: (i) Show that for any two numbers a_1 and a_2, $a_1 a_2 \leq (a_1^2 + a_2^2)/2$.

(ii) Let a_1 and a_2 be nonnegative numbers that have square roots. (This curious assumption is necessary because we do not yet know — and it is not true in the absence of Axiom IX — that all nonnegative numbers have square roots.) Show that

$$\sqrt{a_1 a_2} \leq \frac{1}{2}(a_1 + a_2). \tag{1}$$

(iii) Let a_1, a_2, a_3, a_4 be nonnegative numbers with fourth roots. Use (i) and (ii) to show that

$$\sqrt{a_1 a_2} \, \sqrt{a_3 a_4} \leq \frac{1}{4}\left(a_1^2 + a_2^2 + a_3^2 + a_4^2\right),$$

$$(a_1 a_2 a_3 a_4)^{1/4} \leq \frac{1}{4}(a_1 + a_2 + a_3 + a_4). \tag{2}$$

(The left sides of (1) and (2) are called the **geometric means** of the numbers, and the right sides are the **arithmetic means**.)   ●

PROBLEM 16: Solve the inequality $|x| > 1$.   ●

PROBLEM 17: Solve the inequality $|2x - 4| < \varepsilon$, where ε is a given positive number.   ●

PROBLEM 18: Solve the inequality $x \cdot x - 3x + 2 > 0$.   ●

IV

THE NATURAL NUMBERS

The preceding chapters provide the admittedly tedious underpinning for the elementary algebra of any ordered field. Now we will examine in some detail the important subset consisting of the natural numbers: 1, 2, 3, ... Although the natural numbers are as basic as fingers and toes, the student has very likely never seen *proofs* of their basic properties. We now proceed to develop the mathematical basis for arithmetic. Be forewarned that arithmetic is frequently harder than algebra.

The set $\mathbb{N}$ of **natural numbers** is the smallest subset of $\mathbb{R}$ with the following two properties:

(i) 1 is in the set.   (1)
(ii) $n + 1$ is in the set whenever n is.

The two properties (1) of a set are called **inductive properties**. If T is the intersection of any family of sets S, each of which has the two inductive properties, then T also obviously has these properties. If T is the intersection of *all* sets S with the inductive properties, then T is the smallest such set; that is, $T = \mathbb{N}$. The purpose of these remarks is to show that there is a smallest set with properties (1); that is, to show that $\mathbb{N}$ is sensibly defined.

The usual symbols for the first few natural numbers are defined as follows:

$$2 = 1 + 1, \quad 3 = 2 + 1, \quad 4 = 3 + 1, \quad 5 = 4 + 1,$$

$$6 = 5 + 1, \quad 7 = 6 + 1, \quad 8 = 7 + 1, \quad 9 = 8 + 1.$$

16

The numbers 0, 1, 2, 3, 4, 5, 6, 7, 8, 9 are called **digits**. Fingers and toes are also called digits, a circumstance that is hardly coincidental.

In Chapter 1 we saw examples of fields with two and three elements. The definitions of $\mathbb{N}$ and the digits $0, 1, \ldots, 9$ hold in any field, but obviously the digits do not represent distinct numbers in all fields. In the three-element field, for example, $3 = 0, 4 = 1, \ldots, 9 = 0$. We now, and henceforth, assume also the order axioms, so that it is easy to show that

$$0 < 1 < 2 < 3 < 4 < 5 < 6 < 7 < 8 < 9, \tag{2}$$

which in particular means that the ten digits are distinct numbers.

PROBLEM 1: Prove the inequalities (2).   •

Now let us compute the addition table for digits. Let t stand, temporarily, for $9 + 1$. The first row of the addition table, $0 + 0 = 0, 0 + 1 = 1, 0 + 2 = 2, \ldots$, is immediate from the fact that 0 is an additive identity. The second row simply consists of the definitions of $2, 3, \ldots, 9, t$:

$$1 + 1 = 2, \quad 1 + 2 = 3, \quad \ldots, \quad 1 + 8 = 9, \quad 1 + 9 = t.$$

Now the third row, of sums $2 + d$: $2 + 0 = 2$ and $2 + 1 = 3$ are clear. The next sum is

$$2 + 2 = 2 + (1 + 1) = (2 + 1) + 1 = 3 + 1 = 4.$$

Here we used just the definition of 2 ($2 = 1 + 1$), the definition of 3 ($3 = 2 + 1$), the definition of 4 ($4 = 3 + 1$), and the associative law. Each new sum is computed using the previous one, until finally we get to

$$2 + 8 = 2 + 7 + 1 = 9 + 1 = t,$$

and

$$2 + 9 = 2 + 8 + 1 = t + 1.$$

We define t^n for $n \in \mathbb{N}$ as follows: $t^2 = t \cdot t$, $t^3 = t \cdot t^2$, and so forth, and agree that if $d_0, d_1, d_2, \ldots$ are digits, then

$$d_1 d_0 = d_1 t + d_0,$$
$$d_2 d_1 d_0 = d_2 t^2 + d_1 t + d_0,$$
$$d_3 d_2 d_1 d_0 = d_3 t^3 + d_2 t^2 + d_1 t + d_0,$$

and so on. For example, $10 = 1t + 0 = t$. We earlier used juxtaposition to indicate multiplication, so that $d_1 d_0$ would mean the product of d_1 and d_0. We now modify that earlier agreement to exclude digits, and we stipulate that juxtaposition of digits is reserved for the familiar decimal notation just described.

We can add any two natural numbers written decimally in the usual way, using our addition table for digits. For example, $37 + 68 = 105$:

$$37 + 68 = (3t + 7) + (6t + 8)$$
$$= 9t + (7 + 8)$$
$$= 9t + t + 5$$
$$= (9 + 1)t + 5$$
$$= t^2 + 0t + 5$$
$$= 105.$$

The multiplication table for $1 \cdot 1$ up to $9 \cdot 9$ can be calculated from the addition table and the distributive law. For example,

$$2 \cdot 2 = 2 \cdot (1 + 1) = 2 \cdot 1 + 2 \cdot 1 = 2 + 2 = 4,$$
$$2 \cdot 3 = 2 \cdot (2 + 1) = 2 \cdot 2 + 2 \cdot 1 = 4 + 2 = 6, \tag{3}$$

and so forth.

PROBLEM 2: Calculate the multiplication table for $2 \cdot d$ and $3 \cdot d$ for $d = 0, 1, \ldots, 9$. •

The so-called **Theorem of Induction** is usually stated something like this: If $P(n)$ denotes a proposition that depends on n, and $P(1)$ is true, and $P(n + 1)$ is true whenever $P(n)$ is true, then $P(k)$ is true for all natural numbers k. This is simply a paraphrase of our definition of $\mathbb{N}$. If we let S be the set of all natural numbers k such that $P(k)$ is true, then we have assumed that $1 \in S$ and $n + 1 \in S$ whenever $n \in S$; this implies that $S \supset \mathbb{N}$, and so $P(k)$ is true for all $k \in \mathbb{N}$. We will also accept as part of our common logic the idea of **inductive** (or **recursive**) definition. For example, we define x^n for any $n \in \mathbb{N}$ with the following inductive scheme: $x^1 = x$, and $x^{n+1} = x \cdot x^n$ for all $n \in \mathbb{N}$. We have already made use of this definition for t^n in the decimal notation.

Now we use the theorem of induction to prove some of the basic arithmetic theorems about the natural numbers.

PROPOSITION 1: *Every natural number is positive.*

Proof: We have already seen (Proposition 2 of the last chapter) that $1 > 0$. This is the first step of the inductive proof. Now make the inductive assumption: let n be any natural number, and assume that $n > 0$. Then

$$n + 1 > 0 + 1 = 1 > 0,$$

and so $n + 1 > 0$ whenever $n > 0$. Therefore, all natural numbers are positive. ∎

We define the **integers** to be the natural numbers, their negatives, and 0. Thus an integer is positive if and only if it is a natural number, and an integer is negative if and only if it is the negative of a natural number. An integer is neither positive nor negative if and only if it is 0.

PROPOSITION 2: *The sum of two natural numbers is a natural number.*

Proof: Let m be any fixed natural number. Let S be the set of all natural numbers n such that $m + n \in \mathbb{N}$. Clearly $1 \in S$, since $m + 1 \in \mathbb{N}$ if $m \in \mathbb{N}$. Assume $n \in S$, that is, $m + n \in \mathbb{N}$. Then

$$m + (n + 1) = (m + n) + 1,$$

and $(m + n) + 1 \in \mathbb{N}$ because $m + n \in \mathbb{N}$. Therefore, $n + 1 \in S$ if $n \in S$, and $S = \mathbb{N}$. ■

PROBLEM 3: (i) The sum of two integers is an integer.
(ii) The product of two natural numbers is a natural number.
(iii) The product of two integers is an integer. •

PROPOSITION 3: *If k is a natural number and $k \neq 1$, then $k = p + 1$ for some $p \in \mathbb{N}$. That is, each natural number other than 1 has an immediate predecessor among the natural numbers.*

Proof: Let S be the set consisting of the number 1 together with all natural numbers n such that $n = p + 1$ for some natural number p. Certainly, 1 is in S since we put it there. If $n \in S$, then n is a natural number, so that $n + 1$ is a natural number and $n + 1$ is an element of S by definition of S. Since $n + 1 \in S$ whenever $n \in S$, $S = \mathbb{N}$. ■

COROLLARY 1: *If $n \in \mathbb{N}$ and $n \neq 1$, then $n - 1 \in \mathbb{N}$.* ■

PROBLEM 4: (i) If $n \in \mathbb{N}$, $n \geq 1$.
(ii) There is no natural number n such that $0 < n < 1$. •

PROPOSITION 4: *For all m, $n \in \mathbb{N}$:*

(i) *If $m > n$, then $m \geq n + 1$.*
(ii) *If $m < n + 1$, then $m \leq n$.*
(iii) *$m > n$ if and only if $m - n \in \mathbb{N}$.*
(iv) *$m > n$ if and only if $m = n + k$ for some $k \in \mathbb{N}$.*

Proof: (i) Assume $m > n$, so that $m - n > 0$. We know from Problem 3 that the sum of two integers (here, m and $-n$) is an integer, so that $m - n$ is a positive integer, and therefore a natural number. Hence $m - n \geq 1$, and $m \geq n + 1$.
(ii) Assume $m < n + 1$, so that $n + 1 - m \in \mathbb{N}$ by part (i). Therefore, $n - m + 1 = 1$ or $n - m + 1 > 1$; that is, $n = m$ or $n > m$, which is the same as $m \leq n$.
(iii) If $m - n \in \mathbb{N}$, then $m - n > 0$ and so $m > n$. To prove the other implication, assume that $m > n$, or $m - n > 0$. Then $m - n$ is an integer, and positive, so that $m - n \in \mathbb{N}$.
(iv) This is just a restatement of (iii). ■

PROBLEM 5: For every integer k, there is no integer n such that $k < n < k+1$. Hint: First assume $k \geq 0$ and use Proposition 4(iv) and Problem 4.   •

We will henceforth use without fuss the standard elementary properties of natural numbers like those listed in Proposition 4, as well as the obvious extensions to the negative integers.

The following is a deep and important property of the natural numbers.

PROPOSITION 5 (The Well-Ordering Principle): *Every nonempty subset of* $\mathbb{N}$ *has a smallest element.*

Proof: Let $T \subset \mathbb{N}$, and assume $T \neq \varnothing$. Let S be the set of all natural numbers each of which is less than or equal to all the elements of T. Clearly $1 \in S$, since $1 \leq n$ for *all* $n \in \mathbb{N}$. We know that $S \neq \mathbb{N}$, since if m is any element of T (and we assumed that T has an element), then $m + 1$ is not in S. Since $1 \in S$ and $S \neq \mathbb{N}$, there is some $n \in S$ so that $n + 1 \notin S$; that is, there is some n so that $n \leq k$ for all $k \in T$ but $n + 1 \nleq k$ (i.e., $k < n + 1$) for some $k \in T$. Fix such an element n of S and let k be a corresponding element of T, so that

$$n \leq k < n + 1.$$

There is no natural number strictly between n and $n + 1$ by Problem 4, and so $k = n$. Hence n belongs to both S and T. Since $n \in S$, n is less than or equal to all elements of T. Since $n \in T$, n is the smallest element of T.   ∎

PROBLEM 6: If T is a nonempty family of natural numbers, and there is $M \in \mathbb{N}$ such that $n \leq M$ for all $n \in T$, then T has a largest element.   •

PROBLEM 7: Show that if S is a set of natural numbers containing some given number N, and $n + 1 \in S$ whenever $n \in S$, then S contains all natural numbers $n \geq N$. Hint: Consider the set T that consists of S together with all natural numbers less than N.   •

Inductive proofs are common throughout mathematics, and although the idea is very simple, the details can be tricky. Consider the following sentence: For all $k \in \mathbb{N}$, $k > k + 2$. The inductive step in the "proof" of this sentence is easy — if $n > n + 2$ for some n, then $n + 1 > (n + 1) + 2$, and the sentence holds for $n + 1$ whenever it holds for n. A proof by induction has to have a starting place — usually $n = 1$ — and that is what is missing here.

Here is a slightly more interesting but equally false sentence. In any finite set of nonzero numbers, all the numbers have the same sign. The "proof" goes like this: If a set has only one number, and that number is not zero, then all the elements of the set have the same sign. Now suppose that all the elements in any set of n nonzero numbers have the same sign. Consider any set of $n + 1$ nonzero numbers. Take one number out, so that the remaining n numbers all have the same sign. Now take another element out of the set and put the first one back. You again have a set of n nonzero numbers, and so they all have the same sign. The first number you took out has the same sign as the other n, and so all $n + 1$ numbers have the same sign. The statement is true for any set of $n + 1$ numbers whenever it is true for all sets of n numbers, and so the statement is true for any finite set of nonzero numbers. (The terms "finite set" and "set with n elements" used here have not yet been defined, but we will take care of that in the next chapter.)

PROBLEM 8: What is wrong with the preceding "proof"? N.B. Say what is wrong with the *proof*, not the statement. •

The following familiar result is a consequence of the theorem of induction.

PROPOSITION 6 (The Binomial Theorem): *For any two numbers a and b, and any $n \in \mathbb{N}$,*

$$(a + b)^n = a^n + na^{n-1}b + \cdots + \frac{n!}{(n - k)!k!}a^{n-k}b^k + \cdots + nab^{n-1} + b^n. \quad (4)$$

Proof: The formula (4) has $n + 1$ terms, and the coefficient of $a^{n-k}b^k$ for $k = 0, 1, \ldots, n$ is $n!/(n - k)!k!$, where $n!$ (n factorial) is defined inductively by

$$0! = 1, \qquad 1! = 1, \qquad (n + 1)! = (n + 1) \cdot n!.$$

When $n = 1$, there are two terms in (4) and the coefficient formula gives 1 for $k = 0$ and 1 for $k = 1$:

$$\frac{1!}{(1 - 0)!0!} = 1 \qquad \text{and} \qquad \frac{1!}{(1 - 1)!1!} = 1.$$

Assume that (4) is correct for some n, and calculate the coefficient of $a^{n+1-k}b^k$ in $(a + b)^{n+1}$. The two terms involving $a^{n+1-k}b^k$ arise as a times the term

$$\frac{n!}{(n - k)!k!}a^{n-k}b^k, \quad (5)$$

and b times the term

$$\frac{n!}{(n - (k - 1))!(k - 1)!}a^{n-(k-1)}b^{k-1}. \quad (6)$$

The verification that the coefficients in (5) and (6) add up correctly is the following problem. ∎

PROBLEM 9: Show that the sum of the coefficients in (5) and (6) is $(n + 1)!/(n + 1 - k)!k!$, which is the coefficient of $a^{n+1-k}b^k$ in $(a + b)^{n+1}$ according to (4). •

PROBLEM 10: Prove by induction that the following identities and inequalities hold for all natural numbers n.

(i) $1 + 2 + \cdots + n = \frac{1}{2}n(n + 1)$.

(ii) $1^2 + 2^2 + \cdots + n^2 = \frac{1}{6}n(n + 1)(2n + 1)$.

(iii) $1^3 + 2^3 + \cdots + n^3 = ?$ Hint: From (i) and (ii) you might guess that there is a fourth degree polynomial for this sum.

(iv) $\dfrac{1}{1 \cdot 2} + \dfrac{1}{2 \cdot 3} + \dfrac{1}{3 \cdot 4} + \cdots + \dfrac{1}{n(n + 1)} = \dfrac{n}{n + 1}$.

(v) $1 + 3 + 5 + \cdots + (2n - 1) = n^2$.

(vi) $1 + x + x^2 + \cdots + x^n = \dfrac{1}{1 - x} - \dfrac{x^{n+1}}{1 - x}$.

(vii) $(1 + p)^n \geq 1 + np$ if $p \geq 0$.

(viii) $2n + 1 \leq 2^n$ if $n \geq 3$.

(ix) $n^2 \leq 2^n$ if $n \geq 4$.

(x) $2^{n-1} \leq n!$ for all n.   •

PROBLEM 11: If N is a fixed natural number and S is a set such that $1 \in S$ and $n + 1 \in S$ whenever $n \in S$ and $n < N$, then S contains all natural numbers $k \leq N$.   •

PROBLEM 12: Prove the **Strong Theorem of Induction**: If $1 \in S$ and $n+1 \in S$ whenever $1, 2, \ldots, n \in S$, then $S \supset \mathbb{N}$. Show that the strong theorem of induction implies the following usual (weak) theorem of induction: If $1 \in S$ and $n + 1 \in S$ whenever $n \in S$, then $S \supset \mathbb{N}$. Hint: Be prepared to think about this for a while — perhaps a long while.   •

PROBLEM 13: Prove the laws of exponents for positive integer exponents: for all $x \in \mathbb{R}$,

(i) if $m, n \in \mathbb{N}$, and $x \in \mathbb{R}$, $x^m \cdot x^n = x^{m+n}$;

(ii) if $m, n \in \mathbb{N}$ and $x \in \mathbb{R}$, $(x^m)^n = x^{(mn)}$.   •

PROBLEM 14: Define $x^0 = 1$ for all x, and $x^{-n} = (x^n)^{-1}$ for every $x \neq 0$ and every negative integer $-n$. Finish up the laws of exponents to include all integer exponents, and division (x^m / x^n). Hint: You had better show that $x^{-n} \neq 0$ if n is a positive integer and $x \neq 0$.   •

PROBLEM 15: In the five-element field of Problem 17, Chapter 2, identify the natural numbers $1, 2, 3, 4$; that is, say which is 1, which is 2, which is 3, and which is 4.   •

PROBLEM 16: Calculate the row of $n + 1$ binomial coefficients for $n = 0, 1, 2, 3, 4, 5, 6$, and arrange these rows in a triangular array, with 1 at the top, 1, 1 in the next row, 1, 2, 1 in the third row, and so forth. This is called **Pascal's triangle**. Observe the marvelous property it exhibits, and prove it by induction.   •

V

FINITE AND INFINITE SETS

To describe what "finite" and "infinite" mean we first need to formalize the idea of a one-to-one correspondence, and this is done using the terminology of functions. A **function** f is a set of ordered pairs, (x, y), with the property that no two distinct pairs in f have the same first element. The set of first elements of pairs in f is called the **domain** of f and is denoted $\mathcal{D}(f)$. The set of second elements of pairs in f is called the **range** of f. If $x \in \mathcal{D}(f)$, then there is exactly one y such that $(x, y) \in f$; this y is denoted $f(x)$. We say f is a **function on A onto B** to mean that $A = \mathcal{D}(f)$ and B is the range of f. If we only know that the range of f is a subset of B, we say f is a function **on A to B**, or **on A into B**. The function f is **one-to-one** provided $f(x) = f(y)$ implies $x = y$; that is, every element of the range is $f(x)$ for exactly one x in the domain. A one-to-one function effects a **one-to-one correspondence** between its domain and range. If f is a one-to-one function on A onto B, then the set of pairs (y, x), for $(x, y) \in f$, is also a function. This **inverse function**, denoted f^{-1}, is a function on B onto A, and $f^{-1}(f(x)) = x$ for all $x \in A$ and $f(f^{-1}(y)) = y$ for all $y \in B$.

If k is a natural number, we will denote by $\mathbb{N}_k$ the set of natural numbers less than or equal to k:

$$\mathbb{N}_k = \{n \in \mathbb{N} \, : \, n \leq k\}.$$

The sets $\mathbb{N}_k$ are the prototypical finite sets. We say a set A is **finite** if and only if there is a one-to-one function φ on A onto $\mathbb{N}_k$ for some k. This is clearly equivalent

23

to saying there is a one-to-one function ψ (with $\psi = \varphi^{-1}$), on some $\mathbb{N}_k$ onto A. A set that is not empty and not finite is called **infinite**. If there is a one-to-one function on A onto $\mathbb{N}$, or on $\mathbb{N}$ onto A, then A is a **countable** set.

Now we need to prove some basic facts like these: a countable set is not finite, and vice versa, and there are infinite sets that are not countable. We also want to say what it means for a finite set to have n elements. The English suggests that a set should not have exactly three elements, say, and also have exactly five elements. That is, if there is a one-to-one function on A onto $\mathbb{N}_n$ and also a one-to-one function on A onto $\mathbb{N}_k$, then we had better have $n = k$. It suffices to show there is no one-to-one function on $\mathbb{N}_n$ onto $\mathbb{N}_k$ unless $n = k$.

PROPOSITION 1: *If there is a one-to-one function on $\mathbb{N}_n$ onto $\mathbb{N}_k$, then $n = k$.*

Proof: Let S be the set of numbers n for which the statement is true; that is, let S be all $n \in \mathbb{N}$ such that whenever there is a one-to-one function on $\mathbb{N}_n$ onto some $\mathbb{N}_k$, then $n = k$. To show $1 \in S$, let φ be a one-to-one function on $\mathbb{N}_1 = \{1\}$ onto some $\mathbb{N}_k$. If $k > 1$, then $\varphi^{-1}(1) \in \mathbb{N}_1$ and $\varphi^{-1}(2) \in \mathbb{N}_1$, so that $\varphi^{-1}(1) = \varphi^{-1}(2) = 1$, which is a contradiction. Now suppose that $n \in S$. Let φ be any one-to-one function on $\mathbb{N}_{n+1}$ onto some $\mathbb{N}_k$. If $\varphi(n + 1) = k$, then φ restricted to $\mathbb{N}_n$ is a one-to-one function on $\mathbb{N}_n$ onto $\mathbb{N}_{k-1}$. Hence $n = k - 1$, and $n + 1 = k$ as was to be shown. If $\varphi(n + 1) \neq k$, then $\varphi(i) = k$ for some $i \leq n$. Let φ_0 be the function on $\mathbb{N}_{n+1}$ that interchanges the values of φ at i and $n + 1$; that is, $\varphi_0 = \varphi$ except at i and $n + 1$, and $\varphi_0(n + 1) = \varphi(i) = k$, $\varphi_0(i) = \varphi(n + 1)$. Then φ_0 is a one-to-one function on $\mathbb{N}_{n+1}$ onto $\mathbb{N}_k$ with $\varphi_0(n + 1) = k$, and we have seen that this implies $n + 1 = k$. ∎

Now we can legitimately make the following definition. **A set A has n elements** if and only if there is a one-to-one function φ on A onto $\mathbb{N}_n$. Consequently, each set with n elements is finite, and each finite set has n elements for exactly one $n \in \mathbb{N}$.

If A is a set of n elements, then we can exhibit the elements A:

$$A = \{a_1, a_2, \ldots, a_n\}.$$

Here we have simply used little "a" as the name of a one-to-one function on $\mathbb{N}_n$ onto A, writing $a_1, a_2, \ldots$ for function values instead of the usual $a(1), a(2), \ldots$.

PROBLEM 1: (i) If A has n elements and $a \notin A$, then $A \cup \{a\}$ has $n + 1$ elements.

(ii) If A has n elements and $n > 1$, and $a \in A$, then $A - \{a\}$ has $n - 1$ elements.

(iii) If A has n elements, B has m elements, and $A \cap B = \varnothing$, then $A \cup B$ has $n + m$ elements. Hint: Prove this by induction on m; the case $m = 1$ is part (i). If the statement is true for every second set with m elements, and B has $m + 1$ elements, and $b \in B$, then $B_0 = B - \{b\}$ has m elements by (ii), so that $A \cup B_0$ has $m + n$ elements. Et cetera. •

PROPOSITION 2: *Any nonempty subset of a set with n elements is finite and has n or fewer elements; that is, if $\varnothing \neq A \subset B$, and B has n elements, then A has k elements for some $k \leq n$.*

Proof: If A is a nonempty subset of B, and B has one element, then clearly $A = B$ and A has one element. Suppose the assertion holds for all sets B with n elements. Let B have $n + 1$ elements and assume $\varnothing \neq A \subset B$. If $A = B$, then we are done. If $A \neq B$, let $a \in B - A$. Then $B_0 = B - \{a\}$ has n elements by Problem 1, and $A \subset B_0$, so that A has n or fewer elements.   ■

PROBLEM 2: If A has m elements and B has n elements, then $A \cup B$ has $m + n$ or fewer elements. Hint: $B - A$ has fewer than n elements, and $A \cap (B - A) = \varnothing$.   •

Recall that a set A is **countable** if and only if there is a one-to-one function φ on A onto $\mathbb{N}$. Equivalently, A is countable if and only if there is a 1–1 function on $\mathbb{N}$ onto A.

PROPOSITION 3: *A countable set is not finite. A finite set is not countable.*

We will prove the following statement, which is equivalent (see Problem 3): There is no 1–1 function on $\mathbb{N}$ onto some $\mathbb{N}_n$.

Proof: The proof is by induction on the n in $\mathbb{N}_n$. First, suppose there is a 1–1 function φ on $\mathbb{N}$ onto $\mathbb{N}_1 = \{1\}$. Then $\varphi(1) = \varphi(2) = \varphi(3) = \cdots = 1$, so that φ is not 1–1, and this contradiction shows the statement holds for $n = 1$. For the inductive step, suppose that n is a number such that there is no 1–1 function on $\mathbb{N}$ onto $\mathbb{N}_n$. Assume, to reach a contradiction, that there is a 1–1 function φ on $\mathbb{N}$ onto $\mathbb{N}_{n+1}$. Let $\varphi(k) = n + 1$. We define a new function φ_0 on $\mathbb{N}$ as follows: $\varphi_0(i) = \varphi(i)$ if $i < k$; if $i \geq k$, let $\varphi_0(i) = \varphi(i + 1)$. Clearly, φ_0 is 1–1 on $\mathbb{N}$ onto $\mathbb{N}_n$, so that we have a contradiction.   ■

PROBLEM 3: Show that the statement proved above is equivalent to the statement of Proposition 3.   •

Let A be a finite set of *numbers*, say with n elements, so that we can exhibit the numbers in A:

$$A = \{a_1, a_2, \ldots, a_n\}.$$

We would now like to be able to say that A has a largest element, and a smallest element, and that the elements of A can be arranged in order:

$$a_1 < a_2 < \cdots < a_n.$$

We attack these questions next.

PROPOSITION 4: *A finite set of numbers has a largest element and a smallest element.*

Proof: If A has one element, then that element is obviously largest and smallest. Suppose that every set with n numbers has a largest number. Let A be any set with $n+1$ numbers: $A = \{a_1, a_2, \ldots, a_n, a_{n+1}\}$. Let a_i be the largest number in of the n-element set $\{a_1, a_2, \ldots, a_n\}$. If $a_i > a_{n+1}$, then a_i is the largest element of A, and if $a_{n+1} > a_i$, a_{n+1} is the largest element of A. The existence of a smallest element can be proved the same way, or for variety, as suggested in the following problem. ∎

PROBLEM 4: Use the existence of a largest number in every finite set of numbers to show the existence of a smallest number. Hint: For any set A, let $A^* = \{-a : a \in A\}$. Show the largest element of A^* is the negative of the smallest element of A. •

PROPOSITION 5: *The elements of any finite set of numbers can be arranged in order; that is, if A is a set of n numbers, then there is a 1–1 function a on $\mathbb{N}_n$ onto A such that*

$$a_1 < a_2 < a_3 < \cdots < a_n.$$

Proof: If A has one element, this is clear. Suppose the statement holds for all sets of n numbers, and let A be a set with $n+1$ numbers. Let a be the largest number in A. Then $A_0 = A - \{a\}$ has n numbers, and we can list them as $a_1, a_2, \ldots, a_n$ with $a_1 < a_2 < \cdots < a_n$. Define a_{n+1} to be a. Since a is the largest element, we have

$$a_1 < a_2 < \cdots < a_n < a_{n+1},$$

as desired. ∎

The set $\mathbb{N}$ has the interesting property that it can be put in a 1–1 correspondence with a proper subset of itself. There are, for example, as many even natural numbers as there are in the whole set of natural numbers. The function φ defined by $\varphi(n) = 2n$ is 1–1 on $\mathbb{N}$ onto the even numbers. This property turns out to be characteristic of infinite sets. We will show this after first showing that every infinite set has a countable subset.

PROPOSITION 6: *If A is an infinite set, then there is a 1–1 function φ on $\mathbb{N}$ into A. (N.B. "into," not "onto.")*

Proof: Here we use the idea of an inductive or recursive definition. Let A be an infinite set, so that by definition $A \neq \emptyset$. Let $a \in A$, and define $\varphi(1) = a$. If $A - \{a\} = \emptyset$, then A has one element and is therefore finite and not infinite. Let $b \in A - \{a\}$, and define $\varphi(2) = b$. Now suppose that $\varphi(1), \varphi(2), \ldots, \varphi(n)$ have been defined so that φ is a 1–1 function on $\mathbb{N}_n$ into A. If φ is onto A, then A is a finite set with n elements, contrary to our assumption. Since φ is not onto A, we can let c be any element of $A - \{\varphi(1), \ldots, \varphi(n)\}$, and define $\varphi(n + 1) = c$. Thus, we have defined $\varphi(1)$, and for every $n \in \mathbb{N}$, we defined $\varphi(n + 1)$ in terms of $\varphi(1), \ldots, \varphi(n)$; the function φ is therefore defined inductively on $\mathbb{N}$, into A. To see that φ is one-to-one, consider any two distinct numbers n and k. If n is the larger of the two, then $\varphi(n)$ was explicitly defined so as to be distinct from $\varphi(k)$, since $\varphi(n) \in A - \{\varphi(1), \ldots, \varphi(n-1)\}$. ∎

PROBLEM 5: Show that if A is any infinite set, then there is a 1–1 function on A onto a proper subset of A. Hint: Let φ be one-to-one on $\mathbb{N}$ into A. Define $\psi(x) = x$ for x in A but not in the range of φ. Define ψ on the range of φ to be one-to-one onto a proper subset, using for example the existence of a one–one correspondence between $\mathbb{N}$ and the even numbers. •

PROBLEM 6: A set is finite if and only if it cannot be put in a one-to-one correspondence with a proper subset of itself. Hint: A proper subset of a finite set has fewer elements; that is, if $B \subset A$, $B \neq A$, A has n elements, and B has k elements, then $k < n$. •

A set is **uncountable**, or **uncountably infinite**, provided it is infinite but not countable.

PROPOSITION 7: *There is an uncountable set.*

Proof: Let S be the set of all sequences $x_1, x_2, \ldots$ such that each x_i is either 0 or 1. Suppose this set is countable, and let $s_1, s_2, s_3, \ldots$ be an enumeration of all such sequences; that is, each s_i is a distinct sequence all the terms of which are zero or one. If the terms of s_1 are $s_{11}, s_{12}, s_{13}, \ldots$, and so forth, then we can display *all* such sequences as follows:

$$s_{11}, s_{12}, s_{13}, s_{14}, \ldots$$
$$s_{21}, s_{22}, s_{23}, s_{24}, \ldots$$
$$s_{31}, s_{32}, s_{33}, s_{34}, \ldots$$
$$\cdots.$$

Now we obtain a contradiction by exhibiting a sequence that is different from all of those displayed. Let $t_1 = 0$ if $s_{11} = 1$ and let $t_1 = 1$ if $s_{11} = 0$; that is, let t_1 be different from s_{11}. Let t_2 be different from s_{22}. In general, define t_n to be whichever of 0 and 1 that s_{nn} is not. Then the sequence t differs from all of the sequences s_i, because t differs from s_i in the ith place. ∎

We will be able to show later that the real numbers can all be displayed as sequences of digits using the decimal representation. There are an uncountable number of such sequences by the same argument as that just used, and so $\mathbb{R}$ is uncountable. Similarly, the numbers in $[0, 1]$ can all be exhibited as decimals, so that this set also is uncountable.

PROBLEM 7: (i) Every nonempty subset of a finite set is finite.
(ii) Every nonempty subset of a countable set is finite or countable. •

PROBLEM 8: If A is finite or countable, and B is countable, then $A \cup B$ is countable. •

From Problem 8 it of course follows that any finite union of countable sets is countable. In fact, it is a remarkable result of Cantor that countable unions of countable sets are countable. The idea is this: Let S_1 be the first countable set, with elements $s_{11}, s_{12}, s_{13}, \ldots$, and similarly for $S_2 = \{s_{21}, s_{22}, \ldots\}$, $S_3 = \{s_{31}, s_{32}, \ldots\}$, and so forth. All the elements of $\bigcup\{S_n : n \in \mathbb{N}\}$ are displayed in the proof of Proposition 7. All these elements can be counted by counting along diagonals; thus, the sequence

$$s_{11}, s_{21}, s_{12}, s_{31}, s_{22}, s_{13}, s_{41}, s_{32}, s_{23}, s_{14}, \ldots$$

counts each s_{ij}. Now elements s_{ij} need not be distinct, so that the counting function indicated may not be one-to-one. We therefore need the following proposition.

PROPOSITION 8: *If φ is a function — not necessarily one-to-one — on $\mathbb{N}$ onto a nonempty set A, then A is finite or countable.*

Proof: Let A be nonempty and not finite, and assume that φ is a function on $\mathbb{N}$ onto A. Define a one-to-one function ψ on $\mathbb{N}$ onto A as follows: let $\psi(1) = \varphi(1)$. If $\varphi(2) \neq \varphi(1)$, let $\psi(2) = \varphi(2)$; if $\varphi(2) = \varphi(1)$, then consider the set $S_1 = \{k \in \mathbb{N} : \varphi(k) \neq \varphi(1)\}$. If $S_1 = \varnothing$, then A has only one element, contrary to our assumption. Let k_1 be the smallest number in S_1, and define $\psi(2) = \varphi(k_1)$. Assume that $\psi(1), \psi(2), \ldots, \psi(n)$ have been defined so that ψ is one-to-one on $\mathbb{N}_n$ into A, and the elements $\psi(1), \ldots, \psi(n)$ include all the elements $\varphi(1), \ldots, \varphi(n)$. Let

$$S_n = \{k \in \mathbb{N} : \varphi(k) \neq \psi(i) \text{ for } i = 1, \ldots, n\}.$$

Again, S_n must be nonempty since A is not finite, and we define $\psi(n+1) = \varphi(k_n)$, where k_n is the smallest number in S_n. This scheme provides an inductive definition of ψ on $\mathbb{N}$ into A, and ψ is one-to-one by its definition. The function ψ is onto A, since each $a \in A$ is $\varphi(j)$ for some j, and $\varphi(j)$ is certainly included among the numbers $\psi(1), \ldots, \psi(j)$.   ■

Finally, we will formalize the function that counts the following array diagonally:

$$s_{11}, s_{12}, s_{13}, s_{14}, \ldots$$
$$s_{21}, s_{22}, s_{23}, s_{24}, \ldots$$
$$s_{31}, s_{32}, s_{33}, s_{34}, \ldots$$
$$s_{41}, s_{42}, s_{43}, s_{44}, \ldots.$$

As indicated, we count these elements as follows:

$$s_{11}, s_{21}, s_{12}, s_{31}, s_{22}, s_{13}, s_{41}, s_{32}, s_{23}, s_{14}, \ldots.$$

We define a one-to-one function $\varphi(i, j)$ on pairs (i, j) onto $\mathbb{N}$. The nth diagonal (the elements (i, j) with $i + j = n + 1$) has n elements, and so there are $1 + 2 + \cdots + n =$

$\frac{1}{2}n(n + 1)$ elements in the first n diagonals. If s_{ij} is in the nth diagonal, so that $n + 1 = i + j$, then the index j is the number of s_{ij} beyond the number in the first $n = i + j - 1$ diagonals. There are $\frac{1}{2}(i + j - 2)(i + j - 1)$ elements in the first $n - 1$ diagonals, and so

$$\varphi(i, j) = \frac{1}{2}(i + j - 2)(i + j - 1) + j.$$

PROBLEM 9: Show that $\varphi(i, j)$ just defined is one-to-one on $\mathbb{N} \times \mathbb{N}$ onto $\mathbb{N}$. Hint: Show first that $\varphi(i, j) > \varphi(r, s)$ if $i + j > r + s$; that is, if $n > k$, $s < k$, and $j \geq 1$, then

$$\frac{1}{2}(n - 2)(n - 1) + j > \frac{1}{2}(k - 2)(k - 1) + s.$$

It suffices to show this for $n = k + 1$.   ●

We define the **rational numbers** to be the numbers of the form m/n for integers m and n.

PROBLEM 10: Prove that the set of rational numbers is countable.   ●

VI

LONG DIVISION AND PRIME FACTORIZATION

I n this chapter we develop a few more facts about integers. These facts are well known to every student, and classes with limited time are urged to omit a detailed study of this chapter. The point of including these theorems is to buttress our claim that there are no inadvertent lacunae in our logical development. The student should at least read the propositions and problems, for although the facts are well known, their proofs are not.

We end this chapter with a proof that there is no rational number whose square is 2. This makes clear the need for our final axiom.

We start with the *bête noire* of grammar school mathematics — division, or, more deferentially, long division.

PROPOSITION 1: *For every pair n and d (numerator and denominator) of natural numbers there is a unique pair q and r (quotient and remainder) of nonnegative integers such that*

$$q \leq \frac{n}{d} < q + 1, \tag{1}$$

and

$$n = qd + r, \tag{2}$$

and

$$0 \leq r < d. \tag{3}$$

In other words, for any n, d ∈ ℕ,

$$\frac{n}{d} = q + \frac{r}{d}$$

for a unique integer quotient q and integer remainder r < d.

Proof: Since $d \in \mathbb{N}$, $d \geq 1$ and consequently $n \cdot 1 \leq nd$ and $n/d \leq n$. The point of this apparent banality is to show there are natural numbers (e.g., $n + 1$) that are strictly larger than n/d. (More on this later.) The set of natural numbers that are strictly larger than n/d is nonempty, and so let s be the smallest such number, and let $q = s - 1$. Then $q = 0$ or $q \in \mathbb{N}$, and

$$s - 1 = q \leq \frac{n}{d} < q + 1 = s. \tag{4}$$

Let $r = n - qd$. From (4), $n \geq qd$, and since qd is a natural number or zero, $n - qd$ is zero or a natural number. Now we have (1) and (2). To show that $r < d$, use the second inequality in (4):

$$\frac{n}{d} < q + 1,$$

and so

$$r = n - qd < d.$$

The final statement of Proposition 1, that q and r are unique, is the content of the following problem. ∎

PROBLEM 1: Let n and d be two natural numbers. Show that if

$$n = q_1 d + r_1$$

and

$$n = q_2 d + r_2,$$

where q_1, r_1, q_2, r_2 are natural numbers or zero, and $0 \leq r_i < d$ for $i = 1, 2$, then $q_1 = q_2$ and $r_1 = r_2$. Hint: If $q_1 = q_2$, then obviously $r_1 = r_2$, so assume, to be specific, that $q_1 > q_2$ and hence $q_1 - q_2 \geq 1$. Reach a contradiction by showing that $r_2 - r_1 < d$, and $r_2 - r_1 = (q_1 - q_2)d > d$. •

There are a couple of things we can notice here about Proposition 1. First, we showed that the number n/d can be bracketed between a unique pair of natural numbers. Why, while we were at it, did we not show that every nonnegative number x can be so bracketed? Notice that part of the proof was the demonstration that there are natural numbers larger than n/d. We proved this using the fact that $d \in \mathbb{N}$. Can we not always assume that there is a natural number bigger than any x? The answer is no — not yet, and this is another very convincing argument that we need another axiom. There are ordered fields — that is, systems $(\mathbb{F}, +, \cdot, <)$

satisfying all our axioms so far — in which there are numbers bigger than all natural numbers (see Problem 5). Recall that in all these discussions the word "number" simply refers to any element of any set satisfying the axioms. We will need the final axiom, the Completeness Axiom, to show that $\mathbb{N}$ is not a bounded set.

Now we show that every natural number has the decimal representation we introduced in the last chapter.

PROPOSITION 2: *If* $n \in \mathbb{N}$, *then there are digits* d_0, d_1, ..., d_m *such that* $d_m \neq 0$ *and*

$$n = d_m \ldots d_1 d_0 = d_m t^m + \cdots + d_1 t + d_0,$$

and this representation is unique.

Proof: For any natural number n there is a nonnegative integer m so that $t^m \leq n < t^{m+1}$ (see Problem 2). A complete proof of the proposition depends on a messy induction on m, and the essential idea is apt to get lost in the details. We will therefore prove the result for all n such that $t^3 \leq n < t^4$. Assume, then, that $t^3 \leq n < t^4$, and use the long division proposition with numerator n and denominator t^3 to write

$$n = q_3 t^3 + r_3,$$

with $0 \leq r_3 < t^3$ and $q_3 \in \mathbb{N}$. Since $n < t^4$, $q_3 < t$, and hence q_3 is one of the digits 1, 2, ..., 9 ($n \geq t^3$, so that $q_3 \neq 0$). The same proposition with numerator r_3 and denominator t^2 gives

$$r_3 = q_2 t^2 + r_2,$$

with $0 \leq r^3 < t^2$, and q_2 one of the digits 0, 1, ..., 9. Finally, with r_2 as the numerator and t the denominator,

$$r_2 = q_1 t + r_1,$$

with $0 \leq r_1 < t$ and $q_1 < t$. Putting these equations together, we get

$$\begin{aligned} n &= q_3 t^3 + r_3 \\ &= q_3 t^3 + q_2 t^2 + r_2 \\ &= q_3 t^3 + q_2 t^2 + q_1 t + r_1, \end{aligned}$$

and all of q_3, q_2, q_1, r_1 are digits, with $q_3 \neq 0$.   ∎

PROBLEM 2: Show that if n is any natural number, there is a unique integer $m \in \mathbb{N} \cup \{0\}$ such that $t^m \leq n < t^{m+1}$. Hint: Use induction on n as follows: If $n < t^k$, then $n + 1 \leq 2n < t^{k+1}$. Hence for every n, there is k such that $n < t^k$. Let $m + 1$ be the smallest of these numbers k. There are lots of details missing from this skeleton.   •

PROBLEM 3: Show that if $n \in \mathbb{N}$ and $t^3 \leq n < t^4$, then the representation $n = d_3 d_2 d_1 d_0$ of Proposition 2 is unique. •

PROBLEM 4: Use Proposition 2, for $t^3 \leq n < t^4$, to prove Proposition 2 for n such that $t^4 \leq n < t^5$. This step is the essence of the general inductive proof. •

PROBLEM 5: Here is an example of an ordered field $(\mathbb{F}, +, \cdot, >)$ in which $\mathbb{N}$ is a bounded set, that is, in which there are numbers larger than all natural numbers. Let $\mathbb{F}$ consist of all rational functions $P(x)/Q(x)$, with common factors removed, and further normalized by making the leading coefficient of $Q(x)$ equal 1. The elements of this field — the numbers — look like this:

$$\frac{P(x)}{Q(x)} = \frac{a_n x^n + \cdots + a_1 x + a_0}{x^m + b_{m-1} x^{m-1} + \cdots + b_1 x + b_0}.$$

Addition and multiplication in $\mathbb{R}$ are the usual algebraic operations, and so they satisfy the algebraic axioms. Define $P(x)/Q(x) > 0$ if $a_n > 0$. Define $P_1(x)/Q_1(x) > P_2(x)/Q_2(x)$ to mean $P_1(x)/Q_1(x) - P_2(x)/Q_2(x) > 0$. Thus, for example,

$$\frac{3x^2 - 200x - 1000}{x^3 - x} > 0,$$

and you can check that

$$\frac{x^3 + 1}{x + 5} > \frac{2x^5 + 3}{x^4 - 1},$$

$$\frac{2x^3 + 1}{x + 5} > \frac{x^4 + 3}{x^2 - 1}.$$

Show that the order axioms are satisfied. Specify the elements of $\mathbb{N}$. Show that $x/1 > P(x)/Q(x)$ for all $P(x)/Q(x) \in \mathbb{N}$. •

To see how to get at decimal representations like $\frac{1}{4} = .25$, $\frac{1}{100} = .01$, and so on, we consider any number x such that $0 < x < 1$. Then $0 < 10x < 10$, and there is a unique digit d such that

$$d_1 \leq 10x < d_1 + 1.$$

It follows that

$$\frac{d_1}{10} < x < \frac{d_1}{10} + \frac{1}{10},$$

and hence

$$0 \leq x - \frac{d_1}{10} < \frac{1}{10}.$$

Similarly, since

$$0 \leq 100x - 10d_1 < 10,$$

there is a unique digit d_2 such that

$$d_2 \leq 100x - 10d_1 < d_2 + 1,$$

$$0 \leq x - \frac{d_1}{10} - \frac{d_2}{100} < \frac{1}{100}.$$

Continuing, we find, for every n, digits $d_1, d_2, \ldots, d_n$ such that

$$0 \leq x - \frac{d}{10} - \frac{d_2}{100} - \cdots - \frac{d_n}{10^n} < \frac{1}{10^n}.$$

For any digits $d_1, \ldots, d_n$, we define

$$.d_1 d_2 \ldots d_n = \frac{d_1}{10} + \frac{d_2}{100} + \cdots + \frac{d_n}{10^n},$$

so that for any $x \in [0, 1)$ and any n we have digits $d_1, \ldots d_n$ such that

$$0 \leq x - .d_1 d_2 \ldots d_n < \frac{1}{10^n}.$$

If all d_j are eventually zero, then we have a decimal representation for x. It can also happen that the process does not terminate; for example, if $x = \frac{1}{3}$, all $d_j = 3$. Although we can agree here that $.333\ldots$ is another way of writing $\frac{1}{3}$, we have no way to interpret an arbitrary nonterminating decimal until we discuss limits of sequences.

Next we show that natural numbers can be factored into prime factors, and that 2 has no rational square root.

A natural number p is **prime** provided $p > 1$ and p has no divisors other than 1 and p; that is, there are no natural numbers r and s with $1 < r < p$ and $1 < s < p$ such that $p = rs$.

PROPOSITION 3: *Every natural number bigger than 1 is prime or is a product of primes.*

Proof: For brevity we will interpret "n is a product of primes" to include the case where n is a prime; i.e., a prime is a "product" of one prime. If there are natural numbers larger than one that are not products of primes, then there is a smallest such number, n. Since n is not a prime, it has nontrivial factors: $n = rs$ for some r and s with $1 < r < n$ and $1 < s < n$. Both r and s are products of primes because n is the smallest number that is not, so that $n = rs$ is a product of primes, and we have a contradiction. ∎

To show that the factorization of any natural number into primes is unique, we first need the following result.

PROPOSITION 4: *If p is a prime and p divides a product mn of natural numbers, then p divides m or p divides n.*

Proof: (This clever and difficult proof is due to Kenneth Rogers.) Suppose there are primes p for which the proposition fails. We will — temporarily — call such primes p bad primes, and if p divides mn but p divides neither m nor n, we will call m and n bad factors for p. Now let p be the smallest bad prime, and let m be the smallest bad factor for p, and let n be its companion bad factor. That is, p divides mn and p divides neither m nor n, and if p divides $m'n'$ for some m', $n' \in \mathbb{N}$ and p divides neither m' nor n', then $m' \geq m$ and $n' \geq m$. We will derive a contradiction, thereby proving the theorem, by finding a bad factor m' for p with $m' < m$.

First, note that m cannot be larger than p, for then p would not divide $m - p$, but would divide $(m - p)n$, and so $m - p$ would be a bad factor for p smaller than m. Of course $m = p$ and $m = 1$ are not possible, and so if m is the smallest bad factor for p, then $1 < m < p$. By Proposition 3, m is a product of primes, and we let q be one of these prime factors of m, so that $m = qm'$. Notice that $q \leq m$, so that $q < p$. Since p divides mn, there is $k \in \mathbb{N}$ such that

$$kp = mn = qm'n.$$

From this last equation we see that q divides kp. Since $q < p$, and p is the smallest bad prime, q must divide k or p. Of course, q cannot divide p because p is prime, and so q divides k; let

$$k = k'q.$$

Now we have

$$k'qp = qm'n,$$

$$k'p = m'n,$$

and p divides $m'n$, with $m' < m$. This contradicts the assumption that m was the smallest bad factor for p, since p clearly does not divide m', since $m' < m < p$, and p still does not divide n. ∎

PROBLEM 6: If p divides a product $q_1 \cdots q_m$ of primes, then $p = q_i$ for some i. Hint: Use the preceding proposition and induction on m.   •

PROBLEM 7: If $p_1, \ldots, p_n$ and $q_1, \ldots, q_m$ are primes, and $p_1 \cdots p_n$ divides $q_1 \cdots q_m$, then each p_i is some q_j; that is, there is a one-to-one function φ on $\mathbb{N}_n$ into $\mathbb{N}_m$ such that $p_i = q_{\varphi(i)}$ for each $i \leq n$.   •

PROPOSITION 5: *The factorization of any natural number into primes is unique; that is, if $N = p_1 \cdots p_n = q_1 \cdots q_m$ for primes p_i and q_j, then $n = m$ and there is a one-to-one function φ on $\mathbb{N}_n$ onto $\mathbb{N}_n$ such that $p_i = q_{\varphi(i)}$ for each $i \leq n$.*

Proof: This is immediate from Problem 7. ∎

Finally, we show that 2 is not the square of any rational number m/n. Suppose there were natural numbers m and n such that $(m/n)^2 = 2$. Let us assume that m

is the smallest possible numerator, and that n is its companion. Then

$$m^2 = 2n^2,$$

so that 2 divides m^2, and by Proposition 4, 2 divides m. Let $m = 2r$, so that we have

$$m^2 = 4r^2 = 2n^2,$$
$$2r^2 = n^2,$$

and 2 divides n^2 and, therefore, n. Let $n = 2s$, so that

$$2r^2 = (2s)^2 = 4s^2,$$
$$r^2 = 2s^2,$$
$$2 = (r/s)^2.$$

Since $r < m$, and m was assumed to be the smallest possible numerator, we have a contradiction.

PROBLEM 8: Show that 3 is not the square of any rational number. Generalize.   •

We will show later that every positive number has a square root, cube root, and in general, an nth root. This will follow quite easily from the properties of continuous functions ($x^2, x^3, \ldots, x^n$ are continuous functions), but we obviously need another axiom first — the rational numbers satisfy all our axioms so far, and in this ordered field, 2, 3, and many other numbers do not have a square root.

Define a natural number to be **even** if it is divisible by 2, and **odd** otherwise.

PROBLEM 9: Show that a natural number n is odd if and only if it has the form $n = 2q + 1$ for some $q \in \mathbb{N} \cup \{0\}$.   •

PROBLEM 10: (i) A natural number n is even if and only if $n + 1$ is odd.
(ii) The sum of two even numbers is even.
(iii) The sum of two odd numbers is even.
(iv) The sum of an even and an odd number is odd.
(v) The product of two odd numbers is odd.   •

PROBLEM 11: Show that 5 divides the natural number $d_n d_{n-1} \ldots d_1 d_0$, where the d_i are digits, if and only if $d_0 = 0$ or $d_0 = 5$.   •

PROBLEM 12: Show that 3 divides $d_n d_{n-1} \ldots d_1 d_0$ if and only if 3 divides $d_0 + d_1 + \cdots + d_n$.   •

PROBLEM 13: Show there is no rational number m/n so that $(m/n)^2 = 6$.   •

PROBLEM 14: (i) Show that, for every $n \geq 1$,

$$10^n = 9 \cdot 10^{n-1} + 9 \cdot 10^{n-2} + \cdots + 9 \cdot 10 + 9 + 1.$$

(ii) Show that if d_0, d_1, d_2, d_3, d_4 are digits, then 9 divides $d_4d_3d_2d_1d_0$ if and only if 9 divides $d_0 + d_1 + d_2 + d_3 + d_4$. (The result holds of course for any number of digits, but the notation gets in the way in the general case.)   •

PROBLEM 15: The numbers d_1d_0 that are divisible by 11 have the property that $d_0 - d_1 = 0$. The numbers $d_2d_1d_0$ that are divisible by 11 (namely, 110, 121, 132, ...) have the property that $d_0 - d_1 + d_2$ is 0 or 11. Prove a general result. Hint: $d_3d_2d_1d_0 = d_3(11 - 1)^3 + d_2(11 - 1)^2 + d_1(11 - 1) + d_0$.   •

PROBLEM 16: Show that if $k \in \mathbb{N}$ and k has a square root, $\sqrt{k}$, and j is a natural number such that $j < \sqrt{k} < j + 1$, then $\sqrt{k}$ is not rational. Hint: If $\sqrt{k} = m/n$ and n is the smallest possible denominator, then $\sqrt{k} = (kn - jm)/(m - jn)$, and $m - jn$ is a smaller positive integer.   •

PROBLEM 17: Show there is no largest prime. Hint: If p is the largest prime, then $p! + 1$ has prime factors. How big are they?   •

PROBLEM 18: Notice that $2^2 - 1$, $2^3 - 1$, $2^5 - 1$ are prime, but $2^4 - 1$ is not. Show that $2^n - 1$ is not prime unless n is prime. Hint: First check that

$$a^n - b^n = (a - b)\left(a^{n-1} + a^{n-2}b + \cdots + ab^{n-2} + b^{n-1}\right).$$ •

VII

THE COMPLETENESS AXIOM
AND SEQUENCES

Everything we have done so far is algebra — very elementary algebra. We now introduce our final axiom and commence the vastly more interesting study of analysis. First, we need a few definitions.

A set S of numbers is **bounded above** provided there is a number B such that $x \leq B$ for all $x \in S$. A set is **bounded below** provided there is a number smaller than all its elements, and a set is **bounded** if it is both bounded above and bounded below. If $B \geq x$ for all $x \in S$, then B is an **upper bound** for S, and if $B \leq x$ for all $x \in S$, then B is a **lower bound** for S.

AXIOM IX (The Completeness Axiom): *If a set $S \subset \mathbb{R}$ is nonempty and bounded above, then there is a smallest upper bound for S.*

The smallest upper bound for a set is obviously unique, since there cannot be two smallest elements in the set of upper bounds. We denote the smallest upper bound for S by **sup** S. Here "sup" is short for **supremum**, which we also use in certain literary contexts. The definition of sup S can be rephrased in the following working version: sup $S = B$ if and only if B is an upper bound for S, and for every $\varepsilon > 0$, there is $x \in S$ such that $B - \varepsilon < x \leq B$. The existence of such x is just the statement that $B - \varepsilon$ is *not* an upper bound for S, no matter how small ε is, as long as ε is positive. If S happens to have a largest element x_0, then clearly sup $S = x_0$, but in this case we would not need the extra nomenclature. The more usual and interesting situation is that in which S is bounded above but has no maximum element. In this case, $\sup S$ is a number whose existence or nonexistence might well be moot. The first problem illustrates this idea by using the existence of $\sup S$ for suitable S to show that 2 has a square root. Recall that the rational numbers — the numbers of the form m/n for integers m and n — satisfy all of Axioms I through VIII, but we showed that in this field there is no number whose square is 2.

PROBLEM 1: Let $S = \{x : 0 \le x \text{ and } x^2 < 2\}$. Show that S is nonempty and bounded above, and let $r = \sup S$. To show that $r^2 = 2$, first assume that $r^2 < 2$, and reach a contradiction by showing that $(r + e/10)^2 < 2$, where $e = 2 - r^2$. Thus $(r + e/10) \in S$, contradicting the definition of r as an upper bound for S. To show that $r^2 > 2$ is also untenable, assume that $e = r^2 - 2 > 0$, and show that $(r - e/4)^2 > 2$. This implies (why?) the contradiction that $x^2 > 2$ for some $x \in S$.   •

The preceding problem was given to emphasize that the existence of many numbers — indeed *most* numbers — depends exactly on the Completeness Axiom. We will be able to show quite easily and quite generally that all positive numbers have square roots, cube roots, and so forth, after we develop the properties of continuous functions, but those properties themselves depend on the Completeness Axiom.

PROBLEM 2: If $S \ne \varnothing$ and S is bounded below, then $S^* = \{-x : x \in S\}$ is bounded above. If $B = \sup S^*$, then $-B$ is the largest lower bound for S.   •

Problem 2 shows that nonempty sets that are bounded below have a largest lower bound. This obviously unique largest lower bound for S is denoted **inf** S, where "inf" is an abbreviation for **infimum**. If S is not bounded above, we write $\sup S = \infty$, and if S is not bounded below, we write $\inf S = -\infty$. Here "∞" and "$-\infty$" are not the names of numbers, and these symbols should not be construed as the basis for some new religion.

Now we proceed to the critical result that there are arbitrarily large natural numbers, and arbitrarily small positive numbers of the form $1/n$ for $n \in \mathbb{N}$. Recall that this essential property of the real numbers does not hold in all ordered fields, since we showed that in the field of rational functions

$$\frac{a_n x^n + \cdots + a_0}{x^m + \cdots + b_0}$$

the "number" $\dfrac{x}{1}$ is larger than every "natural number" $\dfrac{m}{1}$. The Completeness Axiom thus rules out some ordered fields that are too small (like the rationals), and rules out some ordered fields that are too big (like the rational functions), and leaves one ordered field, the reals, that is just right. From now on, we assume all nine axioms, and that all sets of numbers are subsets of $\mathbb{R}$, the field of real numbers.

PROPOSITION 1: *(i)* $\mathbb{N}$ *is not bounded above.*

(ii) For every number $\varepsilon > 0$, there is a natural number n so that $1/n < \varepsilon$.

Proof: Assume that $\mathbb{N}$ is bounded above, and let $B = \sup \mathbb{N}$. Then $B - 1/2$ is not an upper bound for $\mathbb{N}$, so that there is $n \in \mathbb{N}$ with

$$B - \frac{1}{2} < n \leq B.$$

It follows that

$$B < B + \frac{1}{2} < n + 1,$$

which, since $n + 1 \in \mathbb{N}$, contradicts the fact that B is an upper bound for $\mathbb{N}$. Part (ii) is contained in the following problem. ∎

PROBLEM 3: (i) For every $\varepsilon > 0$, there is $n \in \mathbb{N}$ so $1/n < \varepsilon$.
(ii) Show that for any given $\varepsilon > 0$, there is $N \in \mathbb{N}$ so that

$$N/\left(2N^2 - 1\right) < \varepsilon.$$

(iii) Show that if $n \in \mathbb{N}$, $n \geq N$, and $N/\left(2N^2 - 1\right) < \varepsilon$, then

$$n/\left(2n^2 - 1\right) < \varepsilon.$$

Hint: You can use differentiation (or witchcraft) to discover that $n/\left(2n^2 - 1\right)$ is decreasing, but obviously you cannot use differentiation (or witchcraft) to prove it, since we have not yet proved anything about either differentiation or witchcraft. •

A **sequence** is a row of numbers like this:

$$x_1, x_2, x_3, \ldots, x_n, \ldots.$$

A sequence has a first number, a second number, and an nth number for every $n \in \mathbb{N}$. We can formalize this by saying that a sequence, x, is a function on $\mathbb{N}$, and we denote the values of this sequence with subscripts; thus x_n, rather than $x(n)$, is the value of the sequence at n. We will use $\{x_n\}$ and other familiar notations to indicate sequences. Frequently, we define $\{x_n\}$ by giving a formula for the nth term: for example,

$$x_n = \frac{n}{2n^2 - 1},$$

it being understood that the domain of this function (i.e., the relevant set of n's) is the set $\mathbb{N}$ of all natural numbers.

We say that the sequence $\{x_n\}$ **converges to** ℓ, written $x_n \longrightarrow \ell$ as $n \longrightarrow \infty$, if for each positive number ε there is a number N such that $|x_n - \ell| < \varepsilon$ for all natural numbers $n \geq N$. The notation "$n \longrightarrow \infty$" in "$x_n \longrightarrow \ell$ as $n \longrightarrow \infty$" is largely honorary, since we know how n runs. Therefore, we will generally write just "$x_n \longrightarrow \ell$." A sequence $\{x_n\}$ **converges** if $x_n \longrightarrow \ell$ for some ℓ; otherwise, $\{x_n\}$ **diverges**.

It is immediate that $x_n \longrightarrow \ell$ if and only if $x_n - \ell \longrightarrow 0$ if and only if $|x_n - \ell| \longrightarrow 0$.

PROBLEM 4: Write out the defining condition for both: (i) $x_n - \ell \longrightarrow 0$; and (ii) $|x_n - \ell| \longrightarrow 0$. Verify that $|x_n - \ell| < \varepsilon$ is equivalent to $\ell - \varepsilon < x_n < \ell + \varepsilon$. •

PROBLEM 5: (i) Write out an ℓ, ε, N condition for "$\{x_n\}$ diverges." Hint: $\{x_n\}$ diverges if and only if for every ℓ, $\{x_n\}$ does not converge to ℓ.

(ii) Show that if $x_n = n$ for all n, $\{x_n\}$ diverges.

(iii) Show that if $x = (-1)^n$ for all n, $\{x_n\}$ diverges. •

If a sequence converges, then it has exactly one limit.

PROPOSITION 2: *If $x_n \longrightarrow \ell$ as $n \longrightarrow \infty$ and $x_n \longrightarrow m$ as $n \longrightarrow \infty$, then* $\ell = m$.

Proof: If $x_n \longrightarrow \ell$, $x_n \longrightarrow m$, and $\ell \neq m$, then, to be specific, assume $\ell < m$ and let $\varepsilon = (m - \ell)/2 > 0$. Then $\ell + \varepsilon = (\ell + m)/2 = m - \varepsilon$. Since $x_n \longrightarrow \ell$ as $n \longrightarrow \infty$, there is N_1 such that $x_n < \ell + \varepsilon$ if $n \geq N_1$. (Here we need just part of the information in $x_n \longrightarrow \ell$.) Since $x_n \longrightarrow m$ as $n \longrightarrow \infty$, there is N_2 such that $m - \varepsilon < x_n$ if $n \geq N_2$. If N is the larger of N_1 and N_2, then both these inequalities hold for $n \geq N$, and in particular for $n = N$, so that we have the absurdity

$$\frac{\ell + m}{2} = m - \varepsilon < x_N < \ell + \varepsilon = \frac{\ell + m}{2}.$$ ∎

Since limits are unique we can write $\lim_{n \to \infty} x_n = \ell$ if $x_n \longrightarrow \ell$. When convenient we will omit the "$n \longrightarrow \infty$" from underneath "lim," and we will write simply $\lim x_n = \ell$.

PROBLEM 6: (i) Show that $1/n \longrightarrow 0$.

(ii) Show that $1/2^n \longrightarrow 0$. Hint: To show $\{2^n : n \in \mathbb{N}\}$ is not bounded above, assume that $B = \sup\{2^n : n \in \mathbb{N}\}$, so that $B/2$ is not an upper bound if $B \neq 0$. •

The convergence of a sequence has nothing to do with the first million terms, or the first ten-to-the-million terms. The tail is the only interesting part of a sequence.

PROBLEM 7: If $x_n \longrightarrow \ell$ and $x_n - y_n \longrightarrow 0$, then $y_n \longrightarrow \ell$. If there is N such that $x_n = y_n$ for all $n \geq N$, and $x_n \longrightarrow \ell$, then $y_n \longrightarrow \ell$. •

PROBLEM 8: If $x_n \longrightarrow \ell \neq 0$, then there is N so that $x_n \neq 0$ for all $n \geq N$. •

We will say that a **sequence** $\{x_n\}$ **is bounded** (or **bounded above**, or **bounded below**), provided the *set* $\{x_n : n \in \mathbb{N}\}$ is bounded (or bounded above, or bounded below).

PROPOSITION 3: *A convergent sequence is bounded.*

Proof: Assume $x_n \longrightarrow x_0$ and let $\varepsilon = 1$. Corresponding to this ε there is some N so that

$$x_0 - 1 < x_n < x_0 + 1$$

for all $n \geq N$. The first N terms have a largest element and a smallest element, and so obviously form a bounded set. The rest of the x_n lie between $x_0 - 1$ and $x_0 + 1$, and so the set of all x_n is bounded. ∎

PROPOSITION 4: *If $x_n \longrightarrow x_0$ and $y_n \longrightarrow y_0$, then $x_n + y_n \longrightarrow x_0 + y_0$ and $x_n y_n \longrightarrow x_0 y_0$.*

Proof: Suppose $x_n \longrightarrow x_0$ and $y_n \longrightarrow y_0$. Let us think of x_n as the nth approximation to x_0, and let $e_n = x_n - x_0$ be the error in this approximation. Similarly, let $e'_n = y_n - y_0$, so that $e_n \longrightarrow 0$ and $e'_n \longrightarrow 0$. When you add the two approximations, $x_n + y_n$, the errors at worst add up; thus,

$$(x_n + y_n) - (x_0 + y_0) = e_n + e'_n,$$
$$|(x_n + y_n) - (x_0 + y_0)| = |e_n + e'_n| \leq |e_n| + |e'_n|.$$

To show that the error between $x_n + y_n$ and the limit $x_0 + y_0$ is less than ε beyond some Nth term, we have only to guarantee that both errors $|e_n|$ and $|e'_n|$ are less than $\varepsilon/2$ beyond the Nth term. So, formally, let ε be any positive number. Since $e_n \longrightarrow 0$, there is N_1 such that $|e_n| = |e_n - 0| < \varepsilon/2$ if $n \geq N_1$. Similarly, there is N_2 such that $|e'_n| < \varepsilon/2$ if $n \geq N_2$. If N is the larger of N_1 and N_2, then both inequalities hold if $n \geq N$; namely,

$$|(x_n + y_n) - (x_0 + y_0)| \leq |e_n| + |e'_n|$$
$$< \frac{\varepsilon}{2} + \frac{\varepsilon}{2} = \varepsilon$$

if $n \geq N$, and hence $(x_n + y_n) \longrightarrow (x_0 + y_0)$.

We prove the product limit in the same way, using the fact that $\{e'_n\}$ is a bounded sequence by Proposition 3. Assume that $|e'_n| < B$ for all n. Let $\varepsilon > 0$. If $x_n = x_0 + e_n$, $y_n = y_0 + e'_n$, then

$$x_n y_n = (x_0 + e_n)\left(y_0 + e'_n\right)$$
$$= x_0 y_0 + e_n y_0 + e'_n x_0 + e_n e'_n,$$
$$|x_n y_n - x_0 y_0| \leq |e_n y_0| + |e'_n x_0| + |e_n e'_n|.$$

We have to show that the right side is less than ε if n is large enough. Pick N_1 such that $|e_n| < \varepsilon/(3|y_0|)$ if $n \geq N_1$; pick N_2 such that $|e'_n| < \varepsilon/(3|x_0|)$ if $n \geq N_2$; and pick N_3 such that $|e_n| < \varepsilon/(3B)$ if $n \geq N_3$. If $x_0 = 0$, so that $\varepsilon/(3|x_0|)$ makes no sense, let $N_1 = 1$, and similarly let $N_2 = 1$ if $y_0 = 0$. Let N be the largest of N_1, N_2, N_3, so that if $n \geq N$, all three inequalities hold. Thus if $n \geq N$,

$$|x_n y_n - x_0 y_0| \leq |e_n| \, |y_0| + |e_n'| \, |x_0| + |e_n| \, |e_n'|$$

$$< \frac{\varepsilon}{3|y_0|} \cdot |y_0| + \frac{\varepsilon}{3|x_0|} |x_0| + \frac{\varepsilon}{3B} B$$

$$= \varepsilon. \qquad\blacksquare$$

PROBLEM 9: Write out in detail the proof of the following special case of Proposition 4: If $x_n \longrightarrow x_0$ and c is any number, then $cx_n \longrightarrow cx_0$. Hint: Consider two cases, $c = 0$ and $c \neq 0$. Show as a corollary of this and Proposition 4 that if $x_n \longrightarrow x_0$ and $y_n \longrightarrow y_0$, then $x_n - y_n \longrightarrow x_0 - y_0$. •

Now we know that addition, subtraction, and multiplication behave in the predictable way as far as sequences are concerned. Next we show that limits of quotients are quotients of limits.

PROPOSITION 5: *If* $x_n \longrightarrow x_0$ *and* $x_0 \neq 0$, *then* $x_n^{-1} \longrightarrow x_0^{-1}$.

Proof: Of course many of the terms x_n may be zero, so that many of the terms x_n^{-1} may be undefined. However, as we have seen, it is only the tail of the sequence $\{x_n^{-1}\}$ we are interested in. Since we know from Problem 8 that there is N such that $x_n \neq 0$ for $n \geq N$, x_n^{-1} exists for $n \geq N$, and $\{x_n^{-1}\}$ has a well-defined tail. In the following argument, we will consider only those n for which x_n^{-1} exists.

Assume that $x_n \longrightarrow x_0 \neq 0$ and let $\varepsilon > 0$. Let $x_n \neq 0$ for $n \geq N$. If we let $e_n = x_n - x_0$, then, for $n \geq N$,

$$\frac{1}{x_n} - \frac{1}{x_0} = \frac{1}{x_0 + e_n} - \frac{1}{x_0}$$

$$= \frac{-e_n}{x_0(x_0 + e_n)}.$$

Pick N such that $|e_n| < |x_0|/2$ if $n \geq N_1$ and, consequently, $|x_0 + e_n| > |x_0|/2$ if $n \geq N_1$. Now pick N_2 such that if $n \geq N_2$,

$$|e_n| < \varepsilon \frac{|x_0|^2}{2}.$$

Let N_3 be the largest of N, N_1, and N_2, so that x_n^{-1} exists and e_n satisfies both of the foregoing inequalities if $n \geq N_3$. Thus,

$$\left| \frac{1}{x_n} - \frac{1}{x_0} \right| = |e_n| \cdot \frac{1}{|x_0| \, |x_0 + e_n|}$$

$$< \frac{\varepsilon |x_0|^2}{2} \frac{1}{|x_0| \frac{|x_0|}{2}} = \varepsilon. \qquad\blacksquare$$

To show that the limit of a sum, difference, or product is the sum, difference, or product of the limits, we did not need to make any assumptions about the limits x_0 and y_0. We saw in Proposition 5 that things are different for division, because of its aversion to zero.

PROBLEM 10: State and prove, using Propositions 4 and 5, a theorem for $\lim(x_n/y_n)$.   •

A **subsequence** of a sequence $\{x_n\}$ is a sequence $\{y_k\}$ the terms of which are some of the terms of $\{x_n\}$, chosen in the order in which they occur in $\{x_n\}$. That is, $y_1 = x_{n_1}$ for some n_1, $y_2 = x_{n_2}$ for some $n_2 > n_1$, and so on. Formally, $\{y_k\}$ is a subsequence of $\{x_n\}$ provided there is a strictly increasing sequence $\{n_k\}$ of natural numbers such that $y_k = x_{n_k}$ for each k. Notice that $\{x_n\}$ is a subsequence of itself, since $n_k = k$ is a possibility. We will systematically use the notation $\{x_{n_k}\}$ to denote a subsequence of $\{x_n\}$, so that in this notation $\{n_k\}$ is always assumed to be strictly increasing.

PROBLEM 11: Show that $\{x_n\}$ converges if and only if every subsequence $\{x_{n_k}\}$ converges.   •

PROBLEM 12: If $\{x_n\}$ is a sequence with no largest or smallest term, and $b = \inf\{x_n : n \in \mathbb{N}\}$, $B = \sup\{x_n : n \in \mathbb{N}\}$, then there are subsequences $\{x_{n_k}\}$, $\{x_{m_k}\}$ such that $x_{n_k} \longrightarrow b$, $x_{m_k} \longrightarrow B$. Hint: Find a *subsequence*, not just a sequence in the set $\{x_n : n \in \mathbb{N}\}$.   •

An interval is by definition one of the following kinds of set: $[a, b]$, (a, b), $[a, b)$, $(a, b]$, $[a, \infty)$, (a, ∞), $(-\infty, b]$, $(-\infty, b)$, $(-\infty, \infty)$. We will find the characterization given in the following proposition useful in the next chapter.

PROPOSITION 6: *I is an interval if and only if I is a nonempty set with the following property: if x, y, z are any numbers such that $x \in I$, $z \in I$, and $x < y < z$, then $y \in I$.*

Proof: It is obvious that all intervals have the given property, and so we assume that I is such a set, and show that I has one of the above forms. We consider the case that I is bounded above but not below, and we leave one of the remaining cases as a problem. Since I is bounded above, there is a smallest upper bound, and we let $b = \sup I$. We will show that $(-\infty, b) \subset I$. Let $y < b$. Since I is not bounded below, y is not a lower bound for I, and there exists a number $x \in I$ with $x < y$. Since $b = \sup I$, and $y < b$, y is not an upper bound for I, so that there is $z \in I$ with $y < z$. Now we have, for any $y < b$, numbers x and z in I with $x < y < z$. It follows that $y \in I$ for every $y < b$, and so $(-\infty, b) \subset I$. Clearly, no number larger than b is in I, so that $I = (-\infty, b)$ or $I = (-\infty, b]$ depending on whether or not $b \in I$.   ∎

PROBLEM 13: Prove Proposition 6 in the case that I is bounded above and below.   •

PROBLEM 14: Assume $\{x_n\}$, $\{y_n\}$, $\{z_n\}$ are sequences such that $x_n \leq y_n \leq z_n$ for all n, and $x_n \longrightarrow \ell$, $z_n \longrightarrow \ell$. Show that $y_n \longrightarrow \ell$. Deduce, as a special case, that if $z_n \longrightarrow 0$ and $|y_n| \leq z_n$ for all n, then $y_n \longrightarrow 0$.   •

PROBLEM 15: If $x_n \longrightarrow 0$ and $\{y_n\}$ is bounded, then $x_n y_n \longrightarrow 0$.   •

PROBLEM 16: (i) If $x_n \leq y_n$ for all n and $x_n \longrightarrow x_0$, $y_n \longrightarrow y_0$, then $x_0 \leq y_0$.
(ii) If $a \leq x_n \leq b$ for all n, and $x_n \longrightarrow x_0$, then $a \leq x_0 \leq b$. Hint: Do not work very hard on (ii); just explain why (ii) follows immediately from (i).   •

In the sequel, we will frequently use the preceding result without comment: if a sequence in $[a, b]$ converges, then the limit is in $[a, b]$.

PROBLEM 17: If $x_n \longrightarrow x_0$, then $|x_n| \longrightarrow |x_0|$.   •

PROBLEM 18: (i) Show that if $e_n \longrightarrow 0$, and $a_n = (e_1 + \cdots + e_n)/n$ for each n, then $a_n \longrightarrow 0$. Hint: If $|e_n| < \varepsilon/2$ for $n \geq N$, then $a_n = (e_1 + \cdots + e_N)/n + (e_{N+1} + \cdots + e_n)/n$ and $\left|(e_{N+1} + \cdots + e_n)/n\right| < \varepsilon/2$.
(ii) Show that if $x_n \longrightarrow \ell$ and $a_n = (x_1 + \cdots + x_n)/n$ for each n, then $a_n \longrightarrow \ell$. Hint: Let $x_n = \ell + e_n$, so that $e_n \longrightarrow 0$.   •

We will say that $x_n \longrightarrow +\infty$ if and only if for each number M (presumably a large number) there is N such that $x_n \geq M$ for all $n \geq N$. We also write $\lim x_n = +\infty$ in this case. The statements $x_n \longrightarrow -\infty$ or $\lim x_n = -\infty$ are defined similarly. If $x_n \longrightarrow \infty$ or $x_n \longrightarrow -\infty$, then $\{x_n\}$ is obviously divergent, since convergent sequences are bounded.

PROBLEM 19: Show that if $\{x_n\}$ is not bounded above, then $x_{n_k} \longrightarrow +\infty$ for some subsequence $\{x_{n_k}\}$.   •

PROBLEM 20: Let A and B be any two bounded nonempty sets of numbers, and let $A + B = \{a + b : a \in A, b \in B\}$. Show that $\sup(A+B) = \sup A + \sup B$. •

PROBLEM 21: Let $\{x_n\}$, $\{y_n\}$ be two bounded sequences with $\sup_n x_n = A$, $\sup_n y_n = B$. Show that $\sup_n(x_n + y_n) \leq A + B$, and give an example where $<$ holds.   •

PROBLEM 22: Show that $\left\{\left(1 + \dfrac{1}{n}\right)^n\right\}$ is an increasing sequence, that is, show that $\left(1 + \dfrac{1}{n+1}\right)^{n+1} \geq \left(1 + \dfrac{1}{n}\right)^n$ for all $n \in \mathbb{N}$. Hint: Write out the binomial expansions for n and for $n + 1$, and compare terms.   •

PROBLEM 23: Give an example of two divergent sequences $\{x_n\}$, $\{y_n\}$ such that $\{x_n + y_n\}$ converges and $\{x_n y_n\}$ converges.   •

PROBLEM 24: Show that for every $x \in \mathbb{R}$, $x^n/n! \longrightarrow 0$. Hint: Let N be a natural number bigger than $2|x|$ and consider what happens from the Nth term on. •

VIII

THREE HEAVY THEOREMS ON SEQUENCES

I n this chapter we prove three very basic theorems. The first two are equivalent to the completeness axiom, and the third is equivalent to the completeness axiom in fields where $\mathbb{N}$ is unbounded. (Recall that we needed the completeness axiom to *prove* that $\mathbb{N}$ was unbounded.) The assertion that the first two statements are equivalent to the completeness axiom means that either statement could have been used in place of the completeness axiom, and the resulting theory would have been the same; exactly the same statements would be true.

The three theorems we prove here are the following:

(1) Every bounded increasing sequence converges.
(2) Every bounded sequence has a convergent subsequence.
(3) If the differences $x_m - x_n$ tend to zero as n and m increase, then $\{x_n\}$ converges.

In the next chapter, which should be viewed as an optional excursion from our central theme, we will show that these statements could be used in place of the completeness axiom. Now we return to the proof of the three results, for which we need some definitions.

A sequence $\{x_n\}$ is **increasing** if $x_{n+1} \geq x_n$ for all n, and **strictly increasing** if $x_{n+1} > x_n$ for all n. Similar definitions hold for **decreasing** and **strictly decreasing** sequences. A sequence is called **monotone** if it is either increasing or decreasing, and **strictly monotone** if it is strictly increasing or decreasing. If $\{x_n\}$ is an increasing sequence, then $x_m \geq x_n$ whenever $m \geq n$. This can be proved by an easy induction, but from now on we will hold such simple truths to be self-evident.

PROPOSITION 1: *Every bounded increasing sequence converges.*

Proof: Let $\{x_n\}$ be a bounded increasing sequence, and let $\ell = \sup\{x_n : n \in \mathbb{N}\}$. If $\varepsilon > 0$, then $\ell - \varepsilon$ is not an upper bound for the sequence, and so there is N so that $\ell - \varepsilon < x_N \leq \ell$. If $n \geq N$, then

$$\ell - \varepsilon < x_N \leq x_n \leq \ell,$$

and consequently,

$$|x_n - \ell| < \varepsilon$$

if $n \geq N$.   ∎

PROBLEM 1: Let $\{x_n\}$ be a decreasing sequence that is bounded below. Show that $\{x_n\}$ converges in two ways:

(i) Mimic the proof of Proposition 1 to show that $x_n \longrightarrow \inf\{x_n : n \in \mathbb{N}\}$.
(ii) Let B be a lower bound for $\{x_n : n \in \mathbb{N}\}$. Show that $\{B - x_n\}$ is increasing and bounded above, and use Proposition 1.   •

PROBLEM 2: If $\{x_n\}$ is an increasing sequence, and $x_{n_k} \longrightarrow \ell$ for some subsequence $\{x_{n_k}\}$, then $x_n \longrightarrow \ell$.   •

PROPOSITION 2 (The Bolzano–Weierstrass Theorem): *Every bounded sequence has a convergent subsequence.*

Proof: Let $\{x_n\}$ be a bounded sequence. We will define inductively two sequences, $\{b_n\}$ and $\{B_n\}$, where b_0 is a lower bound and B_0 is an upper bound for $\{x_n\}$, each interval $[b_{n+1}, B_{n+1}]$ is a half of the interval $[b_n, B_n]$, and each $[b_n, B_n]$ contains infinitely many terms x_j. The b's and B's will squeeze toward a limit ℓ, and since each $[b_k, B_k]$ contains infinitely many terms x_j, we can choose a subsequence $\{x_{n_k}\}$ such that each $x_{n_k} \in [b_k, B_k]$, and hence $x_{n_k} \longrightarrow \ell$.

The sequences $\{b_n\}, \{B_n\}$ are defined as follows. Let b_0, B_0 be any numbers such that $b_0 \leq x_n \leq B_0$ for all n. Let m_0 be the midpoint $(b_0 + B_0)/2$ of $[b_0, B_0]$. If there are infinitely many indices j such that $b_0 \leq x_j \leq m_0$, then let $[b_1, B_1] = [b_0, m_0]$; that is, let $[b_1, B_1]$ be the left half of $[b_0, B_0]$. Otherwise, there must be infinitely many indices j such that x_j lies in the right half of $[b_0, B_0]$, and we let $[b_1, B_1] = [m_0, B_0]$. In either case, $[b_1, B_1]$ is a half of $[b_0, B_0]$ and $[b_1, B_1]$ contains points x_j for infinitely many j. The next step is to let $[b_2, B_2]$ be a half of $[b_1, B_1]$ containing infinitely many terms x_j, and the inductive step is to let $[b_{n+1}, B_{n+1}]$ be a half of $[b_n, B_n]$ containing infinitely many terms x_j. It is clear that $\{b_n\}$ is an increasing sequence bounded above by B_0, and that $\{B_n\}$ is a decreasing sequence bounded below by b_0. Therefore, $\{b_n\}$ and $\{B_n\}$ converge by Proposition 1 and Problem 1. Since $B_n - b_n = (B_0 - b_0)/2^n$, $\lim b_n = \lim B_n$.

We let ℓ be the common limit. To define a subsequence that converges to ℓ, let $x_{n_1} \in [b_1, B_1]$. Pick $n_2 > n_1$ such that $x_{n_2} \in [b_2, B_2]$; this is possible since $[b_2, B_2]$ contains x_j for infinitely many j, and so there is such a j larger than n_1. The inductive step is to pick $n_{k+1} > n_k$ such that $x_{n_{k+1}} \in [b_{k+1}, B_{k+1}]$. Again this is possible because $[b_{k+1}, B_{k+1}]$ has infinitely many terms of the sequence. Since $b_k \leq x_{n_k} \leq B_k$ for all k, and $b_k \longrightarrow \ell$, $B_k \longrightarrow \ell$, we have $x_{n_k} \longrightarrow \ell$.   ∎

PROBLEM 3: Define an interval $[b, B]$ and a sequence $\{x_n\}$ such that $x_j \in [b, B]$ for infinitely many j, but $\{x_j : j \in \mathbb{N}\} \cap [b, B]$ is a finite set. Hint: This is English, not mathematics. We distinguish between infinitely many terms x_n and infinitely many numbers x_n.   •

PROBLEM 4: (i) Give an example of a bounded sequence that has a subsequence that converges to zero and a subsequence that converges to one.

(ii) Give an example of a sequence that has subsequences converging to every natural number.   •

Notice the device used to find a limit of a subsequence in the proof of the Bolzano–Weierstrass Theorem. We located the entire sequence in an initial interval, $[b_0, B_0]$, and determined a half of this interval, $[b_1, B_1]$, containing infinitely many terms of the sequence. Then we halved $[b_1, B_1]$ to find $[b_2, B_2]$ containing infinitely many terms, and so on. The intervals mash down on a single point, and this point serves as the limit for some subsequence. The essence of this argument is contained in the next problem.

PROBLEM 5 (The Nested Intervals Theorem): Let $\{[b_n, B_n]\}$ be any sequence of nested intervals such that $[b_1, B_1] \supset [b_2, B_2] \supset [b_3, B_3] \supset \cdots$. If $B_n - b_n \longrightarrow 0$, then the intersection of all intervals $[b_n, B_n]$ is a single point.   •

PROBLEM 6: Define a sequence $\{x_n\}$ in $[0, 1]$ as follows: let $x_0 = 0$, $x_1 = \dfrac{1}{10}$, $x_2 = \dfrac{2}{10}, \ldots, x_9 = \dfrac{9}{10}$, $x_{10} = 0$, $x_{11} = \dfrac{1}{100}$, $x_{12} = \dfrac{2}{100}, \ldots, x_{109} = \dfrac{99}{100}$, $x_{110} = 0$, $x_{111} = \dfrac{1}{1000}$, $x_{112} = \dfrac{2}{1000}, \ldots$. In other words, the sequence walks from 0 to .9 in steps of $1/10$, then goes back to 0 and walks to .99 in steps of $1/100$, and then goes back to 0 and walks to .999 in steps of $1/1000$, and so on. What numbers are limits of subsequences of this sequence? Is it possible to have a sequence $\{x_n\}$ such that every real number is the limit of some subsequence? If so, describe such a sequence, and if not, say why not.   •

PROBLEM 7: Let r be a one-to-one function on $\mathbb{N}$ onto all the rational numbers in $[0, 1]$. That is, $r_1, r_2, r_3, \ldots$ is a sequence that lists all the rationals in $[0, 1]$, and it lists each one just once. Define a subsequence that converges to $1/2$. Hint: The emphasis here is on "sub."   •

A sequence $\{x_n\}$ is called a **Cauchy sequence** if and only if for each $\varepsilon > 0$ there is a number N such that $|x_n - x_m| < \varepsilon$ whenever $n \geq N$ and $m \geq N$. All convergent sequences are Cauchy sequences, since if $x_n \longrightarrow x_0$ and $\varepsilon > 0$, then there is N so that $|x_k - x_0| < \varepsilon/2$ for all $k \geq N$, and hence $|x_n - x_m| < \varepsilon$ if $n \geq N$ and $m \geq N$. The converse is true, and it provides an invaluable way of determining that a sequence converges when we have no handle on what the limit is.

PROPOSITION 3 (Cauchy's Theorem): *Every Cauchy sequence converges.*

Proof: The proof has two steps. First, every Cauchy sequence is bounded, and so there is a convergent subsequence by the Bolzano–Weierstrass Theorem. Second, if a Cauchy sequence has a convergent subsequence, then the sequence itself converges to the same limit.

Let $\{x_n\}$ be a Cauchy sequence. Corresponding to $\varepsilon = 1$ there is N such that $|x_n - x_m| < 1$ if $n \geq N$ and $m \geq N$. In particular, $|x_n - x_N| < 1$ if $n \geq N$, and so all terms $x_N, x_{N+1}, x_{N+2}, \ldots$ lie in the interval $(x_N - 1, x_N + 1)$. The first $N - 1$ terms have a largest number B and a smallest number b. Hence all x_n are less than or equal to the larger of B and $x_N + 1$. Similarly, all x_n are greater than or equal to the smaller of b and $x_N - 1$.

Since $\{x_n\}$ is bounded, there is a convergent subsequence $\{x_{n_k}\}$, and we let $x_{n_k} \longrightarrow \ell$. To show that $x_n \longrightarrow \ell$, let $\varepsilon > 0$. There is K so that $|x_{n_k} - \ell| < \varepsilon/2$ if $k \geq K$. Since $\{x_n\}$ is Cauchy, there is N such that $|x_i - x_j| < \varepsilon/2$ if $i, j \geq N$. We can take $N \geq K$, so that $n_N \geq N \geq K$ and $|x_{n_N} - \ell| < \varepsilon/2$. If $i \geq N$, then $|x_i - \ell| \leq |x_i - x_{n_N}| + |x_{n_N} - \ell| < \varepsilon/2 + \varepsilon/2 = \varepsilon$. ∎

PROBLEM 8: Let $\{x_n\}$ be any sequence such that $|x_n| < 1/2^n$ for all n. Let $s_n = x_1 + \cdots + x_n$ for all n. Show that $\{s_n\}$ converges. Hint: We know nothing yet about infinite sums, and so future facts like

$$\frac{1}{2^N} + \frac{1}{2^{N+1}} + \cdots = \frac{1}{2^{N-1}}$$

are *not* available for this problem. Notice that with the information given about the $\{x_n\}$, there is no way to determine a limit of $\{s_n\}$, even though we do know there must be a limit. ●

Cauchy's Theorem becomes the definition of completeness in abstract metric spaces. In such a space one knows what sequences converge, but "increasing" probably makes no sense since the space need not be ordered. Similarly, "bounded set" and "bounded sequence" are not generally relevant ideas, since any metric can be replaced by another metric that gives the same convergent sequences, but that makes all sets and all sequences bounded. We remark, however, that the convergence of Cauchy sequences in a field does not imply that $1/n \longrightarrow 0$. Of course, "natural number" makes no sense in a general metric space.

PROBLEM 9: Define a sequence of rational approximations to $\sqrt{2}$ by the following inductive scheme. Let $x_1 = 1$, so that $x_1^2 \leq 2$. Let $x_2 = x_1 + d_1/10 = 1.d_1$, where d_1 is the largest nonnegative integer such that $(x_1 + d_1/10)^2 \leq 2$. Having defined $x_1, x_2, \ldots, x_n$, let $x_{n+1} = x_n + d_n/10^n$, where d_n is the largest nonnegative integer such that $(x_n + d_n/10^n)^2 \leq 2$. Show the following:

(i) $d_n \le 9$ for all n, so that $1.d_1d_2d_3\ldots$ is a decimal expression.
(ii) For all n, $x_n \le x_{n+1} \le 2$.
(iii) $x_m - x_n \le 1/10^{n-1}$ if $m \ge n$; so $\{x_n\}$ is a Cauchy sequence.
(iv) For all n, $2 - x_{n+1}^2 < 5/10^n$. Hint: Use $x_{n+1} \le 2$ and

$$2 - x_{n+1}^2 < \left[x_n + (d_n + 1)/10^n\right]^2 - x_{n+1}^2.$$

All the foregoing results are independent of the completeness axiom; the next result is not.

(v) Show that $\{x_n\}$ converges, say $x_n \longrightarrow x_0$, and $x_0^2 = 2$.   •

PROBLEM 10: Show that every sequence has an increasing subsequence or a decreasing subsequence. Hint: Assume that $\{x_n\}$ has no increasing subsequence, and show that $\{x_n\}$ has a decreasing subsequence as follows. Since there is no increasing subsequence, there is N_1 such that x_{N_1} is strictly larger than all subsequent terms. The tail sequence of terms from $N_1 + 1$ on clearly has no increasing subsequence, so that there is $N_2 > N_1$ such that x_{N_2} is strictly larger than all subsequent terms, and of course $x_{N_1} > x_{N_2}$. Et cetera. Notice that the proof proves more than the statement of the problem; state a stronger result that is proved. •

PROBLEM 11: Use Problem 10 to prove the Bolzano–Weierstrass Theorem. •

PROBLEM 12: Define a sequence $\{r_n\}$ as follows: $r_1 = 2$ and $r_{n+1} = \left(r_n + 3/r_n\right)/2$ for $n \ge 1$.

(i) Use a calculator to find r_2, r_3, r_4, r_5 to eight significant figures.
(ii) Show that $r_n^2 \ge 3$ for all n. Hint: $r_{n+1}^2 = \left(r_n^2 + 6 + 9/r_n^2\right)/4$, so that $r_{n+1}^2 \ge 3$ if $r_n^2 + 9/r_n^2 \ge 6$.
(iii) Show that $r_{n+1}/r_n \le 1$ for all n, so that $\{r_n\}$ is decreasing and bounded below (by what?).
(iv) Let $r_n \longrightarrow r$, and show $r^2 = 3$.
(v) Generalize.

Note: If we had already studied differentiation, it would be appropriate to point out that $\{r_n\}$ is the sequence of Newton's method approximations to the solution of the equation $x^2 - 3 = 0$. Unfortunately we do not have this background, and so the reader should ignore the preceding sentence. •

PROBLEM 13: (i) Verify directly from the definition, and the fact that Cauchy sequences are bounded, that if $\{x_n\}$ and $\{y_n\}$ are Cauchy sequences, then $\{x_n + y_n\}$ and $\{x_ny_n\}$ are Cauchy sequences.
(ii) Assume in addition that $y_n \ge 1/2$ for all n, and show that $\{x_n/y_n\}$ is Cauchy. •

PROBLEM 14: Let $x_n = n/(n+1)$ and let $\varepsilon > 0$. Find N such that $|x_n - x_m| < \varepsilon$ if $n \ge N$ and $m \ge N$. •

PROBLEM 15: Let $s_n = 1 - 1/2 + 1/3 - 1/4 + \cdots + (-1)^{n+1}1/n$. Show that $\{s_n\}$ is a Cauchy sequence. Generalize to the case $s_n = x_1 - x_2 + x_3 - \cdots + (-1)^{n+1}x_n$, where $x_n \geq 0$ for all n, and .... •

PROBLEM 16: Let $0 < x_1 < 1$ and define $\{x_n\}$ inductively by $x_{n+1} = 1 - \sqrt{1 - x_n}$. Show $\{x_n\}$ converges, and find the limit. •

IX

ALTERNATIVE COMPLETENESS AXIOMS

A s we indicated earlier, two of the three principal theorems proved in the last chapter could have been used in place of the Completeness Axiom, and Cauchy's Theorem together with the assumption that $\mathbb{N}$ is unbounded could also be used for completeness. Our nominal aim in this book is to start with explicit axioms for the real numbers, and to prove from them everything a beginning young analyst should know. In fact, that is too ambitious a project, and if it were possible, the result would be tedious beyond belief — something akin to a computer program. Our real aim is therefore to develop enough of the theory so that any reasonable mathematician could to prove any of the myriad details that are omitted. A comparison of axiom systems is therefore not strictly within the purview of this book, and the reader who is dedicated to our main plan should feel free to omit this chapter. The interested reader, however, may gain from this chapter a better understanding of the fundamental idea of completeness in analysis, and a better appreciation of the three powerful theorems of the last chapter.

Our plan is the following. We will temporarily unfrock the Completeness Axiom, while retaining all of Axioms I through VIII, *and* all the algebraic theorems of ordered fields that we proved from those axioms. We then show that if we *assume* that every bounded increasing sequence converges, then we can *prove* that every bounded nonempty set has a smallest upper bound. We then show that the Bolzano–Weierstrass condition (every bounded sequence has a convergent subsequence) implies that every bounded increasing sequence in an ordered field converges, which in turn implies our Completeness Axiom. Both of these proofs use the fact that $\mathbb{N}$ is not a bounded set. An ordered field in which $\mathbb{N}$ is not bounded (and hence $1/n \longrightarrow 0$) is called **Archimedean**. These two new completeness axioms imply that $\mathbb{N}$ is unbounded, and so we prove that first. Finally, we show that if we assume that every Cauchy sequence converges, and we assume that $\mathbb{N}$ is unbounded (this does not follow from the assumption that Cauchy sequences converge), then we can prove the Bolzano–Weierstrass condition, and hence the Completeness Axiom.

PROPOSITION 1: *If Axioms I–VIII hold and every bounded increasing sequence converges, or if every bounded sequence has a convergent subsequence, then $\mathbb{N}$ is not bounded.*

Proof: Assume that every bounded increasing sequence converges, and assume that $\mathbb{N}$ is a bounded set. If $x_n = n$ for all $n \in \mathbb{N}$, then $\{x_n\} = \{n\}$ is a bounded increasing sequence, and hence converges. Let $x_n \longrightarrow \ell$. For $\varepsilon = 1/4$ there is $N \in \mathbb{N}$ such that

$$\ell - \frac{1}{4} < n < \ell + \frac{1}{4}$$

if $n \geq N$. In particular, this says that both N and $N + 1$ lie in an interval of length $1/2$, which is absurd.

The second part of the proof is the following problem.   ■

PROBLEM 1: Prove that if every bounded sequence has a convergent subsequence, and Axioms I–VIII hold, then $\mathbb{N}$ is unbounded.   •

Now we will show that the convergence of bounded increasing sequences implies the Completeness Axiom and, consequently, is equivalent to the Completeness Axiom.

PROPOSITION 2: *If Axioms I through VIII hold, and every bounded increasing sequence converges, then every bounded nonempty set has a smallest upper bound.*

Proof: Assume that every bounded increasing sequence converges, and consequently that $\mathbb{N}$ is unbounded. Let S be a nonempty bounded set. Let $x_0 \in S$, and let B_0 be an upper bound for S. Let $m_0 = (x_0 + B_0)/2$ be the midpoint between x_0 and B_0. If m_0 is an upper bound for S, let $m_0 = B_1$ and let $x_1 = x_0$. Otherwise, pick $x_1 \in S$ with $x_1 > m_0$, and let $B_1 = B_0$. Thus we have x_0 and $x_1 \in S$, B_0 and B_1 are upper bounds for S, $x_0 \leq x_1 \leq B_1 \leq B_0$, and $B_1 - x_1 \leq (B_0 - x_0)/2$. Define x_2 and B_2 the same way: let $m_1 = (x_1 + B_1)/2$, and let $B_2 = m_1$ if m_1 is an upper bound for S, and otherwise pick $x_2 \in S$ with $x_2 > m_1$, and let $B_2 = B_1$. Now we have $x_0 \leq x_1 \leq x_2 \leq B_2 \leq B_1 \leq B_0$ and $B_2 - x_2 \leq (B_0 - x_0)/4$. The inductive procedure is clear, and so in this way we define inductively an increasing sequence $\{x_n\}$ in S, and a decreasing sequence $\{B_n\}$ of upper bounds. The sequence $\{x_n\}$ is bounded by B_0, and so converges; let $x_n \longrightarrow B$. Since $\mathbb{N}$ is unbounded, $1/n \longrightarrow 0$, and hence $1/2^n \longrightarrow 0$. Therefore, $B_n \longrightarrow B$ since $B_n - x_n \leq (B_0 - x_0)/2^n$. For each $x \in S$, $x \leq B_n$ for all n, since all B_n are upper bounds for S. Therefore, for each $x \in S$, $x \leq \lim B_n = B$, and B is an upper bound for S. To show that B is the smallest upper bound, let $\varepsilon > 0$. Since $x_n \longrightarrow B$, there is some n such that $B - \varepsilon < x_n$, and $B - \varepsilon$ is not an upper bound for S for any $\varepsilon > 0$. ∎

The proof that the Bolzano–Weierstrass condition also implies the Completeness Axiom is straightforward, and it is the content of Problem 2.

PROBLEM 2: Assume Axioms I–VIII and assume that every bounded sequence has a convergent subsequence, and show that every bounded nonempty set has a smallest upper bound. Hint: It is sufficient, by Proposition 1, to show that every bounded increasing sequence converges. ●

Finally, we will show that the convergence of Cauchy sequences, together with the Archimedean property, implies the Completeness Axiom.

PROPOSITION 3: *If Axioms I through VIII hold, every Cauchy sequence converges, and* $\mathbb{N}$ *is unbounded, then every bounded sequence has a convergent subsequence.*

Proof: Let $\{x_n\}$ be a bounded sequence, with $a_0 \leq x_n \leq b_0$ for all n. One of the halves of $[a_0, b_0]$ contains infinitely many terms of the sequence, and we let $[a_1, b_1]$ be such a half. One of the halves of $[a_1, b_1]$ in turn contains infinitely many terms of the sequence, and we label one such half $[a_2, b_2]$. Thus,

$$a_0 \leq a_1 \leq a_2 \leq b_2 \leq b_1 \leq b_0$$

and $b_2 - a_2 = (b_0 - a_0)/4$. We define inductively a nest of intervals $[a_n, b_n]$ such that $b_n - a_n = (b_0 - a_0)/2^n$ and each $[a_n, b_n]$ contains infinitely many terms of the sequence. Thus we can choose a subsequence $\{x_{n_k}\}$ so that $x_{n_k} \in [a_k, b_k]$ and $a_k \leq x_{n_k} \leq b_k$. Since both $\{a_k\}$ and $\{b_k\}$ are Cauchy sequences and $b_k - a_k \longrightarrow 0$ (see Problem 3), both $\{a_k\}$ and $\{b_k\}$ converge, and

$$\lim a_k = \lim b_k = \lim x_{n_k}.$$

Thus $\{x_n\}$ has a convergent subsequence $\{x_{n_k}\}$. ∎

PROBLEM 3: Explain why $\{a_k\}$ and $\{b_k\}$ are Cauchy sequences, and why $b_k - a_k \longrightarrow 0$. Where is each of the hypotheses of Proposition 3 used? ●

X

CONTINUOUS FUNCTIONS

I n this chapter we define continuity for functions, and we develop some basic properties of continuous functions. Our functions here are real-valued functions, and we will treat mostly functions of one variable. However, many of these results extend immediately to functions of two or more variables, and we arrange the definitions so we can state those facts. It is also true that properties of functions of two variables have immediate consequences for functions of one variable. For example, addition is a continuous function of two variables, and it follows from this that the sum of two continuous functions of one variable is again continuous. We will say more on this later, but first we need some definitions.

We will use $\mathbb{R}^2$ to denote the plane $\mathbb{R} \times \mathbb{R}$, $\mathbb{R}^3$ to denote three-space $\mathbb{R} \times \mathbb{R} \times \mathbb{R}$, and so on. The space of all n-tuples is denoted $\mathbb{R}^n$. We say f is a **function of one variable** if $\mathcal{D}(f) \subset \mathbb{R}$, where $\mathcal{D}(f)$ denotes the domain of f, and similarly for functions of two, three, or n variables. If nothing is said about f, it can be assumed that f is a function of one variable.

A function f of one variable is **continuous at** x_0 provided $x_0 \in \mathcal{D}(f)$ and $f(x_n) \longrightarrow f(x_0)$ whenever $\{x_n\}$ is a sequence in $\mathcal{D}(f)$ and $x_n \longrightarrow x_0$. Similarly, a function f of two variables is continuous at (x_0, y_0) provided $(x_0, y_0) \in \mathcal{D}(f)$ and $f(x_n, y_n) \longrightarrow f(x_0, y_0)$ whenever $\{(x_n, y_n)\}$ is a sequence in $\mathcal{D}(f)$ and $x_n \longrightarrow x_0$, $y_n \longrightarrow y_0$. We will write $(x_n, y_n) \longrightarrow (x_0, y_0)$ to indicate $x_n \longrightarrow x_0$ and $y_n \longrightarrow y_0$, with similar conventions in other spaces $\mathbb{R}^n$.

Notice that our definition of continuity for f at x_0 makes no demands on the domain of f except that x_0 be in the domain. This is somewhat different from the usual calculus definition, where x_0 is generally required to be an element of an

55

interval contained in $\mathcal{D}(f)$. Our present definition is more flexible, and it admits more direct generalization.

PROBLEM 1: Show that if x_0 is an isolated point of $\mathcal{D}(f)$ — that is, if there is some $\delta > 0$ such that no other point of $\mathcal{D}(f)$ lies in $(x_0 - \delta, x_0 + \delta)$ — then f is continuous at x_0.   •

PROBLEM 2: (i) Write out an operational "definition" for "f is not continuous at x_0." Hint: Since the condition for continuity starts out "$x_0 \in \mathcal{D}(f)$ and for every sequence $\{x_n\}$ in $\mathcal{D}(f), \dots$," the negation should start "$x_0 \notin \mathcal{D}(f)$ or for some. sequence $\{x_n\}$ in $\mathcal{D}(f), \dots$." The part after the "or" is the important part.

(ii) Use your condition from part (i) to show that if $f(x) = 0$ for $x < 0$ and $f(x) = 1/2$ for $x \geq 0$, then f is continuous at x_0 if and only if $x_0 \neq 0$. Hint: There are two things to prove here: (I) if $x_0 \neq 0$, then f is continuous at x_0; (II) f is not continuous at 0, using part (i).   •

PROBLEM 3: Assume that f is continuous at x_0 and $f(x_0) > 0$. Show there is $\delta > 0$ such that $f(x) > 0$ for all $x \in \mathcal{D}(f) \cap (x_0 - \delta, x_0 + \delta)$. Hint: Prove the contrapositive statement, namely, if there is no such $\delta > 0$, then f is not continuous at x_0. Cf. Problem 2(i).   •

PROPOSITION 1: *Define four functions A, S, M, D of two variables as follows: $A(x, y) = x + y$, $S(x, y) = x - y$, $M(x, y) = xy$, $D(x, y) = x/y$. The functions A, S, and M are continuous on $\mathbb{R}^2$, and the function D is continuous on $\{(x_0, y_0) : y_0 \neq 0\}$.*

Proof: In Chapter 7 we showed that if $x_n \longrightarrow x_0$ and $y_n \longrightarrow y_0$, then

$$x_n \pm y_n \longrightarrow x_0 \pm y_0,$$

$$x_n y_n \longrightarrow x_0 y_0,$$

and, if $y_0 \neq 0$,

$$x_n/y_n \longrightarrow x_0/y_0.$$

These are precisely the statements that A, S, and M are continuous at the arbitrary point (x_0, y_0), and that D is continuous at any point (x_0, y_0) with $y_0 \neq 0$.   ■

A function either is or is not continuous at each point of its domain. If f is continuous at every point of its domain, we say simply that f if **continuous**. If S is a subset of $\mathcal{D}(f)$ on which f is continuous, we say f is **continuous on** S.

Given two functions f and g, we form new functions $f + g$, $f - g$, fg, and f/g in the natural way: $(f + g)(x) = f(x) + g(x)$, $(f \cdot g)(x) = f(x)g(x)$, and so forth.

PROPOSITION 2: *If f and g are continuous at x_0, then $f + g$, $f - g$, and fg are continuous at x_0, and f/g is continuous at x_0 if $g(x_0) \neq 0$.*

Proof: To show that $f + g$ is continuous at x_0, let $\{x_n\}$ be a sequence in $\mathcal{D}(f) \cap \mathcal{D}(g)$ such that $x_n \longrightarrow x_0$. Then $f(x_n) \longrightarrow f(x_0)$ and $g(x_n) \longrightarrow g(x_0)$, since f and g are continuous at x_0. Consequently $f(x_n) + g(x_n) \longrightarrow f(x_0) + g(x_0)$, since addition is continuous. The other cases are proved similarly using only the definition of continuity and Proposition 1. ∎

PROBLEM 4: Write out the proofs that $f - g$ and f/g are continuous at x_0 if f and g are, and assuming that $g(x_0) \neq 0$ in the latter case. •

The constant functions and the function $f(x) = x$ are obviously continuous. This leads quickly (see Problem 5) to the fact that polynomials are continuous (everywhere) and rational functions are continuous wherever the denominator is nonzero.

PROBLEM 5: Prove by induction that x^n is continuous for every n, and then, again by induction, that polynomials are continuous. •

Most of the explicit functions we consider are made up as composites of two or more simpler functions. We define the **composition** of f and g, denoted $f \circ g$, by $(f \circ g)(x) = f(g(x))$. Thus $f \circ g$ is defined for all $x \in \mathcal{D}(g)$ such that $g(x) \in \mathcal{D}(f)$.

PROBLEM 6: If g is continuous at x_0 and f is continuous at $g(x_0)$, then $f \circ g$ is continuous at x_0. •

Some of the most useful properties of continuous functions attach to those functions that are continuous on an interval. We will prove some of these familiar results next. An interval is by definition a set of one of the following types: (a, b), $[a, b]$, $[a, b)$, $(a, b]$, $(-\infty, b]$, $(-\infty, b)$, $[a, \infty)$, (a, ∞), $(-\infty, \infty)$. Recall that we showed earlier (Chapter 7, Proposition 6) that intervals can also be characterized as follows: a set I is an interval if and only if I contains every point that is intermediate between two of its points; that is, if x, y, z are any three numbers with x and z in I and $x < y < z$, then $y \in I$. We will use this characterization to show next that continuous functions map intervals onto intervals.

The notation $f[S]$ will denote the set $\{f(x) : x \in S\}$. Thus the square bracket notation changes f from a mapping of points onto points into the corresponding mapping of sets onto sets.

PROPOSITION 3 (Intermediate Value Theorum): *If f is continuous and non-constant on an interval I, then $f[I]$ is an interval.*

Proof: We consider any two points $f(a)$ and $f(b)$ in $f[I]$, with $f(a) < f(b)$. To be specific, assume that $a < b$. We must show that if $f(a) < y < f(b)$, then $y \in f[I]$; that is, we must show that there is some $c \in I$ with $f(c) = y$. Let $m = (a + b)/2$, so that m is the midpoint of $[a, b]$. If $f(m) \geq y$, then let $[a_1, b_1] = [a, m]$, so that $[a_1, b_1]$ is the left half of the interval $[a, b]$ and $f(a_1) < y \leq f(b_1)$. If $f(m) < y$, we let $[a_1, b_1] = [m, b]$, so that $[a_1, b_1]$ is the right half of $[a, b]$ and again $f(a_1) < y \leq f(b_1)$. We continue this halving process, and define inductively a nest of intervals, $[a, b] \supset [a_1, b_1] \supset [a_2, b_2] \supset \cdots$, all of which are subsets of I, such that $f(a_n) < y \leq f(b_n)$ for all n. Since $b_n - a_n \longrightarrow 0$, the Nested Intervals Theorem (Chapter 8, Problem 5) says that there is exactly one number c in all $[a_n, b_n]$, and $a_n \longrightarrow c$, $b_n \longrightarrow c$. Clearly $c \in I$, and so f is continuous at c. Consequently, $f(a_n) \longrightarrow f(c)$ and $f(b_n) \longrightarrow f(c)$. Since $f(a_n) < y$ for all n, $f(c) \leq y$. Since $f(b_n) \geq y$ for all n, $f(c) \geq y$. Therefore, $f(c) = y$, and $y \in f[I]$. Thus $f[I]$ contains any point between two of its points, and so $f[I]$ is an interval. Notice that the point c we found must be between a and b.   ■

PROBLEM 7: Suppose f is continuous on $[a, b]$, $f(a) < y < f(b)$, and $f(m) = y$, where $m = (a + b)/2$. Is the point c in the proof of Proposition 3 necessarily equal to m? Is $c < m$ possible? Is $c > m$ possible?   ●

A function f is **bounded above on** S provided the set $f[S]$ is bounded above, and similarly for f is **bounded below on** S. If f is bounded above and below on S, then f is **bounded on** S. The Intermediate Value Theorem then says, for example, that if f is continuous on an interval I and f is bounded below on I but not bounded above on I, then $f[I]$ has the form $[a, \infty)$ or (a, ∞) for some a.

PROBLEM 8: Describe a continuous function f on $I = (0, 1)$ such that: (i) $f[I] = [0, 1]$; (ii) $f[I] = (0, \infty)$.   ●

We will say that a function f is **increasing on** S (**strictly increasing on** S) provided $f(x_2) \geq f(x_1) (f(x_2) > f(x_1))$ whenever $x_1 < x_2$ and $x_1, x_2 \in S$. Similar definitions hold for **decreasing** and **strictly decreasing**. A function that is either increasing or decreasing is called **monotonic** and **strictly monotonic** if it is strictly increasing or decreasing. A function that is strictly increasing or strictly decreasing on a set S is obviously one-to-one on S, and hence it has an inverse function, f^{-1}, defined on $f[S]$. That is, $f^{-1}(y) = x$ means $f(x) = y$, and f^{-1} is a one-to-one function on $f[S]$ onto S.

PROPOSITION 4: *If $n \in \mathbb{N}$, the function $f(x) = x^n$ is strictly increasing on $[0, \infty)$ onto $[0, \infty)$.*

Proof: If $0 \leq x_1 < x_2$, then from Chapter 3 we know that

$$x_1^2 \leq x_1 x_2 < x_2^2,$$

so that x^2 is strictly increasing on $[0, \infty)$. The same argument will show that if x^n is strictly increasing on $[0, \infty)$, then so is x^{n+1}, which completes the inductive proof that x^n is one-to-one on $[0, \infty)$ for all $n \in \mathbb{N}$. Clearly the set $\{x^n : x \geq 0\}$ is bounded below by zero. The set is not bounded above since $x^n > x$ if $x > 1$. Therefore, the continuous function x^n maps the interval $[0, \infty)$ onto the interval $[0, \infty)$, and the mapping is one-to-one.   ■

We will denote the mapping inverse to x^n by $x^{1/n}$. Notice that we have shown that every nonnegative number has a unique nonnegative nth root.

PROBLEM 9: Show that if $x \geq 0$ and $m, n \in \mathbb{N}$, then $(x^m)^{1/n} = \left(x^{1/n}\right)^m$. Hint: This is purely a formal verification once you know that all the roots exist.

Justify each of the following equations and explain the proof that lurks therein:

$$\left[\left(x^{1/n}\right)^{m}\right]^{n} = \left(x^{1/n}\right)^{(mn)} = \left[\left(x^{1/n}\right)^{n}\right]^{m} = x^{m}. \qquad \bullet$$

Now we want to define $x^{m/n}$ by $x^{m/n} = [x^{m}]^{1/n} = [x^{1/n}]^{m}$ for every rational number m/n, using Problem 9. However, we still have to worry about whether $x^{2/3} = x^{4/6}$, since these two expressions have different definitions.

PROBLEM 10: Show that if $m, n, k, \ell \in \mathbb{N}$ and $m/n = k/\ell$, then $[x^{m}]^{1/n} = [x^{k}]^{1/\ell}$. Hint: Let $A = [x^{m}]^{1/n}$ and $B = [x^{k}]^{1/\ell}$ and show that $A^{kn} = B^{\ell m}$ if $kn = \ell m$ (i.e., if $m/n = k/\ell$).   •

Now using Problems 9 and 10 we can unambiguously define $x^{m/n} = [x^{m}]^{1/n}$ for $x \geq 0$ and $m, n \in \mathbb{N}$. The usual laws for positive rational exponents are easy to check, and we will assume these laws have been verified. As usual we define $x^{0} = 1$ if $x > 0$, and $x^{-r} = 1/x^{r}$ for r a positive rational and $x > 0$. Notice that although negative numbers have some roots — for example, negative numbers have cube roots — we define x^{r} only for $x > 0$ and r any rational, with of course the agreement that $0^{r} = 0$ if $r > 0$. We will extend the definition to include irrational exponents later. The continuity of the functions $x^{m/n}$ will follow from Proposition 6.

Continuous functions map intervals onto intervals, and if the domain is a closed bounded interval, $[a, b]$, we can say more.

PROPOSITION 5: *If f is continuous on $[a, b]$, then f assumes a maximum value d on $[a, b]$, and f assumes a minimum value c on $[a, b]$, and consequently $f[a, b] = [c, d]$.*

Proof: First we show that f is bounded. Suppose on the contrary that f is not bounded above. Then for each $n \in \mathbb{N}$, there is $x_{n} \in [a, b]$ such that $f(x_{n}) \geq n$. Let $\{x_{n_{k}}\}$ be a convergent subsequence of $\{x_{n}\}$, with $x_{n_{k}} \longrightarrow x_{0}$. Clearly $x_{0} \in [a, b]$, and $f(x_{n_{k}}) \geq n_{k} \geq k$ for all k, since $\{n_{k}\}$ is a strictly increasing sequence. Since f is continuous at x_{0}, $f(x_{n_{k}}) \longrightarrow f(x_{0})$. The convergent sequence $\{f(x_{n_{k}})\}$ must be bounded, but that contradicts $f(x_{n_{k}}) \geq k$ for all k. The proof that f is bounded below is similar (see Problem 11).

Since f is bounded above on $[a, b]$ we let $d = \sup\{f(x) : x \in [a, b]\}$ and show that $d = f(x_{0})$ for some $x_{0} \in [a, b]$. For each n, $d - 1/n$ is not an upper bound for $f[a, b]$, and so we pick $y_{n} \in [a, b]$ with

$$d - \frac{1}{n} < f(y_{n}) \leq d. \qquad (1)$$

Let $\{y_{n_{k}}\}$ be a convergent subsequence, with $y_{n_{k}} \longrightarrow y_{0} \in [a, b]$. Then $f(y_{n_{k}}) \longrightarrow f(y_{0})$, and from (1), $f(y_{n_{k}}) \longrightarrow d$, so that $f(y_{0}) = d$. Thus the sup, d, of the function values is actually a maximum value. The existence of a minimum value $c = f(y_{1})$ is proved similarly. Since $f[a, b]$ is an interval, it must be the interval $[c, d]$.   ■

PROBLEM 11: Prove that if f is continuous on $[a, b]$, then f is bounded below, and if $c = \inf\{f(x) : x \in [a, b]\}$, then $c = f(z)$ for some $z \in [a, b]$. Do this two ways: (i) Mimic the proof of Proposition 5. (ii) Use the fact that $-f$ has a maximum value by Proposition 5 to show that f has a minimum value. •

PROBLEM 12: Assume that f is continuous on $[a, b]$, and let d be the maximum value of f on $[a, b]$. Show there is a minimum number $x_0 \in [a, b]$ such that $f(x_0) = d$. •

The inverse of a strictly increasing function is also strictly increasing; bigger numbers correspond to bigger numbers no matter which direction you consider the mapping. Less obvious is the fact that the inverse of a strictly increasing continuous function is continuous, provided the correspondence is from one interval to another.

PROPOSITION 6: *If f is a continuous strictly increasing or strictly decreasing function on an interval I onto an interval J, then f^{-1} is continuous on J.*

Proof: We will assume that f is strictly increasing on I onto J. Let $y_0 \in J$, and consider first the case that y_0 is an interior point of J. Let $f^{-1}(y_0) = x_0$, so that $f(x_0) = y_0$, and x_0 is necessarily an interior point of I. Let $[a_1, b_1]$ be an interval of radius r centered at x_0, with r so small that $[a_1, b_1] \subset I$. Let $[a_n, b_n]$ be the interval centered at x_0 with radius r/n, so that $a_n \longrightarrow x_0$ and $b_n \longrightarrow x_0$. Let $c_n = f(a_n), d_n = f(b_n)$, so that $c_n < y_0 < d_n$. Then $c_n \longrightarrow y_0, d_n \longrightarrow y_0$ because f is continuous at x_0. To show that f^{-1} is continuous at y_0, let $\{y_k\}$ be a sequence J such that $y_k \longrightarrow y_0$ and let $\varepsilon > 0$. We must show that $f^{-1}(y_k) \longrightarrow f^{-1}(y_0) = x_0$. Pick N such that $r/N < \varepsilon$, and hence $(a_N, b_N) \subset (x_0 - \varepsilon, x_0 + \varepsilon)$. Pick K such that if $k \geq K$, $y_k \in (c_N, d_N)$; that is, let K correspond to $\varepsilon' = \min\{y_0 - c_N, d_N - y_0\}$. If $y_k \in (c_N, d_N)$, then $f^{-1}(y_k) \in (a_N, b_N)$ and $|f^{-1}(y_k) - f^{-1}(y_0)| < \varepsilon$. Thus $f^{-1}(y_k) \longrightarrow f^{-1}(y_0)$, and f^{-1} is continuous at y_0. The case where y_0 is an endpoint of J is the following problem. ■

PROBLEM 13: Prove Proposition 6 in the case where y_0 is the upper endpoint of J. •

PROBLEM 14: Prove that x^r is continuous on $[0, \infty)$ for every rational $r > 0$. •

PROBLEM 15: Define a strictly increasing continuous function such that f^{-1} is not continuous. Hint: The domain of f cannot be an interval by Proposition 6. Try a domain like $[0, 1) \cup \{2\}$, with $f(x) = x$ on $[0, 1)$. Any value of $f(2)$ makes f continuous at 2 (why?), so just find one that makes f^{-1} discontinuous at $f(2)$. •

PROBLEM 16: Define f on $(0, 1)$ as follows: If x is irrational, then $f(x) = 0$. If $x = m/n$ for natural numbers m and n with no common factors, then $f(m/n) = 1/n$. Where is f continuous and where is f discontinuous? Hint: If x_0 is rational, then there is certainly a sequence $\{x_n\}$ of irrational points such that $x_n \longrightarrow x_0$, and that answers the question about rational points. If x_0 is irrational and

$x_n \longrightarrow x_0$, then some of the x_n may be rational. What can you say about their denominators?   •

PROBLEM 17: In Proposition 6, to ensure that the function f was one-to-one on an interval I, we assumed that f was strictly increasing or strictly decreasing on I. Show that there is no other way for a continuous function to be one-to-one on an interval; that is, show that a continuous one-to-one function on an interval is necessarily strictly monotone. Hint: Assume the contrary, and first show that there must be points a, b, c in I with $a < b < c$ and $f(a) < f(b)$ and $f(b) > f(c)$ (or $f(a) > f(b)$ and $f(b) < f(c)$). In the first case, f must map $[a, b]$ onto an interval containing $\left[f(a), f(b)\right]$ and must map $[b, c]$ onto an interval containing $\left[f(c), f(b)\right]$. Points just to the left of $f(b)$ will be covered twice.   •

The result of Problem 17 above gives us a formally stronger version of Proposition 6.

PROPOSITION 6′: *If f is continuous and one-to-one on an interval I onto the interval J, then f^{-1} is continuous on J.*

PROBLEM 18: Give an example of a function f that is continuous and one-to-one on a set S, and f is neither increasing nor decreasing.   •

PROBLEM 19: If f is continuous on S, then $|f|$ is continuous on S, but not conversely.   •

PROBLEM 20: Let f be defined on (a, ∞) and let $w(\varepsilon) = \sup \left\{|f(x_1) - f(x_2)| : x_1, x_2 \in (a, a + \varepsilon)\right\}$. Show that f can be defined at a so the resulting function is continuous at a if and only if $\lim_{\varepsilon \to 0^+} w(\varepsilon) = 0$.   •

PROBLEM 21: If f and g are continuous, then $h(x) = \max\{f(x), g(x)\}$ is continuous.   •

PROBLEM 22: If $f(x) = a_3 x^3 + a_2 x^2 + a_1 x + a_0$ and $a_3 \neq 0$, then $f(x) = 0$ for some x. Generalize.   •

PROBLEM 23: If f is continuous on $\mathbb{R}$ into the set of rational numbers, then f is constant. Hint: To show there is an irrational number between any two rational numbers, first show that $m\sqrt{2}/n$ is irrational for all $m, n \in \mathbb{N}$.   •

PROBLEM 24: Does a bounded continuous function on $\mathbb{R}$ necessarily have a maximum and minimum?   •

XI

UNIFORM CONTINUITY

Recall that we have a sequential definition for continuity: f is continuous at x_0 provided $f(x_n) \longrightarrow f(x_0)$ whenever $x_n \longrightarrow x_0$. One good reason for giving this particular definition is that it is the way most people actually think about continuity. The more usual calculus-text definition is not so intuitive, and that is why you probably do not remember it exactly. However, the ε–δ definition of the calculus texts is both standard and useful, and so we introduce it here as an equivalent criterion for continuity.

PROPOSITION 1 (The ε–δ Criterion for Continuity): *A function f is continuous at x_0 if and only if $x_0 \in \mathcal{D}(f)$ and for every $\varepsilon > 0$ there is $\delta > 0$ such that if $x \in \mathcal{D}(f)$ and $|x - x_0| < \delta$, then $|f(x) - f(x_0)| < \varepsilon$.*

Proof: Suppose first that f is continuous at x_0; that is, suppose $f(x_n) \longrightarrow f(x_0)$ whenever all $x_n \in \mathcal{D}(f)$ and $x_n \longrightarrow x_0$. Let $\varepsilon > 0$. Suppose that there is for this ε no positive δ as advertised in the proposition. Then for every $\delta > 0$ something goes wrong; namely, there is for each $\delta > 0$ some $x \in \mathcal{D}(f)$ so that $|x - x_0| < \delta$ and $|f(x) - f(x_0)| \geq \varepsilon$. For $\delta = 1/n$, we pick $x_n \in \mathcal{D}(f)$ with $|x_n - x_0| < 1/n$ and $|f(x_n) - f(x_0)| \geq \varepsilon$. Then, clearly, $x_n \longrightarrow x_0$ but $\{f(x_n)\}$ fails to converge to $f(x_0)$. The contradiction shows that the ε–δ criterion is necessary for continuity of f at x_0.

Now assume the ε–δ criterion at $x_0 \in \mathcal{D}(f)$. To show that f is continuous at x_0, let $\{x_n\}$ be a sequence in $\mathcal{D}(f)$ such that $x_n \longrightarrow x_0$. To show that $f(x_n) \longrightarrow f(x_0)$, we let $\varepsilon > 0$ and search for a corresponding N. There is, corresponding to x_0, and this ε, a positive number δ such that $|f(x) - f(x_0)| < \varepsilon$ if $|x - x_0| < \delta$.

Pick N so that $|x_n - x_0| < \delta$ if $n \geq N$. Then $|f(x_n) - f(x_0)| < \varepsilon$ if $n \geq N$, and $f(x_n) \longrightarrow f(x_0)$. ∎

PROBLEM 1: Show that the ε–δ criterion of Proposition 1 could equally well have been written: "for every $\varepsilon > 0$, there is $\delta \in (0, 1)$ such that ...," or "for every $\varepsilon > 0$, there is $\delta \in (0, 10^{-5})$ such that ...." Can the condition also be written with ε restricted — for example, "For every $\varepsilon \in (0, 1)$...," or "For every $\varepsilon \in (0, 10^{-5})$..."? Suppose $x_0 \in \mathcal{D}(f)$ and for every $\varepsilon > 0$ you can find $\delta > 0$ such that $|f(x) - f(x_0)| \leq 100\varepsilon$ whenever $x \in \mathcal{D}(f)$ and $|x - x_0| < \delta$. Show that f is continuous at x_0.   •

PROBLEM 2: Use the ε–δ criterion to show that the following functions are continuous at the given point. (i) x^3 at $x_0 = 2$; (ii) x^3 at $x_0 = 100$; (iii) $1/x$ at $x_0 = 1/100$. Hint: $x^3 - 2^3 = (x - 2)(x^2 + 2x + 4)$. Refer to Problem 1 and let $\varepsilon > 0$. If $|x - 2| < \delta_0 = 1$, then $1 < x < 3$ and $0 < x^2 + 2x + 4 < 19$. Hence if $|x - 2| < \delta \leq 1$, $|x^3 - 2^3| < 19\delta$.   •

PROBLEM 3: Suppose f is continuous at x_0 and $f(x_0) > 0$. Use the ε–δ criterion to show there is some interval $(x_0 - \delta, x_0 + \delta)$ such that $f(x) > 0$ for all points x of $\mathcal{D}(f) \cap (x_0 - \delta, x_0 + \delta)$.   •

PROBLEM 4: Write out the negation of the ε–δ condition that f is continuous at x_0; that is, give an ε–δ condition for "f is not continuous at x_0." Use your condition to show that f is not continuous at 0 if $f(x) = 0$ for $x < 0$ and $f(x) = 1$ for $x \geq 0$.   •

If a given δ "works" for some ε in the ε–δ criterion, then any smaller positive δ will also work. However, Problem 2 shows that the set of δ's that work for a given ε and x_0 will ordinarily depend on x_0 as well as ε. The steeper the curve is at x_0, the smaller δ will have to be for a given value of ε. For some curves of limited steepness, a uniform $\delta > 0$ will work for every given $\varepsilon > 0$ and all x_0.

PROBLEM 5: (i) Let $f(x) = 1/x$ for $x \in [1/2, \infty)$, so that f is continuous on $[1/2, \infty)$. Let $\varepsilon > 0$. Find a $\delta > 0$ (in terms of ε) that will work in the ε–δ criterion uniformly for all $x_0 \in [1/2, \infty)$.

(ii) Let $g(x) = x^2$ for $0 \leq x \leq 2$. For each $\varepsilon > 0$, find a $\delta > 0$ that works uniformly for all $x_0 \in [0, 2]$.   •

Motivated by the kind of behavior indicated in Problem 5, we make the following definition: f is **uniformly continuous on a set** S provided $S \subset \mathcal{D}(f)$ and for every $\varepsilon > 0$ there is a $\delta > 0$ such that, for all x, $x_0 \in S$, if $|x - x_0| < \delta$ then $|f(x) - f(x_0)| < \varepsilon$. If f is merely continuous on S, then given ε and x_0 there is a satisfactory δ depending on both ε and x_0; if f is uniformly continuous on S, then given ε there is a δ that works for all $x_0 \in S$. Of course, uniform continuity on S implies uniform continuity on any subset of S, as well as continuity at each point of S, as the language suggests.

Problem 5 shows that $1/x$ is uniformly continuous on $[1/2, \infty)$ and, therefore, uniformly continuous on $[1, \infty)$ or on $[13, 15]$. However, since the curve $y = 1/x$ becomes increasingly steep near 0, there is no single δ that will work with a given ε and all $x > 0$ (see Problem 6). The function $1/x$ is continuous on $(0, \infty)$, but it is not uniformly continuous on $(0, \infty)$.

The next proposition gives the most useful sufficient condition for uniform continuity.

PROPOSITION 2: *If f is continuous on a closed bounded interval $[a, b]$, then f is uniformly continuous on $[a, b]$.*

Proof: Assume, on the contrary, that f is continuous on $[a, b]$ but f is not uniformly continuous on $[a, b]$. Then there is some $\varepsilon > 0$ for which no δ works uniformly over $[a, b]$; that is, there is some $\varepsilon > 0$ such that, no matter how small δ is, there are points x and x' in $[a, b]$ with $|x - x'| < \delta$ and $|f(x) - f(x')| \geq \varepsilon$. For each $\delta = 1/n$, choose points x_n and x_n' with $|x_n - x_n'| < 1/n$ and $|f(x_n) - f(x_n')| \geq \varepsilon$. Since $\{x_n\}$ is a bounded sequence, there is a convergent subsequence: $x_{n_k} \longrightarrow x_0 \in [a, b]$. Clearly, we also have $x_{n_k}' \longrightarrow x_0$. Since f is continuous at x_0, there is $\delta > 0$ that corresponds to x_0 and $\varepsilon/2$ such that $|x - x_0| < \delta$ implies $|f(x) - f(x_0)| < \varepsilon/2$. There is K such that $|x_{n_k} - x_0| < \delta$ if $k \geq K$ and $|x_{n_k}' - x_0| < \delta$ if $k \geq K$. It follows that if $k \geq K$,

$$|f(x_{n_k}) - f(x_{n_k}')| \leq |f(x_{n_k}) - f(x_0)| + |f(x_0) - f(x_{n_k}')|$$

$$< \frac{\varepsilon}{2} + \frac{\varepsilon}{2} = \varepsilon.$$

But $|f(x_{n_k}) - f(x_{n_k}')| \geq \varepsilon$ for all k, and so we have a contradiction. ∎

PROBLEM 6: In the preceding proof we used the following condition, which is equivalent to "f is not uniformly continuous on S": for some $\varepsilon > 0$ and every $\delta > 0$, there are points $x, x' \in S$ such that $|x - x'| < \delta$ and $|f(x) - f(x')| \geq \varepsilon$. Verify this condition to show that $1/x$ is not uniformly continuous on $(0, \infty)$; that is, produce a culprit value for ε, and show that for this ε and every $\delta, \ldots$, etc. Hint: Any ε will do; take your pick of $\varepsilon = 10^6$, $\varepsilon = 47$, $\varepsilon = 1$, $\varepsilon = 10^{-6}$. •

PROPOSITION 3: *If f is uniformly continuous on a set S, then $\{f(x_n)\}$ is a Cauchy sequence whenever $\{x_n\}$ is a Cauchy sequence in S.*

Proof: Assume that f is uniformly continuous on S, and let $\{x_n\}$ be a Cauchy sequence in S. To show that $\{f(x_n)\}$ is a Cauchy sequence, let $\varepsilon > 0$. For this ε pick a corresponding δ from the uniform continuity assumption. Since $\{x_n\}$ is Cauchy, there is N such that $|x_n - x_m| < \delta$ whenever $n, m \geq N$. Therefore, $|f(x_n) - f(x_m)| < \varepsilon$ whenever $n, m \geq N$, and $\{f(x_n)\}$ is a Cauchy sequence. ∎

Proposition 3 almost has this consequence: if f is uniformly continuous on (a, b), then f can be extended to a continuous function on $[a, b]$. The next problem indicates the missing steps.

PROBLEM 7: Show that if f is uniformly continuous on (a, b), then f can be defined at a and b so that the resulting function is continuous. Hint: If $x_n \longrightarrow b-$, then $\{f(x_n)\}$ has a limit. Show the limit does not depend on the particular sequence $\{x_n\}$ approaching b. Generalize to domains other than intervals (a, b). For example, does $\mathcal{D}(f)$ have to contain an interval on one side of the point the value of which is to be defined?   •

PROBLEM 8: Show that in Proposition 3 it is not enough to assume that f is merely continuous on S; that is, give a set S, a bounded function f that is continuous on S, and a Cauchy sequence $\{x_n\}$ in S such that $\{f(x_n)\}$ is not a Cauchy sequence. Hint: Your function had better not be uniformly continuous, and your sequence $\{x_n\}$ better not converge to a point of $\mathcal{D}(f)$.   •

The converse of Proposition 3 would be that if $\{f(x_n)\}$ is Cauchy whenever $\{x_n\}$ is Cauchy, then f is uniformly continuous. This is not true. However, if f always maps Cauchy sequences into Cauchy sequences, then we can conclude that f is continuous. Uniform continuity of f follows if, in addition, $\mathcal{D}(f)$ is bounded.

PROPOSITION 4: *If $\{f(x_n)\}$ is a Cauchy sequence whenever $\{x_n\}$ is a Cauchy sequence in $\mathcal{D}(f)$, then f is continuous. If in addition $\mathcal{D}(f)$ is bounded, then f is uniformly continuous.*

Proof: To show that f is continuous, let $x_0 \in \mathcal{D}(f)$ and let $\{x_n\}$ be a sequence in $\mathcal{D}(f)$ that converges to x_0. Then the sequence

$$x_1, x_0, x_2, x_0, x_3, x_0, \ldots$$

is a Cauchy sequence, and so the corresponding sequence of function values converges; that is, $f(x_n) \longrightarrow f(x_0)$ (see Problem 9), and f is continuous. Now suppose in addition that $\mathcal{D}(f)$ is a bounded set, and f maps all Cauchy sequences into Cauchy sequences. Suppose f is not uniformly continuous, and ε is a culprit for which no δ works. Then for each $\delta = 1/n$, we pick points x_n and x_n' with $|x_n - x_n'| < 1/n$ and $|f(x_n) - f(x_n')| \geq \varepsilon$. Since $\mathcal{D}(f)$ is bounded, $\{x_n\}$ has a convergent subsequence: $x_{n_k} \longrightarrow x_0$. Hence $x_{n_k}' \longrightarrow x_0$ also. Notice that we do not know that $x_0 \in \mathcal{D}(f)$. The sequence

$$x_{n_1}, x_{n_1}', x_{n_2}, x_{n_2}', x_{n_3}, x_{n_3}', \ldots$$

is a Cauchy sequence, so that

$$f(x_{n_1}), f(x_{n_1}'), f(x_{n_2}), f(x_{n_2}'), \ldots$$

must also be a Cauchy sequence. However, the sequence of function values is not Cauchy since there are arbitrarily large k with $|f(x_{n_k}) - f(x_{n_k}')| \geq \varepsilon$.   ■

PROBLEM 9: Show that the sequence

$$x_1, x_0, x_2, x_0, x_3, x_0, x_4, \ldots$$

is Cauchy if and only if $x_n \longrightarrow x_0$.   •

PROBLEM 10: Show that the assumption that $\mathcal{D}(f)$ be bounded was necessary in the latter part of Proposition 4; that is, give a continuous function f (with $\mathcal{D}(f)$ necessarily unbounded) such that $\{f(x_n)\}$ is Cauchy whenever $\{x_n\}$ is Cauchy, but f is not uniformly continuous. Hint: Try $f(x) = x^2$ on $\mathbb{N} \cup \{n + 1/n : n \in \mathbb{N}\}$ with $\varepsilon = 2$. What are the Cauchy sequences in $\mathcal{D}(f)$? •

PROBLEM 11: Assume f is uniformly continuous and one-to-one on an interval I. We know that f^{-1} is continuous. Is f^{-1} necessarily uniformly continuous? Hint: Try $f(x) = 1/x$ on $[1, \infty)$. What is f^{-1} and what is its domain? •

We know that if f is continuous on a closed bounded interval $[a, b]$, then f is bounded. This is not true if the interval is merely bounded, since, for example, $1/x$ is unbounded on $(0, 1)$. However if f is uniformly continuous on a bounded interval, then f is bounded. In fact, the domain need not be an interval — just a bounded set.

PROPOSITION 5: *If f is uniformly continuous on a bounded set S, then f is bounded.*

Proof: Let $\varepsilon = 1$ and let δ be such that $|f(x) - f(x')| < 1$ if $x, x' \in S$ and $|x - x'| < \delta$. Since S is bounded, S can be covered by a finite number of intervals of length less than δ. In each such interval that intersects S, pick one point of S, and let $\{x_1, x_2, \ldots, x_N\}$ be the resulting finite subset of S. Every point of S is within δ of some x_i, and so for all $x \in S$,

$$|f(x)| \leq 1 + \max\{|f(x_1)|, \ldots, |f(x_N)|\}. \qquad \blacksquare$$

If f is continuous on a bounded closed interval, then f is uniformly continuous. This includes functions that get arbitrarily steep, like $\sqrt{x}$ on $[0, 1]$. It is easy to see that $\sqrt{x}$ is also uniformly continuous on $[1, \infty)$, since

$$\left|\sqrt{x} - \sqrt{x'}\right| = \frac{|x - x'|}{\sqrt{x} + \sqrt{x'}} \leq \frac{1}{2}|x - x'|,$$

and $\delta = \varepsilon$ works for all $x \geq 1$, $x' \geq 1$. The next problem suggests how these two facts can be put together to show, for example, that $\sqrt{x}$ is uniformly continuous on $[0, \infty)$.

PROBLEM 12: If f is uniformly continuous on the interval $[a, b]$ and uniformly continuous on $[b, \infty)$, then f is uniformly continuous on $[a, \infty)$. Generalize. Do the sets that meet at b have to be intervals? Can you use $(a, b]$ or $(-\infty, b]$ instead of $[a, b]$? •

In Problem 7, you were invited to show that if f is uniformly continuous on a set S, then f can be defined on points that are limits of sequences in S in such a way that the extended function is continuous. We will prove this next, and then use this fact to extend the definition of x^r, for fixed x, from rational numbers, $r = m/n$, to all numbers.

PROPOSITION 6: *Let f be uniformly continuous on S, and let S* consist of S together with all limits of sequences in S. There is a unique function f* that is continuous on S* such that f* = f on S.*

Proof: Of course we define $f^*(x) = f(x)$ if $x \in S$. For $x_0 \in S^* - S$, let $\{x_n\}$ be a sequence in S with $x_n \longrightarrow x_0$. Then $\{x_n\}$ is Cauchy, and so $\{f(x_n)\}$ is Cauchy. Let $f(x_n) \longrightarrow \ell$. To show this limit does not depend on the sequence $\{x_n\}$, let $\{y_n\}$ be any other sequence in S such that $y_n \longrightarrow x_0$. Then the mixed sequence,

$$x_1, y_1, x_2, y_2, x_3, y_3, \ldots,$$

is certainly again a Cauchy sequence, which makes the sequence of function values Cauchy; that is,

$$f(x_1), f(y_1), f(x_2), f(y_2), \ldots \tag{1}$$

is Cauchy. If $|f(x_n) - \ell| < \varepsilon$ for all large n, and all terms of (1) are within ε of each other beyond some point, then $|f(y_n) - \ell| < 2\varepsilon$ for all sufficiently large n, and $f(y_n) \longrightarrow \ell$. Now we can define $f^*(x_0)$ unambiguously by $f^*(x_0) = \lim_{n\to\infty} f(x_n)$, where $\{x_n\}$ is *any* sequence in S that converges to x_0. To show that f^* is continuous on S^*, we let $x_0 \in S^*$ and consider a sequence $\{z_n\}$ in S^* with $z_n \longrightarrow x_0$. If the points z_n are all in S, then we of course know that $f^*(z_n) = f(z_n) \longrightarrow f^*(x_0)$. Now suppose that $\{z_n\}$ consists entirely of points of $S^* - S$. For each z_n we look at some sequence in S that converges to z_n, the function values of which therefore converge to $f^*(z_n)$. From such a sequence converging to z_n, we pick $u_n \in S$ with

$$|u_n - z_n| < \frac{1}{n} \quad \text{and} \quad |f(u_n) - f^*(z_n)| < \frac{1}{n}. \tag{2}$$

Then $u_n \longrightarrow x_0$, so that $f(u_n) \longrightarrow f^*(x_0)$. But from (2) we then also have $f^*(z_n) \longrightarrow f^*(z_0)$. Thus $f^*(z_n) \longrightarrow f^*(x_0)$ for sequences $\{z_n\}$ that are wholly in S, or are wholly in $S^* - S$. If $\{z_n\}$ has points in both S and $S^* - S$, then for a given ε, we pick N_1 such that $|f(z_n) - f^*(x_0)| < \varepsilon$ for the S-points z_n if $n \geq N_1$, and we pick N_2 such that $|f^*(z_n) - f^*(x_0)| < \varepsilon$ for the $(S^* - S)$-points with $n \geq N_2$, and we choose the larger of N_1 and N_2 to show $f^*(z_n) \longrightarrow f^*(x_0)$. ∎

PROBLEM 13: Show that the f^* of Proposition 6 is unique. •

Now we apply the foregoing result to the function x^r, for fixed x, as a function on the set S of rationals r. In order to make x look more fixed, we consider a fixed positive number $p > 1$ and let $f(r) = p^r$. If we show that f is uniformly continuous on the positive rationals r in $[0, M]$, then f can be extended to a continuous function $f^*(x) = p^x$ for $x \in [0, M]$. Since M is arbitrary, f can be extended to a continuous function f^* on $[0, \infty)$. The steps are outlined in the following problem.

PROBLEM 14: Let $p > 1$.

(i) If $0 \leq r < s$ and r, s are rationals, then $p^r < p^s$. Hint: $r = k/N$ and $s = \ell/N$ for some k, ℓ, N.

(ii) Show that $p^{1/n} \longrightarrow 1$ as $n \longrightarrow \infty$, and show also that $q^{1/n} \longrightarrow 1$ if $0 < q \leq 1$. Hint: If $p^{1/n} \geq 1 + \varepsilon$ for arbitrarily large values of n, then $p \geq (1 + \varepsilon)^n$ for arbitrarily large n.

(iii) To show that f is uniformly continuous on $\mathbb{Q} \cap [0, M]$, where $\mathbb{Q}$ is the set of rationals, let $0 \leq r < s \leq M$, with $r, s \in \mathbb{Q}$. Then

$$0 \leq p^s - p^r = p^r(p^{s-r} - 1) \leq p^M(p^{s-r} - 1).$$

(iv) With p^x a continuous function on $[0, \infty)$ for all $p \geq 1$, define q^x for $0 < q \leq 1$ by $q^x = [(1/q)^x]^{-1}$, and show that p^x is continuous on $[0, \infty)$ for all $p > 0$.   •

PROBLEM 15: If f is increasing, bounded, and continuous on (a, b), then f is uniformly continuous on (a, b).   •

PROBLEM 16: (i) If f and g are uniformly continuous on S, then $f + g$ is uniformly continuous on S.

(ii) If f is uniformly continuous on S into T, and g is uniformly continuous on T, then $g \circ f$ is uniformly continuous on S.   •

PROBLEM 17: (i) Give an example of functions f and g such that both f and g are uniformly continuous on $\mathbb{R}$, but fg is not.

(ii) Add a hypothesis that will ensure that fg is uniformly continuous.   •

XII

CLOSED SETS; COMPACT SETS; OPEN SETS

A close look at the theorems of Chapters 10 and 11 will reveal that in many instances the hypotheses were stronger than necessary. In particular, many theorems had the hypothesis that the domain of the function in question was a closed interval $[a, b]$. In this chapter we isolate the properties of a closed bounded interval that were actually used in these theorems. In this way we obtain results that are stronger, and more importantly, results that generalize more gracefully to sets in $\mathbb{R}^n$ and other metric spaces. In this chapter all functions are functions of one variable, and all sets are subsets of $\mathbb{R}$. In a later chapter we will explicitly generalize to other spaces $\mathbb{R}^n$.

Consider the theorem that a continuous function on $[a, b]$ has a maximum value. What did we actually use about $[a, b]$? First, we used the fact that every sequence in $[a, b]$ has a convergent subsequence. This is of course also true for any bounded set. Second, we used the fact that the limit of any convergent sequence in $[a, b]$ is again a point of $[a, b]$. This is true for more general sets than closed intervals, and so we now give this useful property a name: a set F is **closed** provided F contains the limit of any convergent sequence in F. A set that is both closed and bounded is called **compact**. With these definitions, the old proof of the maximum theorem yields the following stronger result.

PROPOSITION 1: *If f is continuous on a compact set K, then f has a maximum value on K and a minimum value on K.*

69

PROBLEM 1: Write out the proof of Proposition 1 and check that compactness was all that is necessary. (Cf. Proposition 5, Chapter 10.)   •

The closed intervals $(-\infty, b]$, $[a, b]$, $[a, \infty)$, $(-\infty, \infty)$ are certainly closed sets. The following problem asks you to verify the closedness of some others.

PROBLEM 2: (i) A finite set is closed.
(ii) If $x_n \longrightarrow x_0$ and $F = \{x_n : n \in \mathbb{N}\} \cup \{x_0\}$, then F is closed. Hint: $\{x_n\}$ is not the only convergent sequence in F, and x_0 is not the only limit of such a sequence.
(iii) If $F = \bigcup_{n\in\mathbb{N}}[2n, 2n + 1]$, then F is closed.
(iv) If for every $x \notin F$ there is an open interval $(x - \delta, x + \delta)$ that contains no points of F, then F is closed.   •

It is useful to know what combinations of closed sets are again closed.

PROPOSITION 2: *Any finite union of closed sets is closed. The intersection of any family of closed sets is closed.*

Proof: Assume F_1 and F_2 are closed sets. Let $\{x_n\}$ be a sequence in $F_1 \cup F_2$ with $x_n \longrightarrow x_0$. We must show that $x_0 \in F_1 \cup F_2$. There must be an infinite number of indices n with $x_n \in F_1$, or an infinite number of n with $x_n \in F_2$ (or both). Suppose $x_n \in F_1$ for infinitely many n. Then there is a subsequence $\{x_{n_k}\}$ in F_1, and of course $x_{n_k} \longrightarrow x_0$ as $k \longrightarrow \infty$. Since F_1 is closed, $x_0 \in F_1$ and hence $x_0 \in F_1 \cup F_2$. The same argument shows that if the union of any n closed sets is closed, then the union of any $n + 1$ closed sets is closed, and so any finite union of closed sets is closed.

Now let $\mathcal{F}$ be any family of closed sets. If $\{x_n\}$ is a convergent sequence in $\bigcap \mathcal{F}$, then $\{x_n\}$ is a sequence in each $F \in \mathcal{F}$. Hence the limit of $\{x_n\}$ is in each closed set $F \in \mathcal{F}$, and therefore in $\bigcap \mathcal{F}$.   ∎

PROBLEM 3: Show that an infinite union of closed sets need not be closed. •

It is clear that finite unions and arbitrary intersections of bounded sets are again bounded, so we have the following corollary of Proposition 2.

COROLLARY: *A finite union of compact sets is compact. The intersection of any family of compact sets is compact.*   ∎

Recall that the continuous image of a compact interval $[a, b]$ is another compact interval $[c, d]$, where c and d are the minimum and maximum values of the function on $[a, b]$. If the domain is compact but not an interval, then the image need not be an interval, but the image is still compact.

PROPOSITION 3: *If f is continuous on a compact set K, then $f[K]$ is compact. More euphonically: the continuous image of a compact set is compact.*

Proof: In the first part of Problem 1 you showed that if f is continuous on a compact set K, then f is bounded on K. Therefore, we need only show that $f[K]$ is a closed set. Let $\{y_n\}$ be any sequence in $f[K]$, with $y_n \longrightarrow y_0$. We must show that $y_0 \in f[K]$. For each $y \in f[K]$ there may be many x's in K with $f(x) = y$, but for each y_n we pick one $x_n \in K$ with $f(x_n) = y_n$. Since K is bounded, $\{x_n\}$ has a convergent subsequence: $x_{n_k} \longrightarrow x_0$. Since K is closed, $x_0 \in K$, and f is continuous at x_0. Therefore, $f(x_{n_k}) \longrightarrow f(x_0)$. But $f(x_{n_k}) = y_{n_k} \longrightarrow y_0$, so that $f(x_0) = y_0$, and $y_0 \in f[K]$.   ∎

Continuous functions on compact intervals $[a, b]$ are uniformly continuous (Proposition 2, Chapter 11). That proof also used only the fact that the domain was compact, as the next problem indicates.

PROBLEM 4: Show that if f is continuous on a compact set K, then f is uniformly continuous on K.   ●

PROBLEM 5: Show that if $\mathcal{D}(f)$ is compact but not an interval, then f can be one-to-one and neither increasing nor decreasing.   ●

The following proposition gives a characterization of compact sets that is useful in more general metric spaces.

PROPOSITION 4: *A set K is compact if and only if each sequence in K has a subsequence that converges to a point of K.*

Proof: It is clear that the condition is necessary, since in bounded sets every sequence has a convergent subsequence, and for closed sets the limit must again be in the set. To show the condition is sufficient, assume that every sequence in K has a subsequence converging to a point of K. If K is not bounded, then there is a sequence $\{x_n\}$ in K with $|x_n| \geq n$, and this sequence has no convergent subsequence. If K is not closed, then there is a convergent sequence $\{x_n\}$ in K the limit of which is not in K. Every subsequence of $\{x_n\}$ also has the same limit, which is not in K.   ∎

If f is continuous and one-to-one on a compact set K that is not an interval, then f need not be strictly monotone, and $f[K]$ need not be an interval, but $f[K]$ is compact, and f^{-1} is still continuous on $f[K]$. The following proposition verifies this statement, and illustrates the utility of the ε–δ criterion for continuity.

PROPOSITION 5: *If f is continuous and one-to-one on a compact set K, then f^{-1} is continuous on the (compact) set $f[K]$.*

Proof: Assume f is continuous and one-to-one on a compact set K, but, contrary to the proposition, that f^{-1} is not continuous at some $y_0 \in f[K]$. Then there is some $\varepsilon > 0$ such that for every $\delta > 0$ there is a point $y \in f[K]$ with $|y - y_0| < \delta$ and $|f^{-1}(y) - f^{-1}(y_0)| \geq \varepsilon$. (This is the ε–δ condition for "f^{-1} is *not* continuous at y_0.") For each $\delta = 1/n$ we choose $y_n \in f[K]$ with $|y_n - y_0| < 1/n$ and $|f^{-1}(y_n) - f^{-1}(y_0)| \geq \varepsilon$. Let $x_n = f^{-1}(y_n)$, so that $\{x_n\}$ is a sequence in the compact set K. Let $\{x_{n_k}\}$ be a convergent subsequence, with $x_{n_k} \longrightarrow x_0 \in K$. Since f is continuous at x_0, $f(x_{n_k}) \longrightarrow f(x_0)$. But $f(x_{n_k}) = y_{n_k}$, and $y_{n_k} \longrightarrow y_0$

since $y_n \longrightarrow y_0$. Therefore, $f(x_0) = y_0$, or $f^{-1}(y_0) = x_0$. Now we have

$$\left|f^{-1}(y_{n_k}) - f^{-1}(y_0)\right| \geq \varepsilon$$

for all k, which is the same as

$$|x_{n_k} - x_0| \geq \varepsilon$$

for all k, contradicting $x_{n_k} \longrightarrow x_0$. ∎

We will let A' denote the complement of any set A; that is, $A' = \mathbb{R} - A$. A set $U \subset \mathbb{R}$ is called **open** provided its complement U' is closed. We saw in Problem 2(iv) that if each point x of a set $U = F'$ lies in an open interval $(x - \delta, x + \delta)$ that contains no points of F, then F is closed, and so U is open. This sufficient condition for U to be an open set is also necessary. Suppose U is open, so that $F = U'$ is closed. If there is no interval $(x_0 - \delta, x_0 + \delta)$ around some $x_0 \in U$ such that $(x_0 - \delta, x_0 + \delta)$ contains no points of F, then $(x_0 - 1/n, x_0 + 1/n)$ contains some point x_n of F for each n. Thus $\{x_n\}$ is a sequence in F and $x_n \longrightarrow x_0 \notin F$, which is a contradiction. We will state this criterion for openness as a proposition, the proof of which lies in the foregoing discussion.

PROPOSITION 6: *A set U is open if and only if for each $x \in U$ there is some $\delta > 0$ such that $(x - \delta, x + \delta)$ is a subset of U.*

The whole line is a closed set, and so its complement, $\varnothing$, is an open set. The empty set, $\varnothing$, is also a closed set, since there are no sequences in $\varnothing$ the limits of which are not in $\varnothing$; so $\mathbb{R}$ is an open set. These are the only two sets that are both open and closed, but many sets are neither open nor closed.

PROBLEM 6: Show that a set U is open if and only if U is a union of open intervals. Hint: Half of this is immediate from Proposition 6, but the other half is not quite immediate. Be careful. •

PROBLEM 7: Every nonempty open set (in $\mathbb{R}$) is a finite or countable union of disjoint open intervals. (These intervals are called the **components** of U.) Hint: For $x \in U$, let C_x (the component of x) be the union of all open intervals I such that $x \in I \subset U$. Show that C_x is an open interval, and that if $C_x \cap C_y \neq \varnothing$ for x, $y \in U$, then $C_x = C_y$. •

Properties of closed sets automatically yield properties of open sets via DeMorgan's laws for the set operations. For example, $(A \cap B)' = A' \cup B'$ for all sets A and B. If A and B are open sets, then A' and B', and hence $A' \cup B'$, are closed sets, so that $A \cap B$ is an open set. Similarly, if $\mathcal{U}$ is any family of open sets and $\mathcal{F}$ is the family of all (closed) complements of sets in $\mathcal{U}$, then $\bigcap \mathcal{F}$ is a closed set. Since $\left(\bigcup \mathcal{U}\right)' = \bigcap \mathcal{F}, \bigcup \mathcal{U}$ is an open set. Thus we have the following.

PROPOSITION 7: *Any union of open sets is open. Any finite intersection of open sets is open.*

Our last theorem is a big one — the Heine–Borel Theorem. To state the result we need another definition. A family $\mathcal{U}$ of open sets is an **open cover** or **open covering** of a set S provided each point of S is in one of the open sets U of $\mathcal{U}$. Equivalently, $\mathcal{U}$ is an open cover of S if and only if $S \subset \bigcup \mathcal{U}$. If some finite number $U_1, \ldots, U_n$ of the sets in $\mathcal{U}$ is a cover of S, so that $S \subset U_1 \cup \cdots \cup U_n$, then we say S has a **finite subcover** of $\mathcal{U}$ sets. The following characterization of compactness becomes the definition of compactness in more general settings. Most really good theorems eventually turn into definitions.

PROPOSITION 8 (The Heine–Borel Theorem): *K is a compact set if and only if every open cover of K has a finite subcover; that is, if $K \subset \bigcup \mathcal{U}$ for some family $\mathcal{U}$ of open sets, then there are $U_1, \ldots, U_n \in \mathcal{U}$ such that $K \subset U_1 \cup \ldots \cup U_n$.*

Proof: Assume first that K is compact and that $\mathcal{U}$ is an open cover of K. We must show that some finite number of open sets of the family cover K. We suppose this is not the case and reach a contradiction.

Since K is compact, K is bounded; let $K \subset [a, b]$. Divide $[a, b]$ into the two halves, $\left[a, (a + b)/2\right]$ and $\left[(a + b)/2, b\right]$. If K has no finite subcover, then the part of K in at least one of the halves has no finite subcover. Let $[a_1, b_1]$ be a half of $[a, b]$ such that $K \cap [a_1, b_1]$ has no finite subcover by sets of $\mathcal{U}$. Halve $[a_1, b_1]$ and obtain an interval $[a_2, b_2]$ such that $K \cap [a_2, b_2]$ has no finite subcover by sets of $\mathcal{U}$; continue the process to obtain a sequence $\{[a_n, b_n]\}$ of such intervals. The intervals $[a_n, b_n]$ shrink to a single point x_0; that is, $a_n \longrightarrow x_0$ and $b_n \longrightarrow x_0$. Each $[a_n, b_n]$ obviously contains a point x_n of K and $x_n \longrightarrow x_0$, and hence $x_0 \in K$. Because $x_0 \in K$, there is some open set U of the family $\mathcal{U}$ such that $x_0 \in U$. Since U is open, there is $\delta > 0$ such that $(x_0 - \delta, x_0 + \delta) \subset U$. For large enough N, $[a_N, b_N] \subset (x_0 - \delta, x_0 + \delta) \subset U$. It follows that $K \cap [a_N, b_N]$ does have a finite subcover, namely, a cover by the single set U. The contradiction shows that every open cover of a compact set has a finite subcover.

The proof that only compact sets have finite subcovers for every open cover is easier. Suppose K is a set such that every open cover of K has a finite subcover. We show first that K must be bounded. The family $\mathcal{U}$ of all open sets $(-n, n)$, $n \in \mathbb{N}$, covers $\mathbb{R}$ and therefore covers K. The union of the finite number of these open sets that cover K is some bounded interval $(-N, N)$, and hence K is a bounded set. Now we show that K must be closed. Suppose, on the contrary, that there is a sequence $\{x_n\}$ in K with $x_n \longrightarrow x_0 \notin K$. Let $\mathcal{U}$ be all the open sets $\left(-\infty, x_0 - 1/n\right)$ and $\left(x_0 + 1/n, \infty\right)$. This family covers all of $\mathbb{R}$ except x_0, and therefore it covers K. Any finite subset of $\mathcal{U}$ will fail to cover some interval $\left[x_0 - 1/N, x_0 + 1/N\right]$, which clearly contains points x_n of K. Therefore, K is closed and bounded. ∎

Since the intersection of any family of closed sets is closed, there is a smallest closed set containing any given set S, namely, the intersection of *all* closed sets that contain S. This smallest closed superset of S is called the **closure** of S, and denoted $\overline{S}$. If S is closed, then clearly $\overline{S} = S$. A more "constructive" way of describing $\overline{S}$ would be to just add to S all limits of sequences in S. However, it is not obvious that this set is closed; the set certainly contains all limits of sequences in S, but does it contain all limits of sequences in $\overline{S}$? You decide in the next problem.

PROBLEM 8: Let S be any set, and let S^* be the union of S and all limits of sequences in S. Is $S^* = \overline{S}$; that is, is S^* a closed set?   •

PROBLEM 9: Show that $(A \cup B)^- = \overline{A} \cup \overline{B}$ for any two sets A and B. How about $(A \cap B)^- = \overline{A} \cap \overline{B}$? $(A \cap B)^- \supset \overline{A} \cap \overline{B}$? $(A \cap B)^- \subset \overline{A} \cap \overline{B}$?   •

PROBLEM 10: A point x is a **boundary point** of a set S if and only if every open interval $(x - \delta, x + \delta)$, $\delta > 0$, contains a point of S and a point of $\mathbb{R} - S$. Show that a closed set contains all its boundary points, and that an open set contains none of its boundary points.   •

PROBLEM 11: Show that $\mathbb{R}$ is the only nonempty set that is both open and closed. Hint: If A is open, A is a union of disjoint open intervals by Problem 7. Suppose one of these intervals is bounded or half-bounded — does its endpoint belong to A?   •

PROBLEM 12: Show the set of all rational numbers is neither open nor closed.   •

PROBLEM 13: If F is a closed set and U is an open set, then $F - U$ is closed and $U - F$ is open.   •

PROBLEM 14: (i) Show that for any set A there is a largest open subset of A. (This set is called the **interior** of A and is denoted $A°$.)

(ii) A is open if and only if $A = A°$.

(iii) What can you say about $(A \cup B)°$ and $(A \cap B)°$?

(iv) Show that x is a boundary point of A (Problem 10) if and only if $x \in \overline{A} - A°$.   •

XIII

DERIVATIVES

n this chapter we will introduce the derivative of a function, and show how properties of the derivative influence the behavior of the function. The derivative of f at a, $f'(a)$, will be defined as the limit of the slopes of the segments from $(a, f(a))$ to nearby points $(x, f(x))$. This means we will need a definition for

$$\lim_{x \to a} \frac{f(x) - f(a)}{x - a}. \tag{1}$$

We have already discussed limits of sequences, and now we will introduce the kind of limit for functions that makes sense out of (1).

We say that $f(x) \longrightarrow \ell$ as $x \longrightarrow a$ (read $f(x)$ approaches ℓ as x approaches a) provided f is defined on some deleted interval $\{x : 0 < |x - a| < \delta\}$ around a, and for every sequence $\{x_n\}$ in this deleted interval such that $x_n \longrightarrow a$, $f(x_n) \longrightarrow \ell$. Notice that f need not be defined at $x = a$ (the difference quotients (1) are not), and if f is defined at a, its value there has no bearing on whether the limit exists or what the limit is. The requirement that f be defined for all x near a is similar to the requirement for sequences that x_n be defined for all large enough n. It is clear from our earlier work on limits of sequences that limits are unique if they exist, so we will also use "$\lim_{x \to a} f(x) = \ell$" for "$f(x) \longrightarrow \ell$ as $x \longrightarrow a$."

It is convenient to be able to talk about limits as x approaches a from one side only. Accordingly, we say that $f(x) \longrightarrow \ell$ as $x \longrightarrow a+$, or $\lim_{x \to a+} f(x) = \ell$, provided f is defined on some interval $(a, a + \delta)$, and $f(x_n) \longrightarrow \ell$ whenever $\{x_n\}$ is a sequence in $(a, a + \delta)$ such that $x_n \longrightarrow a$. Left sided limits, $\lim_{x \to a-} f(x) = \ell$, are defined similarly.

PROBLEM 1: Show that $\lim_{x \to a} f(x) = \ell$ if and only if $\lim_{x \to a^+} f(x) = \ell$ and $\lim_{x \to a^-} f(x) = \ell$. Hint: Do not overlook the conditions on the domain, and do not assume that any sequence approaching a lies only on one side of a. •

PROBLEM 2: Show that if $f(x) \longrightarrow \ell$ as $x \longrightarrow a$ and $g(x) \longrightarrow m$ as $x \longrightarrow a$, then

    (i) $f(x) + g(x) \longrightarrow \ell + m$ as $x \longrightarrow a$;
    (ii) $f(x) - g(x) \longrightarrow \ell - m$ as $x \longrightarrow a$;
    (iii) $f(x)g(x) \longrightarrow \ell m$ as $x \longrightarrow a$;
    (iv) $f(x)/g(x) \longrightarrow \ell/m$ as $x \longrightarrow a$ if $m \neq 0$.

Hint: Most of the work has been done earlier in the treatment of sequences. However, you still have to worry about the domains, particularly the domain of f/g. •

PROBLEM 3: If f is defined on some interval around a, then f is continuous at a if and only if $\lim_{x \to a} f(x) = f(a)$. •

PROBLEM 4: If $g(x) \longrightarrow b$ as $x \longrightarrow a$, f is defined on an interval around b, and $f(x) \longrightarrow f(b)$ as $x \longrightarrow b$, then $f(g(x)) \longrightarrow f(b)$ as $x \longrightarrow a$. Show that it is not enough to assume $f(x) \longrightarrow \ell$ as $x \longrightarrow b$ to conclude that $f(g(x)) \longrightarrow \ell$. Hint: Suppose $g(x) = b$ for all x such that $0 < |x - a| < 1$. •

The close relationship between the limit concept and continuity, as indicated in Problem 3, suggests that an ε–δ criterion for the existence of a limit would be useful.

PROPOSITION 1: *For any f, $\lim_{x \to a} f(x) = \ell$ if and only if for every $\varepsilon > 0$ there is $\delta > 0$ such that $f(x)$ is defined and $|f(x) - \ell| < \varepsilon$ whenever $0 < |x - a| < \delta$.*

Proof: Assume the ε–δ condition holds. For $\varepsilon_0 = 1$, there is some $\delta_0 > 0$ so that, in particular, $f(x)$ is defined for $0 < |x - a| < \delta_0$. This condition on $\mathcal{D}(f)$ is part of the requirement for $\lim_{x \to a} f(x) = \ell$. Now let $\{x_n\}$ be any sequence in $\{x : 0 < |x - a| < \delta_0\}$ with $x_n \longrightarrow a$. To show that $f(x_n) \longrightarrow \ell$, we let $\varepsilon > 0$ and let $\delta > 0$ be a corresponding number from the ε–δ condition, so that $|f(x) - \ell| < \varepsilon$ whenever $0 < |x - a| < \delta$. There is N such that $|x_n - a| < \delta$ if $n \geq N$, since $x_n \longrightarrow a$, and $0 < |x_n - a|$ for all n by assumption. Therefore, $|f(x_n) - \ell| < \varepsilon$ if $n \geq N$, and $f(x_n) \longrightarrow \ell$. This completes the proof that the ε–δ condition is sufficient for $\lim_{x \to a} f(x) = \ell$.

Now assume $f(x) \longrightarrow \ell$ as $x \longrightarrow a$. There is δ_0 such that $f(x)$ is defined on $\{x : 0 < |x - a| < \delta_0\}$. Let $\varepsilon > 0$. Assume, contrary to the ε–δ condition, that there is no δ for this ε such that $|f(x) - \ell| < \varepsilon$ if $0 < |x - a| < \delta$. For each $\delta = 1/n$, we can then pick x_n with $0 < |x_n - a| < 1/n$ and $|f(x_n) - \ell| \geq \varepsilon$. Clearly $x_n \longrightarrow a$ but $f(x_n) \not\longrightarrow \ell$, which contradicts our assumption. Therefore, $f(x) \longrightarrow \ell$ as $x \longrightarrow a$ implies the ε–δ condition. ∎

Now we define the **derivative of** f **at** a, denoted $f'(a)$, by

$$f'(a) = \lim_{x \to a} \frac{f(x) - f(a)}{x - a}. \tag{2}$$

The **difference quotient** $(f(x) - f(a))/(x-a)$ is not defined at a, but it is defined for all other x where f is defined. The existence of the limit (2) implies that f is defined on some open interval around a, since $f(a)$ must also be defined for (2) to make sense.

PROPOSITION 2: *If $f'(a)$ exists, f is continuous at a.*

Proof: If $f(x) - f(a)$ does not approach 0 as $x \longrightarrow a$ (the condition for continuity at a), then there is an $\varepsilon > 0$ and a sequence $x_n \longrightarrow a$ with $|f(x_n) - f(a)| \geq \varepsilon$. The difference quotient will be unbounded on this sequence $\{x_n\}$ and, therefore, not convergent. ∎

PROPOSITION 3: *Assume that $f'(a)$ and $g'(a)$ exist.*

(i) *If $h(x) = f(x) \pm g(x)$, then $h'(a) = f'(a) \pm g'(a)$.*
(ii) *If $h(x) = f(x)g(x)$, then $h'(a) = f(a)g'(a) + g(a)f'(a)$.*
(iii) *If $h(x) = f(x)/g(x)$ and $g(a) \neq 0$, then*
$h'(a) = (g(a)f'(a) - f(a)g'(a))/g(a)^2$.

Proof: The proofs of the sum and difference formulas are straightforward. For example, if $h(x) = f(x) + g(x)$, then

$$h'(a) = \lim_{x \to a} \frac{h(x) - h(a)}{x - a}$$
$$= \lim_{x \to a} \frac{(f(x) + g(x)) - (f(a) + g(a))}{x - a}$$
$$= \lim_{x \to a} \left[\frac{f(x) - f(a)}{x - a} + \frac{g(x) - g(a)}{x - a} \right]$$
$$= \lim_{x \to a} \left[\frac{f(x) - f(a)}{x - a} \right] + \lim_{x \to a} \left[\frac{g(x) - g(a)}{x - a} \right]$$
$$= f'(a) + g'(a).$$

If we give rein to our primitive craving for linearity, the preceding calculations practically suggest themselves. This is not true for the product rule, however. Let $h(x) = f(x)g(x)$, so that

$$h'(a) = \lim_{x \to a} \frac{f(x)g(x) - f(a)g(a)}{x - a}. \tag{3}$$

Now if you forget that you know the product rule, it is not clear where one goes from (3) to express $h'(a)$ in terms of $f'(a)$ and $g'(a)$. To motivate the calculations, we turn to the dependent variable notation, which is most natural for those who have physical interpretations for the variables. Let Δf and Δg be the changes in

f and g engendered by the change $\Delta x = x - a$. Thus

$$\Delta f = f(x) - f(a), \qquad \Delta g = g(x) - g(a);$$

$$f'(a) = \lim_{\Delta x \to 0} \frac{\Delta f}{\Delta x}, \qquad g'(a) = \lim_{\Delta x \to 0} \frac{\Delta g}{\Delta x}.$$

Now the difference quotient in (3) looks like this:

$$\frac{h(x) - h(a)}{x - a} = \frac{(f(a) + \Delta f)(g(a) + \Delta g) - f(a)g(a)}{\Delta x}$$

$$= g(a)\frac{\Delta f}{\Delta x} + f(a)\frac{\Delta g}{\Delta x} + \Delta f \frac{\Delta g}{\Delta x}. \tag{4}$$

Taking the limit as $\Delta x \to 0$ (i.e., $x \to a$) on both sides shows that $h'(a)$ exists and has the right value, since the limits of $\Delta f/\Delta x$ and $\Delta g/\Delta x$ exist by hypothesis and $\Delta f \to 0$ as $\Delta x \to 0$ since f is continuous at a; therefore,

$$h'(a) = g(a)f'(a) + f(a)g'(a) + 0 \cdot g'(a). \qquad \blacksquare$$

PROBLEM 5: Write out the proof of the quotient rule, Proposition 3(iii).   •

PROPOSITION 4 (The Chain Rule): *Let $h(x) = f(g(x))$. If $g'(a)$ exists and $f'(g(a))$ exists, then $h'(a) = f'(g(a))g'(a)$.*

Proof: Assume that $f'(g(a))$ and $g'(a)$ exist. Then f is defined on some interval $(g(a) - \varepsilon, g(a) + \varepsilon)$ around $g(a)$. Since g is continuous at a, $g(x) \in (g(a) - \varepsilon, g(a) + \varepsilon)$ for all x in some interval $(a - \delta, a + \delta)$. Thus $f(g(x))$ is defined for all x in $(a - \delta, a + \delta)$. For $y \neq g(a)$ with $g(a) - \varepsilon < y < g(a) + \varepsilon$, define $e(y)$ by

$$\frac{f(y) - f(g(a))}{y - g(a)} - f'(g(a)) = e(y), \tag{5}$$

and let $e(g(a)) = 0$. Then $e(y) \to 0$ as $y \to g(a)$, and so e is continuous at $g(a)$. From (5) it follows that for all $y \in (g(a) - \varepsilon, g(a) + \varepsilon)$ (including $y = g(a)$, where we get $0 = 0$),

$$f(y) - f(g(a)) = f'(g(a))(y - g(a)) + e(y)(y - g(a)). \tag{6}$$

Replace y by $g(x)$ in (6), with $x \in (a - \delta, a + \delta)$ so that $g(x) \in (g(a) - \varepsilon, g(a) + \varepsilon)$, and divide both sides by $x - a$. Then, if $0 < |x - a| < \delta$,

$$\frac{f(g(x)) - f(g(a))}{x - a} = f'(g(a))\frac{(g(x) - g(a))}{x - a} + e(g(x))\frac{(g(x) - g(a))}{x - a}.$$

As $x \to a$, both terms on the right approach limits, so that

$$h'(g(a)) = f'(g(a))g'(a) + 0 \cdot g'(a). \qquad \blacksquare$$

We will use the operator notation $\frac{d}{dx}$ in the usual way. Thus $\frac{d}{dx}f(x) = f'(x)$ wherever $f'(x)$ exists, and it will follow from the next problem that $\frac{d}{dx}x^2 = 2x$.

PROBLEM 6: Show that for all integers n, $\frac{d}{dx}x^n = nx^{n-1}$. Hint: Many integers are negative, and one integer is zero. You can use the binomial theorem for $n \in \mathbb{N}$ if you really enjoy algebra, but it is easier and more elegant to use induction. •

The following theorem on the derivative of an inverse function has obvious geometric content. The graphs of $y = f(x)$ and $y = f^{-1}(x)$ are symmetric about the line $y = x$. If f has a tangent at (a, b), then surely f^{-1} has a tangent at the symmetric point (b, a). The symmetric image of the line through (a, b) with slope m is the line through (b, a) with slope $1/m$ (draw the picture). That is what the following proposition says.

PROPOSITION 5: *If f is continuous and one-to-one on an open interval I around a, $f'(a) = m \neq 0$, and $f(a) = b$, then $(f^{-1})'(b) = 1/m$.*

Proof: Recall that a continuous one-to-one function on an interval I is either strictly increasing or strictly decreasing (Chapter 10, Problem 17). Let us assume to be specific that f is strictly increasing and maps I onto an open interval J around b. We also know that f^{-1} is continuous (Chapter 12, Proposition 5). If we let $x = f^{-1}(y)$, $f(x) = y$, then

$$\left(f^{-1}\right)'(b) = \lim_{y \to b} \frac{f^{-1}(y) - f^{-1}(b)}{y - b}$$

$$= \lim_{y \to b} \frac{f^{-1}(f(x)) - f^{-1}(f(a))}{f(x) - f(a)}$$

$$= \lim_{y \to b} \frac{x - a}{f(x) - f(a)}$$

$$= \lim_{y \to b} \frac{1}{\left(\frac{f(x)-f(a)}{x-a}\right)} = \frac{1}{m}.$$

As $y \longrightarrow b$, $f^{-1}(y) = x \longrightarrow a$, so that the last limit is a consequence of Problem 4: the difference quotient in the denominator is a function of x, which is a function of y, and the appropriate limits exist. Notice that if $y \neq b$, $f^{-1}(y) \neq f^{-1}(b)$, or $x \neq a$, so that the division is legitimate. ∎

PROBLEM 7: Use Proposition 5 to show that if $n \in \mathbb{N}$, $\frac{d}{dx}x^{1/n} = (1/n)x^{1/n-1}$ for $x > 0$. Hint: You know that x^n and $x^{1/n}$ have derivatives at every $x > 0$, so that you can use the chain rule and differentiate both sides of $\left(x^{1/n}\right)^n = x$ to get $n\left(x^{1/n}\right)^{n-1}\frac{d}{dx}\left(x^{1/n}\right) = 1$ for all $x > 0$. •

The connection between the behavior of f and the sign of its derivative rests ultimately on the following simple fact.

PROPOSITION 6: *If $f'(a) > 0$, then there is some $\delta > 0$ such that f is defined on $(a - \delta, a + \delta)$, and $f(x) > f(a)$ if $a < x < a + \delta$ and $f(x) < f(a)$ if $a - \delta < x < a$. Similarly, if $f'(a) < 0$, then $f(x) < f(a)$ on some interval to the right of a and $f(x) > f(a)$ on some interval to the left of a.*

Proof: Assume $f'(a) > 0$. Then corresponding to $\varepsilon = f'(a)$ there is some $\delta > 0$ such that f is defined on $(a - \delta, a + \delta)$, and

$$\left| \frac{f(x) - f(a)}{x - a} - f'(a) \right| < f'(a)$$

if $0 < |x - a| < \delta$. The inequality implies that the difference quotient is positive for $0 < |x - a| < \delta$, and hence that $f(x)$ lies above $f(a)$ to the right of a and below $f(a)$ to the left of a. ∎

Mr. Rolle's claim to fame is a corollary of the following consequence of Proposition 6.

PROBLEM 8: If f is defined on $(a - \delta, a + \delta)$ for some $\delta > 0$ and $f(a) \geq f(x)$ for all $x \in (a - \delta, a + \delta)$, or $f(a) \leq f(x)$ for all $x \in (a - \delta, a + \delta)$, and $f'(a)$ exists, then $f'(a) = 0$. Notice that this is a local theorem; f might be defined on all of $\mathbb{R}$ and have many ups and downs, but if f has a *local* maximum or minimum at a, then $f'(a) = 0$. Rolle's Theorem is this corollary: if f is continuous on $[c, d]$, $f(c) = f(d)$, and f' exists on (c, d), then f has a local maximum or minimum at some $a \in (c, d)$, and thus $f'(a) = 0$.   •

The mean value (or average value) of a function over an interval will be defined in the next chapter as a certain integral. For now we take as a definition that the mean value of $f'(x)$ over $[a, b]$ is $(f(b) - f(a))/(b - a)$; that is, the mean value of the derivative over $[a, b]$ is the constant slope that would get the function from $(a, f(a))$ to $(b, f(b))$. The important Mean Value Theorem says that the mean value of the derivative must be assumed at some interior point of the interval.

PROPOSITION 7 (Mean Value Theorem): *If f is continuous on $[a, b]$ and f' exists on (a, b), then $f'(c) = (f(b) - f(a))/(b - a)$ for some $c \in (a, b)$.*

PROBLEM 9: Prove the Mean Value Theorem. Hint: The vertical distance $v(x)$ between the curve $y = f(x)$ and the line between $(a, f(a))$ and $(b, f(b))$ is $v(x) = f(x) - [f(a) + m(x - a)]$, where $m = (f(b) - f(a))/(b - a)$. Show that v has a relative maximum or minimum at some point c of (a, b), so that $v'(c) = 0$ and $f'(c) = m$.   •

PROBLEM 10: If $f'(x) \geq 0$ on (a, b), then f is increasing on (a, b). If $f'(x) > 0$ on (a, b), then f is strictly increasing on (a, b). If $f'(x) \leq 0$ on (a, b), then f is decreasing on (a, b). If $f'(x) < 0$ on (a, b), then f is strictly decreasing on (a, b). Prove one of the four statements and indicate wherein the difference between "strictly" and "nonstrictly" lies. Deduce as a corollary that if $f'(x) = 0$ on (a, b), then f is constant on (a, b). Hint: Notice that f is both increasing and decreasing.   •

PROBLEM 11: (i) Prove: If f is continuous on $(a - \delta, a]$ and $f'(x) < m$ on $(a - \delta, a)$, then $f(x) > f(a) + m(x - a)$ on $(a - \delta, a)$. State the analogous properties for $f'(x) > m$ on an interval to the right of a.

(ii) Show that if f' is defined on an open interval around a, and $f'(a) = 0$, $f''(a) < 0$, then f has a local maximum at a.   •

A function f is defined to be **convex** on an interval I (any kind) provided

$$f((1 - \lambda)x_1 + \lambda x_2) \le (1 - \lambda)f(x_1) + \lambda f(x_2) \tag{7}$$

for all $x_1, x_2 \in I$ and all $\lambda \in [0, 1]$. As λ goes from 0 to 1, the point $(1 - \lambda)x_1 + \lambda x_2$ goes from x_1 to x_2. The point $(1 - \lambda)x_1 + \lambda x_2$ is the fraction λ of the distance from x_1 to x_2; for example, if $\lambda = \frac{1}{4}$, then $\frac{3}{4}x_1 + \frac{1}{4}x_2$ is $\frac{1}{4}$ the distance from x_1 to x_2. The geometric interpretation of (7) is that the graph of f always lies on or under the segment joining any two of its points. This is what is sometimes called **concave up**. The function x^2 is convex, and the function x is convex; the function $\sqrt{x}$ is not convex.

The **second derivative** of f, denoted f'', is the derivative of f', so that f' is increasing on an interval on which $f''(x) \ge 0$. Some calculus texts make the statements (sometimes, alas, definitions!) that f is convex (concave up) on I if $f''(x) \ge 0$ on I, and f is concave down ($-f$ is convex) on I if $f''(x) \le 0$ on I. A nonnegative second derivative is certainly sufficient for a function to be convex, but hardly necessary. For example, $|x|$ is convex, but there is no first or second derivative at 0. More generally, any continuous function the graph of which consists of line segments, with slopes increasing from left to right, is convex. Of course such a function will not have a derivative at the corners. A convex function need not be everywhere differentiable, but it is necessarily continuous.

PROPOSITION 8: *If f is convex on an open interval I, then f is continuous on I.*

Proof: Assume that f is convex on the open interval I, but not continuous at some point $a \in I$. To be specific, and to simplify both the picture and the notation, we can assume (see Problem 12) that $a = 0$, $f(a) = 0$, and f is convex on an open interval around 0. If f is not continuous at 0, then there is some $\varepsilon > 0$ such that at least one of the following four things happen (see Problem 13):

(i) There is a sequence $\{x_n\}$ in I that decreases to 0 with $f(x_n) \ge \varepsilon$ for all n.
(ii) There is a sequence $\{x_n\}$ in I that increases to 0 with $f(x_n) \ge \varepsilon$ for all n.
(iii) There is a sequence $\{x_n\}$ in I that decreases to 0 with $f(x_n) \le -\varepsilon$ for all n.
(iv) There is a sequence $\{x_n\}$ in I that increases to 0 with $f(x_n) \le -\varepsilon$ for all n.

First, suppose case (i). The points $(0, 0)$ and $(x_1, f(x_1))$ are on the graph, and so the graph must lie under the line $y = (f(x_1)/x_1) x$ for $0 \le x \le x_1$. Since $f(x_1) \ge \varepsilon$, the segment from $(0, 0)$ to $(x_1, f(x_1))$ crosses the line $y = \varepsilon$ at some b with $0 < b \le x_1$. If $x_N < b$, then $f(x_N) \ge \varepsilon$, and $f(x_N)$ lies above the line from $(0, 0)$ to $(x_1, f(x_1))$, which is a contradiction. Case (ii) is handled similarly (see Problem 13).

Now assume case (iii), and let $\{x_n\}$ be a sequence that decreases to 0, with $f(x_n) \le -\varepsilon$ for all n, and $-x_n \in I$ for all n. Since $f(0)$ must lie below the segment from $(-x_n, f(-x_n))$ to $(x_n, f(x_n))$, it follows that $f(-x_n) \ge -f(x_n) \ge \varepsilon$. Since

$f(-x_n) \geq \varepsilon$ for all n, we have case (ii), and thus a contradiction. The argument for case (iv) is the same. ∎

PROBLEM 12: Show that if f is convex on an interval around x_0, and $g(x) = f(x_0 + x) - f(x_0)$, then g is convex on an interval around 0, and $g(0) = 0$. Since g is continuous at 0 if and only if f is continuous at x_0, this shows that the simplifying assumption in the proof of Proposition 8 sacrifices no generality. •

PROBLEM 13: Show that if f is not continuous at a, where a belongs to an open interval I, then there is some $\varepsilon > 0$ such that one of the four possibilities in the proof of Proposition 8 holds. Write out the details of the proof in case (ii). •

PROBLEM 14: Prove that if $f''(x) \geq 0$ on (a, b), then f is convex on (a, b). Hint: Let $x_1, x_2 \in (a, b)$ with $x_1 < x_2$, and let $(f(x_2) - f(x_1))/(x_2 - x_1) = m$ be the slope of the secant line. There is some $c \in (x_1, x_2)$ with $f'(c) = m$. Since f' is increasing, $f'(x) \leq m$ on $[x_1, c]$, and the graph of f lies on or under the secant line at least for $x_1 \leq x \leq c$. If there is some x_0 with $(x_0, f(x_0))$ above the secant line, then $x_0 > c$, and $f'(x) \geq m$ on $[x_0, x_2]$. Show why this prevents the curve from getting back down to $(x_2, f(x_2))$. •

PROBLEM 15: If f' is defined and bounded on (a, b), then f is uniformly continuous on (a, b). •

PROBLEM 16: Derivatives have the intermediate value property; that is, if f' exists on an open interval containing x_1 and x_2 and $f'(x_1) < m < f'(x_2)$, then there is $c \in (x_1, x_2)$ such that $f'(c) = m$. Hint: Show first by replacing $f(x)$ by $f(x) - mx$ that it is sufficient to prove in the special case where $f'(x_1) < 0 < f'(x_2)$ that $f'(c) = 0$ for some c. Show that if $f'(x_1) < 0$ and $f'(x_2) > 0$, then the minimum of f on $[x_1, x_2]$ must be strictly less than both $f(x_1)$ and $f(x_2)$, and thus must occur at some $c \in (x_1, x_2)$. •

PROBLEM 17: Let f, g, h be three functions defined on some interval around a, with $f(a) = h(a)$, and $g(x)$ always between $f(x)$ and $h(x)$, so that in particular $g(a) = f(a) = h(a)$. Prove that if $f'(a) = h'(a)$, then $g'(a) = f'(a) = h'(a)$. •

We will borrow some future facts about the sine and cosine function for the purpose of a present example. If $f(x) = x^2 \sin 1/x$ for $x \neq 0$, and $f(0) = 0$, then $f'(x)$ exists for all x, and $f'(0) = 0$ by Problem 17 since $f(x)$ is mashed between x^2 and $-x^2$. Since $f'(x) = 2x \sin 1/x - \cos 1/x$, there is a sequence $x_n = 1/2\pi n \longrightarrow 0$ with $f'(x_n) \longrightarrow -1$, and there is a sequence $y_n = 1/(2n + 1)\pi \longrightarrow 0$ with $f'(y_n) \longrightarrow 1$. Hence f' is not continuous at 0. Moreover, since f' has the intermediate value property, $f'(x)$ must oscillate between $+1$ and -1 infinitely often as $x \longrightarrow 0$. The next problem shows that this kind of oscillation is the only way a derivative can be discontinuous. Derivatives never have **jump discontinuities**; that is, discontinuities where $\lim_{x \to a-} f'(x)$ exists and $\lim_{x \to a+} f'(x)$ exists and the limits are different. If this happened, there would have to be a corner at $(a, f(a))$, and $f'(a)$ would not exist.

PROBLEM 18: If $f'(x)$ exists on an open interval around a, and $\lim_{x \to a-} f'(x)$ $= m_1$ and $\lim_{x \to a+} f'(x) = m_2$, then $m_1 = m_2 = f'(a)$. Hint: Suppose $m_1 < m_2$, and let $m_1 < m_3 < m_4 < m_2$. Then $f'(x) < m_3$ on some interval $(a - \delta_1, a)$ and $f'(x) > m_4$ on some interval $(a, a + \delta_2)$. The Mean Value Theorem then says that $(f(a) - f(x))/(a - x) < m_3$ for $x \in (a - \delta_1, a)$, and so $f'(a) \leq m_3$. A similar argument on the right shows that $f'(a) \geq m_4$.   ●

PROBLEM 19: (i) If f is convex on an open interval I, and $a, b, c \in I$ with $a < b < c$, then

$$\frac{f(b) - f(a)}{b - a} \leq \frac{f(c) - f(b)}{c - b}.$$

Hint: Let $b = (1 - \lambda)a + \lambda c$, so that

$$f(b) \leq (1 - \lambda)f(a) + \lambda f(c),$$

$$(1 - \lambda)f(b) + \lambda f(b) \leq (1 - \lambda)f(a) + \lambda f(c), \tag{8}$$

$$(1 - \lambda)(f(b) - f(a)) \leq \lambda (f(c) - f(b)).$$

Also,

$$(1 - \lambda)b + \lambda b = (1 - \lambda)a + \lambda c,$$

and so

$$b - a = \frac{\lambda}{1 - \lambda}(c - b). \tag{9}$$

Put (8) and (9) together.

(ii) Show that if $f'(a)$ and $f'(c)$ exist, then $f'(a) \leq f'(c)$, so that f' increases if f is convex and differentiable on I.   ●

PROBLEM 20 (Cauchy's Mean Value Theorem): If f and g are continuous on $[a, b]$, $f'(x)$ and $g'(x)$ exist for $x \in (a, b)$, and $g'(x) \neq 0$ on (a, b), then there is some $c \in (a, b)$ such that $(f(b) - f(a))/(g(b) - g(a)) = f'(c)/g'(c)$. Hint: Think of the curve $y = f(t)$, $x = g(t)$, $a \leq t \leq b$, and the line $\ell(t) = f(a) + m\left[g(t) - g(a)\right]$, where $m = (f(b) - f(a))/(g(b) - g(a))$. The curve and the line meet at $(g(a), f(a))$ and $(g(b), f(b))$, and so the vertical distance $v(t) = f(t) - \ell(t)$ between them must have a maximum or minimum at some $c \in (a, b)$, and $v'(c) = 0$.   ●

PROBLEM 21 (l'Hospital's Rule for 0/0 as $x \longrightarrow a+$): If $\lim_{x \to a+} f(x) = 0$ and $\lim_{x \to a+} g(x) = 0$, $f'(x)$ and $g'(x)$ exist on (a, b) for some $b > a$, and $\lim_{x \to a+} f'(x)/g'(x) = \ell$, then $\lim_{x \to a+} f(x)/g(x) = \ell$. Hint: See Problem 20.   ●

PROBLEM 22: Prove l'Hospital's rule for 0/0 as $x \longrightarrow \infty$. Hint: If $f'(x)/g'(x) \longrightarrow \ell$, there is x_0 such that if $x > x_1 \geq x_0$, then

$$\frac{f(x_1)/g(x_1) - f(x)/g(x_1)}{1 - g(x)/g(x_1)}$$

is within ε of ℓ. Let $x \longrightarrow \infty$, and conclude that $|f(x_1)/g(x_1) - \ell| \leq \varepsilon$ if $x_1 \geq x_0$. Worry some about the possibility $g(x_1) = 0$. ●

PROBLEM 23 (l'Hospital's Rule for ∞/∞ as $x \longrightarrow \infty$): If $f'(x)$, $g'(x)$ exist for $x \geq a$, $f(x) \longrightarrow \infty$ and $g(x) \longrightarrow \infty$ as $x \longrightarrow \infty$, and $f'(x)/g'(x) \longrightarrow \ell$ as $x \longrightarrow \infty$, then $f(x)/g(x) \longrightarrow \ell$ as $x \longrightarrow \infty$. Hint: Pick $x_0 > a$ beyond which $|f'(x)/g'(x) - \ell| < \varepsilon$. Pick $x_1 > x_0$ beyond which $g(x_0)/g(x) < 1$, and write

$$\frac{f(x) - f(x_0)}{g(x) - g(x_0)} = \frac{f(x)/g(x) - f(x_0)/g(x)}{1 - g(x_0)/g(x)}.$$

Pick $x_2 > x_1$ beyond which $g(x_0)/g(x)$ is so close to 0 that

$$(\ell - \varepsilon)\left(1 - \frac{g(x_0)}{g(x)}\right) > \ell - 2\varepsilon,$$

$$(\ell + \varepsilon)\left(1 - \frac{g(x_0)}{g(x)}\right) < \ell + 2\varepsilon.$$

Pick $x_3 > x_2$ beyond which $|f(x_0)/g(x)| < \varepsilon$. Show that $|f(x)/g(x) - \ell| < 3\varepsilon$ if $x \geq x_3$. ●

XIV

THE DARBOUX INTEGRAL

he ordinary integral of the calculus, $\int_a^b f(x)\,dx$, is designed to make sense out of limits of sums of the form $\sum_{i=1}^{n} f(x_i)\Delta x_i$. Sums of this sort have many interpretations in the physical sciences, but area is the preferred mathematical concept. We will study two equivalent definitions for the integral, one due to Riemann and one due to Darboux. We start with the Darboux integral, which is familiar from the calculus and lends itself best to an interpretation as area.

Suppose f is a bounded nonnegative function defined on the interval $[a, b]$. Suppose we want to define what is meant by the area under the graph of f, that is, the area of the region

$$S = \{(x, y) : a \le x \le b,\ 0 \le y \le f(x)\}. \tag{1}$$

Let us start by asking what properties an area function should have. That is, if $A(T)$ denotes the area of a plane region T, what properties should the function A have? We assume of course that $A(T) \ge 0$ for all T, and just three additional properties:

   (i) If R is a rectangle with dimensions ℓ and w, then $A(R) = \ell w$.
   (ii) If S and T are nonoverlapping regions (i.e., they have no disc in common) then $A(S \cup T) = A(S) + A(T)$.
   (iii) If $S \subset T$, then $A(S) \le A(T)$.

A line segment can obviously be contained in rectangles of arbitrarily small area. Therefore, by the monotone property (iii), it follows that $A(L)$ must be zero

85

for any line segment L. From (ii) it then follows that the area of any finite number of line segments must be zero.

If an area function A satisfies the foregoing three conditions, and S is the region (1) under the graph of a *continuous* function f, then we will show that $A(S)$ must be the integral $\int_a^b f$ that we define in this chapter.

A **partition** of an interval $[a, b]$ is a finite set $P = \{x_0, x_1, \ldots, x_n\}$ of points of $[a, b]$ labeled so that

$$a = x_0 < x_1 < x_2 < \cdots < x_n = b.$$

We consider a fixed *bounded* function f defined on $[a, b]$. For a given partition $P = \{x_0, \ldots, x_n\}$ of $[a, b]$, we let

$$m_i = \inf\{f(x) : x_{i-1} < x < x_i\},$$

$$M_i = \sup\{f(x) : x_{i-1} < x < x_i\}.$$

If $m \le f(x) \le M$ for all $x \in [a, b]$, then

$$m \le m_i \le M_i \le M$$

for all i. Define the **lower sum** $L(f, P)$ and the **upper sum** $U(f, P)$ for f and P by

$$L(f, P) = \sum_{i=1}^{n} m_i(x_i - x_{i-1}),$$

$$U(f, P) = \sum_{i=1}^{n} M_i(x_i - x_{i-1}).$$

The lower sum can be interpreted as the total area of a finite number of rectangles under the graph of f (if $f \ge 0$), and $U(f, P)$ as the area of a finite number of rectangles that cover the area under f. Our m_i and M_i are infs and sups over open intervals, whereas the usual calculus text takes infs and sups over closed intervals $[x_{i-1}, x_i]$. We use open intervals because we want to partition $[a, b]$ into *disjoint* sets, since that is the approach used later for the important Lebesgue integral. Our approach appears to ignore a finite number of values of f for any given partition, and thus an area corresponding to a finite number of line segments. Since any finite number of line segments must have area zero, this poses no problem.

For any partition P we have the obvious inequalities

$$m(b - a) \le L(f, P) \le U(f, P) \le M(b - a),$$

where again $m \le f(x) \le M$ on $[a, b]$. The corresponding picture shows that if you add a point to a partition, one lower rectangle is replaced by two nonoverlapping rectangles whose combined area is larger. For example, let $P = \{x_0, x_1, \ldots, x_n\}$ and let Q be a partition with one more point, say x^* between x_0 and x_1:

$$Q = \{x_0, x^*, x_1, x_2, \ldots, x_n\}.$$

Let

$$m_1' = \inf\{f(x) : x_0 < x < x^*\},$$
$$m_1'' = \inf\{f(x) : x^* < x < x_1\}.$$

Then $m_1' \geq m_1$ and $m_1'' \geq m_1$, since m_1' and m_1'' are infs of smaller sets. Therefore,

$$m_1'(x^* - x_0) + m_1''(x_1 - x^*) \geq m_1(x_1 - x_0),$$

and the other terms of $L(f, Q)$ are the same as those of $L(f, P)$. Therefore, $L(f, Q) \geq L(f, P)$.

We say that Q is a **refinement** of P, written $P < Q$ or $Q > P$, if P and Q are both partitions of the same interval $[a, b]$, and every point of P is also a point of Q.

PROPOSITION 1: *If $P < Q$, then $L(f, P) \leq L(f, Q)$ and $U(f, Q) \leq U(f, P)$.*

Proof: If $P < Q$, then we can get Q by dropping a finite number of extra points into P, one at a time. Each extra point makes the lower sum bigger or leaves it unchanged, and similarly makes each upper sum smaller or leaves it unchanged. ∎

Since lower sums represent areas inside the region under f (for $f \geq 0$), and upper sums represent areas covering the region under f, we had better have $L(f, P) \leq U(f, Q)$ for all P and Q. We verify this next.

PROPOSITION 2: *For all partitions P and Q of $[a, b]$, $L(f, P) \leq U(f, Q)$.*

Proof: For any given P and Q, let R be the common refinement consisting of all points in either P or Q. Then $P < R$ and $Q < R$, so that

$$L(f, P) \leq L(f, R) \leq U(f, R) \leq U(f, Q). \qquad ∎$$

If S is again the region (1) under a nonnegative function f, and $A(S)$ is to be the area of S, then we must surely have

$$L(f, P) \leq A(S) \leq U(f, Q)$$

for all partitions P and Q. Therefore, we must have

$$\sup_P L(f, P) \leq A(S) \leq \inf_Q U(f, Q).$$

If the sup and inf above coincide, then there is only one choice for $A(S)$. In this case, we say that f is Darboux **integrable** and denote the integral $\int_a^b f$ or $\int_a^b f(x)\, dx$. The following result is obvious, but it is our constant tool for showing integrability, and so we emphasize it with the title of proposition.

PROPOSITION 3: *The bounded function f is integrable over $[a, b]$ if and only if for each $\varepsilon > 0$ there is a partition P such that $U(f, P) - L(f, P) < \varepsilon$.*

The notation $\int_a^b f(x)\,dx$ is the historical one, suggesting the sum of terms $f(x)\,\Delta x$ for small values Δx. The integral sign is just the old English S, for sum. The variable x in $\int_a^b f(x)\,dx$ is a dummy, and it can be replaced by anything except f or d:

$$\int_a^b f = \int_a^b f(x)\,dx = \int_a^b f(t)\,dt = \int_a^b f(u)\,du.$$

As indicated, $x, t, u, \ldots$ are socially acceptable dummy variables, but $a, b, c, \alpha,$ $Q, T, \ldots$ are not used in polite circles. The use of a dummy variable is helpful in certain "change of variable" theorems, but it is otherwise redundant.

In our discussion here, it was convenient to picture a nonnegative function f, so that $L(f, P)$ and $U(f, P)$ represent areas of rectangles inside and covering the region under f. However, none of the foregoing inequalities or propositions depend on f being nonnegative; therefore, we now and henceforth assume only that f is bounded on $[a, b]$.

PROBLEM 1: If $f(x) \equiv c > 0$ on $[a, b]$, then the region under the graph of f is a rectangle, and we have two definitions for the area: length times width is one definition, and the integral is the other. Are they the same?   •

PROBLEM 2: If f is integrable on $[a, b]$ and $g = f$ except for a finite number of points, then g is integrable and $\int_a^b g = \int_a^b f$. (Observe how much simpler this is using open intervals to define m_i and M_i as opposed to using closed intervals.)   •

PROBLEM 3: Assume f and g are integrable over $[a, b]$, and show the following:

(i) If $f \le g$ on $[a, b]$, then $\int_a^b f \le \int_a^b g$; in particular, if $f \ge 0$, then $\int_a^b f \ge 0$.

(ii) If c is constant, then cf is integrable and $\int_a^b cf = c\int_a^b f$. Hint: Treat $c > 0$ and $c < 0$ separately.   •

Not every function is integrable, and so not every region under a nonnegative bounded function has an area defined by the Darboux integral. For example, let $f(x) = 0$ if x is a rational number in $[0, 1]$ and let $f(x) = 1$ if x is irrational. Every lower sum for f is zero and every upper sum is one, and so f is not integrable. A more sophisticated integral, the Lebesgue integral, does include this function among the honored integrable ones, but no integral regards all functions as integrable, and there will always be regions to which we cannot ascribe an area. For now, we show that continuous functions are integrable.

PROPOSITION 4: *If f is continuous on $[a, b]$, then f is integrable on $[a, b]$.*

Proof: Assume that f is continuous on the compact interval $[a, b]$. We let $\varepsilon > 0$, and show there is a partition P such that $U(f, P) - L(f, P) < \varepsilon$. Since f is continuous on a closed bounded interval, f is uniformly continuous. Therefore, corresponding to the positive number $\varepsilon/(b - a)$ there is $\delta > 0$ such that $|f(x) - f(x')| < \varepsilon/(b - a)$ whenever x and x' are in $[a, b]$ and $|x - x'| < \delta$. Let P be a partition of $[a, b]$ with $x_i - x_{i-1} < \delta$ for all i. Since $|f(x) - f(x')| < \varepsilon/(b - a)$ whenever x and x' are in (x_{i-1}, x_i), it follows that $M_i - m_i \leq \varepsilon/(b - a)$ for each i. Hence

$$U(f, P) - L(f, P) = \sum (M_i - m_i)(x_i - x_{i-1})$$

$$\leq \frac{\varepsilon}{b - a} \sum (x_i - x_{i-1})$$

$$= \frac{\varepsilon}{b - a} \cdot (b - a) = \varepsilon. \qquad \blacksquare$$

PROPOSITION 5: *If* $a < b < c$, *then* f *is integrable over* $[a, c]$ *if and only if* f *is integrable over* $[a, b]$ *and* $[b, c]$, *and then*

$$\int_a^b f + \int_b^c f = \int_a^c f.$$

Proof: Assume that f is integrable over $[a, c]$. Let $\varepsilon > 0$. We show that there is a partition P of $[a, b]$ and a partition Q of $[b, c]$ such that $U(f, P) - L(f, P) < \varepsilon$, so that f is integrable over $[a, b]$, and $U(f, Q) - L(f, Q) < \varepsilon$, so that f is integrable over $[b, c]$.

Since f is integrable over $[a, c]$, there is a partition R of $[a, c]$ such that $U(f, R) - L(f, R) < \varepsilon$. We can assume that b is a point of R, since adding a point would make the upper and lower sums closer together. Let P be the partition of $[a, b]$ consisting of the points of R that are in $[a, b]$, and let Q be the partition of $[b, c]$ consisting of the points of R that are in $[b, c]$. Then

$$U(f, P) + U(f, Q) = U(f, R),$$
$$L(f, P) + L(f, Q) = L(f, R). \qquad (2)$$

Therefore,

$$\left[U(f, P) - L(f, P) \right] + \left[U(f, Q) - L(f, Q) \right] = U(f, R) - L(f, R) < \varepsilon. \quad (3)$$

Since both terms on the left are nonnegative, both are less than ε, and f is integrable over both $[a, b]$ and $[b, c]$. Moreover,

$$L(f, P) \leq \int_a^b f \leq U(f, P),$$

$$L(f, Q) \leq \int_b^c f \leq U(f, Q).$$

Adding, and using (2) and (3), we see that

$$\int_a^c f - \varepsilon < L(f, R) \leq \int_a^b f + \int_b^c f \leq U(f, R) < \int_a^c f + \varepsilon,$$

or equivalently,

$$\left| \int_a^b f + \int_b^c f - \int_a^c f \right| < \varepsilon.$$

Since ε is arbitrary, we have the desired equality. The proof of the "if" implication is the next problem. ∎

PROBLEM 4: Show that if f is integrable over $[a, b]$ and over $[b, c]$, then f is integrable over $[a, c]$, and $\int_a^b f + \int_b^c f = \int_a^c f$. Hint: The equality is automatic once you know that f is integrable over $[a, c]$. •

PROBLEM 5: If f is continuous on (a, b) and bounded, then f is integrable over $[a, b]$. Hint: Let $|f(x)| \leq M$ on $[a, b]$. Let $a < c < d < b$, with c so close to a and d so close to b that $M(c - a) + M(b - d) < \varepsilon$. Let $P = \{x_0, x_1, \ldots, x_n\}$ be a partition of $[c, d]$ with $U(f, P) - L(f, P) < \varepsilon$. Let $Q = \{a, c, x_1, \ldots, d, b\}$, so that Q is a partition of $[a, b]$, and show that $U(f, Q) - L(f, Q) < 2\varepsilon$. •

PROBLEM 6: If f is bounded on $[a, b]$ and has at most a finite number of discontinuities in $[a, b]$, then f is integrable over $[a, b]$. Hint: This follows from Problem 5 and Proposition 5 with a little argument, but no further computation. •

PROBLEM 7: If f is defined on $[a, b]$ and uniformly continuous on (a, b), then f is integrable on $[a, b]$. Hint: See Problem 7 of Chapter 11, and Proposition 4 and Problem 2 in this chapter. Do not forget to show that f is bounded. •

The **average value** or **mean value** of an integrable function f over $[a, b]$ is defined to be $\int_a^b f/(b - a)$. Thus if $f \geq 0$, the mean value of f is the height of a rectangle over $[a, b]$ that has the same area as the region under f.

PROBLEM 8: Show that if f is continuous on $[a, b]$, then f somewhere assumes its mean value; that is, there is some $c \in [a, b]$ such that $f(c)(b - a) = \int_a^b f$. Can you say more — that there is such $c \in (a, b)$? Show the statement is false if f is only assumed to be integrable and not necessarily continuous. •

PROBLEM 9: Let f be integrable over $[a, b]$, and for $x \in (a, b)$, define $F(x) = \int_a^x f$. Show that F is continuous on (a, b). Is F uniformly continuous on (a, b)? •

PROBLEM 10: Show that any monotone function is integrable over $[a, b]$. •

PROBLEM 11: Let $f(x) = 0$ for every irrational number $x \in [0, 1]$, and let $f(m/n) = 1/n$ for every rational number m/n written in lowest terms. (See Problem 16, Chapter 10.) Show that f is integrable over $[0, 1]$. Hint: There are only a finite number of points where $f(x) \geq 1/N$. If these points can be encased in disjoint intervals with total length less than ε, then $U(f, P) < (1/N) \cdot 1 + 1 \cdot \varepsilon$. •

PROBLEM 12: If f is bounded on $[a, b]$ and integrable on $[c, b]$ for every $c \in (a, b)$, then f is integrable on $[a, b]$. (This will show, for example, that $\sin(1/x)$ is integrable on $[0, 1]$ once we define $\sin x$.) Hint: Take c close to a and see Problem 5. •

PROBLEM 13: Define f on $[0, 1]$ as follows: f is 1 on $(1/2, 3/4]$ and -1 on $(3/4, 1]$; f is 1 on $(1/4, 3/8]$ and -1 on $(3/8, 1/2]$, and so forth; that is, f is 1 on the first half of $(1/2^n, 2/2^n]$ and -1 on the second half, for $n = 1, 2, \ldots$. Let $f(0) = 0$. Show that f (which has a countable number of discontinuities) is integrable, and find $\int_0^1 f$.   •

PROBLEM 14: Let $f(0) = 1$ and $f(1/n) = 1$ for $n \in \mathbb{N}$, and $f(x) = 0$ otherwise. Let $F(x) = \int_{-1}^x f$. Show that the integral exists for $x \in [-1, 1]$, and that $F'(0)$ exists even though f is not continuous at 0.   •

XV

THE RIEMANN DEFINITION

To define the Darboux integral of f, we squeeze up with nonoverlapping rectangles that lie under the graph of f, and squeeze down with rectangles over the graph of f. If these two squeezes produce the same number, then that number is the Darboux integral of f. The Riemann definition, which gives exactly the same integral for exactly the same functions, uses a slightly different approach. If $P = \{x_0, \ldots, x_n\}$ is a partition of $[a, b]$ and all the subinterval lengths $\Delta x_i = x_i - x_{i-1}$ are very small, then $\sum_{i=1}^{n} f(c_i)\Delta x_i$ is a pretty good approximation to the area under f, no matter what point $c_i \in (x_{i-1}, x_i)$ is used. The integral is accordingly defined to be the limit of the sums $\sum_{i=1}^{n} f(c_i)\Delta x_i$ as the partitions get finer and finer. Now we need a definition for this kind of limit, since the sums $\sum_{i=1}^{n} f(c_i)\Delta x_i$ are not functions of n, but functions of pairs (P, c), where P is a partition of $[a, b]$, and $\{c_1, \ldots, c_n\}$ is a finite sequence with $c_i \in (x_{i-1}, x_i)$ for each i. Rather than making an *ad hoc* definition for limits of sums $\sum f(c_i)\Delta x_i$, we define a general kind of limit, which will include anything that might reasonably be called a limit in analysis or topology. In particular, our old familiars $\lim_{n \to \infty} x_n$ and $\lim_{x \to a} f(x)$ will be special cases of the general definition.

The idea of limit involves a function the values of which approach a limit, ℓ, as the variable runs through the domain in some specific manner. We start, therefore, with conditions on the domain of a function that allow us to say how the variable runs through the domain. A **directed set** is a nonempty set D on which there is a partial ordering $\prec$ that satisfies certain conditions. A **partial ordering** is a relation $\alpha \prec \beta$ that holds for *some* pairs α and β of elements in D. For D to be a directed set with the ordering $\prec$, the following conditions must be satisfied:

(i) $\alpha < \alpha$ for all $\alpha \in D$.

(ii) If $\alpha < \beta$ and $\beta < \gamma$, then $\alpha < \gamma$ for all $\alpha, \beta, \gamma \in D$.

(iii) For any two elements α and β of D, there is an element $\gamma \in D$ such that $\alpha < \gamma$ and $\beta < \gamma$.

We will use α, β, γ, and so forth, for generic elements of a directed set as a reminder that the elements of D need not be countable (like $\mathbb{N}$) or totally ordered (like $\mathbb{R}$). We say β is *farther out* than α if $\alpha < \beta$, and we write $\beta > \alpha$ to mean the same as $\alpha < \beta$.

The notation $<$ was already used for a relation between partitions, and this was not accidental. The set of all partitions of a given interval is a directed set with the refinement relation $P < Q$.

> **PROBLEM 1:** (i) Show that refinement, $P < Q$, makes the set of all partitions of a fixed interval a directed set.
>
> (ii) If $D = \mathbb{N}$ and $m < n$ means $m \leq n$, then D is a directed set.
>
> (iii) If $D = \{x : 0 < |x - a| < \delta\}$ and $x < y$ means $|y - a| \leq |x - a|$, then D is a directed set.
>
> (iv) If D is all finite subsets of some set A and $F < G$ means $F \subset G$, then D is a directed set. •

The things that have limits are functions on directed sets. We will call a function on a directed set a **net**. If x is a net on the directed set D, we write $\{x_\alpha\}$ to suggest the similarity with sequential behavior. Finally, the limit definition is this: If $\{x_\alpha\}$ is a net on D, then $x_\alpha \longrightarrow \ell$ (x_α converges to ℓ) if and only if for each $\varepsilon > 0$ there is $\alpha_0 \in D$ such that $|x_\alpha - \ell| < \varepsilon$ whenever $\alpha > \alpha_0$. Condition (iii) for directed sets ensures that limits of nets are unique.

> **PROBLEM 2:** If $\{x_\alpha\}$ is a net on D, and $x_\alpha \longrightarrow \ell$ and $x_\alpha \longrightarrow m$, then $\ell = m$. •

Since limits are unique, we can write $\lim_\alpha x_\alpha = \ell$ to mean $x_\alpha \longrightarrow \ell$.

The proofs of the usual limit-of-a-sum (difference, product, quotient) theorems are nearly the same as the proofs for sequences.

PROPOSITION 1: Let $\{x_\alpha\}, \{y_\alpha\}, \{z_\alpha\}$ *be nets on the same directed set D, and assume that $x_\alpha \to \ell$ and $y_\alpha \to m$. Then*

$$x_\alpha \pm y_\alpha \longrightarrow \ell \pm m,$$

$$x_\alpha y_\alpha \longrightarrow \ell m,$$

$$\frac{x_\alpha}{y_\alpha} \longrightarrow \frac{\ell}{m} \text{ if } m \neq 0.$$

If $x_\alpha \leq z_\alpha \leq y_\alpha$ for all α, and $\ell = m$, then $z_\alpha \longrightarrow \ell$.

Proof: We prove the product limit by way of illustration and leave the others as a problem. Assume $x_\alpha \longrightarrow \ell$ and $y_\alpha \longrightarrow m$, or what is the same, that $x_\alpha - \ell \longrightarrow 0$, $y_\alpha - m \longrightarrow 0$. Let $r_\alpha = x_\alpha - \ell$ and $s_\alpha = y_\alpha - m$, so that

$x_\alpha = \ell + r_\alpha, y_\alpha = m + s_\alpha$ with $r_\alpha \longrightarrow 0, s_\alpha \longrightarrow 0$. Then

$$x_\alpha y_\alpha - \ell m = (\ell + r_\alpha)(m + s_\alpha) - \ell m$$

$$= m r_\alpha + \ell s_\alpha + r_\alpha s_\alpha.$$

Let $\varepsilon > 0$. Pick α_1 so that if $\alpha > \alpha_1$, then $|r_\alpha| < 1$. Pick α_2 so that if $\alpha > \alpha_2$, $|r_\alpha m| < \varepsilon/3$. Let α_0 be farther out than both α_1 and α_2, so that if $\alpha > \alpha_0$, both inequalities hold. Using the same technique, find β_0 so that if $\alpha > \beta_0$, $|s_\alpha| < \varepsilon/3$ and $|s_\alpha \ell| < \varepsilon/3$. Let γ_0 be farther out than both α_0 and β_0. If $\alpha > \gamma_0$, then

$$|x_\alpha y_\alpha - \ell m| \leq |m r_\alpha| + |\ell s_\alpha| + |r_\alpha s_\alpha|$$

$$< \frac{\varepsilon}{3} + \frac{\varepsilon}{3} + \frac{\varepsilon}{3} = \varepsilon. \qquad \blacksquare$$

PROBLEM 3: Assume that $x_\alpha \longrightarrow \ell$ and $y_\alpha \longrightarrow m$ and prove that $x_\alpha - y_\alpha \longrightarrow \ell - m$ and $x_\alpha / y_\alpha \longrightarrow \ell/m$ if $m \neq 0$. Hint: If $m \neq 0$, then y_α could still be zero for lots of α's, but you can show that there is some α_0 so that $y_\alpha \neq 0$ if $\alpha > \alpha_0$. The net $\{x_\alpha / y_\alpha\}$ is interpreted to mean the function of α with value equal to x_α / y_α whenever $y_\alpha \neq 0$. Thus $\{x_\alpha / y_\alpha\}$ is generally a net on some tail-like subset of the original directed set D. $\qquad\bullet$

Now we return to Riemann's definition of the integral. Let f be any function — bounded or not — on $[a, b]$. If $P = \{x_0, \ldots, x_n\}$ is a partition of $[a, b]$ and $c_i \in (x_{i-1}, x_i)$ for $i = 1, \ldots, n$, then we define the **Riemann sum for f and P and c** by

$$R(f, P, c) = \sum_{i=1}^{n} f(c_i)(x_i - x_{i-1}). \qquad (1)$$

For a fixed function f on $[a, b]$, $R(f, P, c)$ is a net on the set of pairs (P, c). Any function c with $c_i \in (x_{i-1}, x_i)$ for each i is called a **choice function** for the partition P. We partially order the pairs (P, c) as follows:

$$(P, c) < (P', c') \text{ if and only if } P < P'.$$

In other words, the pairs (P, c) are ordered by refinement of the partition, and the choice function plays no role in the ordering. If c and c' are two choice functions for the same partition P, then $(P, c) > (P, c')$ and $(P, c') > (P, c)$, since $P > P$ for any P. Since the ordering of pairs (P, c) depends only on P, we will write $\lim_P R(f, P, c)$ instead of the more cumbersome $\lim_{(P,c)} R(f, P, c)$.

If $\lim_P R(f, P, c) = I$, we say that f is Riemann integrable over $[a, b]$, with Riemann integral I. We proceed rapidly to show that Riemann integrability and Darboux integrability mean the same thing. Occam's razor ("Entities should not be multiplied unnecessarily") is one of the soundest ideas to come out of thousands of years of philosophical inquiry.

To show that the two definitions of integrability are equivalent, we must first notice that the Darboux integral is defined *only* for bounded functions. Upper and lower sums make no sense for unbounded functions. The Riemann sums, however, are defined for all functions on $[a, b]$, and it is conceivable that we could have the net $\{R(f, P, c)\}$ converging for an unbounded f. We settle this question first.

PROPOSITION 2: *If f is Riemann integrable on $[a, b]$, then f is bounded.*

Proof: Assume that f is integrable over $[a, b]$, and let $\lim_P R(f, P, c) = I$. For $\varepsilon = 1$ there is a partition P_0 of $[a, b]$ such that $|R(f, P, c) - I| < 1$ whenever $P > P_0$, and where c is any choice function for P. In particular, if c and c' are any two choice functions for P_0, then

$$|R(f, P_0, c) - I| < 1 \qquad \text{and} \qquad |R(f, P_0, c') - I| < 1,$$

and so $|R(f, P_0, c) - R(f, P_0, c')| < 2$. Fix any choice function c for $P_0 = \{x_0, \ldots, x_n\}$. Suppose f is not bounded, and to be specific suppose f is not bounded above. Then f is not bounded above on some subinterval (x_{i-1}, x_i), and we may as well assume this subinterval is (x_0, x_1). Define a second choice function c' for P_0 by letting $c'_i = c_i$ for $i = 2, 3, \ldots, n$. For $i = 1$, pick a point $c'_1 \in (x_0, x_1)$ such that $f(c'_1) > N$, where N can be as large as we like. Since $c'_i = c_i$ for $i \geq 2$,

$$R(f, P_0, c') - R(f, P_0, c) = \left(f(c'_1) - f(c_1)\right)(x_1 - x_0)$$
$$> \left(N - f(c_1)\right)(x_1 - x_0).$$

We can take N so large that $R(f, P_0, c') - R(f, P_0, c) > 2$, which contradicts our choice of P_0. ∎

We say that a net $\{x_\alpha\}$ is **increasing** if $x_\alpha \leq x_\beta$ whenever $\alpha < \beta$, with similar agreements for decreasing nets.
A net $\{x_\alpha\}$ is **bounded above** provided there is a number M such that $x_\alpha \leq M$ for all α, and $\{x_\alpha\}$ is **bounded below** if there is M so that $M \leq x_\alpha$ for all α. A net is **bounded** if it is bounded above and below. The upper sums $\{U(f, P)\}$ for a given function form a decreasing net, and the lower sums $\{L(f, P)\}$ are an increasing net.

PROBLEM 4: If $\{x_\alpha\}$ is an increasing bounded net, then $\lim x_\alpha = \sup\{x_\alpha : \alpha \in D\}$. If $\{x_\alpha\}$ is a decreasing bounded net, then $\lim x_\alpha = \inf\{x_\alpha : \alpha \in D\}$. It follows that $\sup_P L(f, P) = \lim_P L(f, P)$ and $\inf_P U(f, P) = \lim_P U(f, P)$ for any bounded function f. •

From Problem 4 we see that f is Darboux integrable if and only if

$$\lim_P L(f, P) = \lim_P U(f, P).$$

For any partition P of $[a, b]$ and any choice function c for P, we have

$$m_i \leq f(c_i) \leq M_i$$

for all i, and hence

$$\sum m_i \Delta x_i \leq \sum f(c_i)\Delta x_i \leq \sum M_i \Delta x_i,$$
$$\text{or} \quad L(f, P) \leq R(f, P, c) \leq U(f, P). \tag{2}$$

If $\{L(f, P)\}$, $\{U(f, P)\}$, and $\{R(f, P, c)\}$ were nets on the same directed set, it would follow immediately from (2) that every Darboux integrable function is Riemann integrable, with the same integral. However, the directed sets are different; in one case the set of partitions is the directed set, and in the other the set of pairs (P, c) is the directed set. Nevertheless, it is true that Darboux integrability implies Riemann integrability, and the proof is simple if not immediate.

PROPOSITION 3: *If f is Darboux integrable on* [a, b] *with integral I, then f is Riemann integrable on* [a, b] *with integral I.*

Proof: Assume f is Darboux integrable with integral I, and let $\varepsilon > 0$. There is a partition P_0 so that $U(f, P) - L(f, P) < \varepsilon$ if $P > P_0$. Therefore,

$$I - \varepsilon < L(f, P) \leq I \leq U(f, P) < I + \varepsilon$$

for all $P > P_0$. Since

$$L(f, P) \leq R(f, P, c) \leq U(f, P)$$

for all P, we have in particular that

$$I - \varepsilon < R(f, P, c) < I + \varepsilon$$

if $P > P_0$, for any choice c, and so $\lim_P R(f, P, c) = I$. ∎

To show that Riemann integrability implies Darboux integrability, we show that for a given partition P there are choices that make the Riemann sum arbitrarily close to $L(f, P)$, and choices such that the Riemann sum is arbitrarily close to $U(f, P)$. If all the Riemann sums for $P > P_0$ are close to the integral I, then so are the upper and lower sums. The details follow.

PROPOSITION 4: *If f is Riemann integrable over* [a, b] *with integral I, then f is Darboux integrable over* [a, b], *and by Proposition 3 the integrals are the same.*

Proof: Assume that $\lim_P R(f, P, c) = I$. Let $\varepsilon > 0$. There is $P_0 = \{x_0, x_1, \ldots, x_n\}$ so that $|R(f, P_0, c) - I| < \varepsilon$ for any choice c for P_0. Let m_i and M_i be as usual for P_0. For each i, choose c_i and c_i' so that

$$M_i - f(c_i) < \frac{\varepsilon}{b - a},$$

$$f(c_i') - m_i < \frac{\varepsilon}{b - a}.$$

Then

$$0 \leq U(f, P_0) - R(f, P_0, c)$$

$$= \sum_{i=1}^{n} (M_i - f(c_i)) \Delta x_i$$

$$< \left[\frac{\varepsilon}{b - a} \right] \sum_{i=1}^{n} \Delta x_i = \varepsilon.$$

Similarly,

$$0 \le R(f, P_0, c_i') - L(f, P_0) < \varepsilon.$$

Since $R(f, P_0, c)$ and $R(f, P_0, c')$ are within ε of each other, $U(f, P_0) - L(f, P_0) < 3\varepsilon$, and f is Darboux integrable. ∎

Now we have just one concept of "integrable," and one notation, $\int_a^b f$, for its integral.

The linearity of *the* integral is easy to show using the (Riemann) limit-of-a-net idea.

PROPOSITION 5: *If f and g are integrable on $[a, b]$, and k is a constant, then kf, $f + g$, and $f - g$ are integrable on $[a, b]$, and*

$$\int_a^b kf = k \int_a^b f,$$

$$\int_a^b f + g = \int_a^b f + \int_a^b g,$$

$$\int_a^b f - g = \int_a^b f - \int_a^b g.$$

Proof: Assume that f and g are integrable, so that $\lim_P R(f, P, c) = \int_a^b f$, $\lim_P R(g, P, c) = \int_a^b g$. Notice that for all pairs (P, c),

$$R(kf, P, c) = kR(f, P, c),$$

$$R(f + g, P, c) = R(f, P, c) + R(g, P, c).$$

Since the nets $\{R(f, P, c)\}$ and $\{R(g, P, c)\}$ converge, the nets $\{R(kf, P, c)\}$ and $\{R(f + g, P, c)\}$ converge by Proposition 1, and

$$\lim_P R(kf, P, c) = k \lim_P R(f, P, c),$$

$$\lim_P R(f + g, P, c) = \lim_P R(f, P, c) + \lim_P R(g, P, c).$$

That is,

$$\int_a^b kf = k \int_a^b f,$$

and

$$\int_a^b f + g = \int_a^b f + \int_a^b g. \quad \blacksquare$$

We will define $\int_a^a f = 0$ for any function f, and $\int_b^a f = - \int_a^b f$ if $a < b$.

Problem 5: Show that $\int_a^b f + \int_b^c f = \int_a^c f$ for all a, b, c, provided f is integrable on an interval containing all of a, b, c. Hint: Here we do not assume $a \leq b \leq c$.   •

Problem 6: Let f be integrable on $[a, b]$, and $F(x) = \int_a^x f$ for $a \leq x \leq b$. If f is continuous at $x_0 \in [a, b]$, then $F'(x_0) = f(x_0)$. Hint: Do this first for $a < x_0 < b$. If $x_0 = a$ or $x_0 = b$, then F is not defined on an interval around x_0 and so $F'(x_0)$ does not exist according to our definition. Show how to define one-sided derivatives $F'(a+)$ and $F'(b-)$ so that the statement holds for $x_0 = a$ or $x_0 = b$.   •

The Fundamental Theorem of Calculus now follows easily from the Riemann definition.

Proposition 6 (The Fundamental Theorem of Calculus): *Suppose f is integrable on $[a, b]$ and there is F on $[a, b]$ such that $F'(x) = f(x)$ for all $x \in [a, b]$, where $F'(a)$ and $F'(b)$ are the one-sided derivatives you defined in Problem 6. (Notice that such a function F does exist if f is continuous.) Then*

$$\int_a^b f = F(b) - F(a).$$

Proof: Since F' exists on $[a, b]$, F is continuous on $[a, b]$. The one-sided continuity of F at a and b follows from the existence of the one-sided derivative in the usual way. Let $\varepsilon > 0$ and let P be a partition such that

$$\left| R(f, P, c) - \int_a^b f \right| < \varepsilon$$

for all choices c for P. If $P = \{x_0, x_1, \ldots, x_n\}$, then the Mean Value Theorem for F holds on each interval $[x_{i-1}, x_i]$; that is, for each i there is $c_i \in (x_{i-1}, x_i)$ such that

$$F(x_i) - F(x_{i-1}) = F'(c_i)(x_i - x_{i-1})$$
$$= f(c_i)(x_i - x_{i-1}).$$

Notice that

$$\sum_{i=1}^n F(x_i) - F(x_{i-1}) = F(b) - F(a).$$

Moreover,

$$\sum_{i=1}^n f(c_i)(x_i - x_{i-1}) = R(f, P, c).$$

Thus we have a choice c for P such that

$$R(f, P, c) = F(b) - F(a),$$

and for *any* choice c for P

$$\left| R(f, P, c) - \int_a^b f \right| < \varepsilon.$$

Hence $|\int_a^b f - (F(b) - F(a))| < \varepsilon$ for every $\varepsilon > 0$, and equality holds. ∎

The Riemann integral is defined only for bounded functions f and bounded domains $[a, b]$. The idea is extended to certain "improper" cases as follows. Suppose f is defined on $(a, b]$ and f is integrable on $[a + \varepsilon, b]$ for all $\varepsilon > 0$ (with $a + \varepsilon < b$), but f is unbounded at a. Then we define

$$\int_a^b f = \lim_{\varepsilon \to 0+} \int_{a+\varepsilon}^b f$$

provided this limit exists. Similarly, if f is unbounded near b, but integrable on every interval $[a, b - \varepsilon]$, we define

$$\int_a^b f = \lim_{\varepsilon \to 0+} \int_a^{b-\varepsilon} f.$$

If these limits exist, we say that f is improperly integrable on $[a, b]$.

The limits $\lim_{x \to \infty} f(x)$ and $\lim_{x \to -\infty} f(x)$ are defined to be limits of nets in the following way. Suppose f is defined for all $x \geq N$ for some N. The set $[N, \infty)$ is a directed set with the usual ordering: $x_1 < x_2$ means $x_1 \leq x_2$. Then if we write "$\lim_{x \to \infty}$" for the usual "$\lim_x$," we see that $\lim_{x \to \infty} f(x) = \ell$ means that for each $\varepsilon > 0$ there is some $x_0 \in [N, \infty)$ such that $|f(x) - \ell| < \varepsilon$ whenever $x \geq x_0$. The limit as $x \longrightarrow -\infty$ is defined similarly.

If f is integrable on every interval $[a, b]$ for $a < b$, then $\int_a^b f$ is a net on the directed set $\{b : b \geq a\}$, and we define

$$\int_a^\infty f = \lim_b \int_a^b f = \lim_{b \to \infty} \int_a^b f.$$

A similar definition holds for $\int_{-\infty}^a f$.

PROBLEM 7: (i) If $f \geq 0$ on $[a, \infty)$ and integrable on every interval $[a, b]$, then f is integrable on $[a, \infty)$ if and only if $\{\int_a^b f : a < b\}$ is bounded.
(ii) Give an example of a nonnegative function f on $[0, \infty)$ such that $\int_0^\infty f$ exists, but $f(x)$ does not approach zero as $x \longrightarrow \infty$. •

PROBLEM 8: (i) Let $f(x)$ be 1 on $[0, 1/2)$ and -1 on $[1/2, 1)$. Divide $[1, 2)$ into four equal intervals, and let f be alternately $+1$ and -1 on these. In general, divide $[n, n + 1)$ into 2^n equal intervals and let f be alternately $+1$ and -1 on these. Show that $|f(x)| \equiv 1$ on $[0, \infty)$ and $\int_0^\infty f = 0$.
(ii) Give an example of f on $[0, \infty)$ such that $|f(x)| \longrightarrow \infty$ as $x \longrightarrow \infty$ and $\int_0^\infty f = 0$. •

PROBLEM 9: If f is continuous on $[a, \infty)$ and $|f|$ is integrable over $[a, \infty)$, then f is integrable over $[a, \infty)$. Hint: Let $f^+(x) = \max\{f(x), 0\}$ and $f^-(x) = \max\{-f(x), 0\}$, so that $f(x) = f^+(x) - f^-(x)$ and $|f(x)| = f^+(x) + f^-(x)$. Show that f^+ and f^- are continuous and that the sets $\{\int_a^b f^+ : b > a\}$, $\{\int_a^b f^- : b > a\}$ are bounded. Cf. Problem 7(i). $\bullet$

PROBLEM 10: Let f and g be continuous functions on $[a, \infty)$ such that $\int_a^\infty f^2$ and $\int_a^\infty g^2$ both exist. Show that $\int_a^\infty fg$ exists. Hint: It is sufficient to show that $\int_a^\infty |fg|$ exists. Notice that $2|f(x)g(x)| \le |f(x)|^2 + |g(x)|^2$ for all x, so that it is sufficient to show that $\int_a^\infty (f^2 + g^2)$ exists. $\bullet$

PROBLEM 11: For each partition $P = \{x_0, x_1, \ldots, x_n\}$ of a fixed interval $[a, b]$, define $\|P\| = \max_i(x_i - x_{i-1})$. ($\|P\|$ is called the **norm** of P.) Define a partial ordering $\oslash$ on pairs (P, c) by $(P, c) \oslash (Q, c')$ if and only if $\|Q\| \le \|P\|$. Show that $R(f, P, c)$ is a net on the directed set of pairs (P, c) with ordering $\oslash$. For f defined on $[a, b]$, say that f is norm-integrable with integral I provided $\lim_{\|P\|} R(f, P, c) = I$. (That is, given $\varepsilon > 0$ there is a pair (P_0, c_0) such that $|R(f, P, c) - I| < \varepsilon$ whenever $\|P\| \le \|P_0\|$.) Show that f is norm-integrable with integral I if and only if $\int_a^b f = I$. Hint: It is easy to show that norm-integrability implies integrability. To show the other implication, let $\varepsilon > 0$ and choose $P_0 = \{x_0, \ldots, x_N\}$ such that $U(f, P_0) - L(f, P_0) < \varepsilon$. Let $P_1 = \{y_0, y_1, \ldots, y_m\}$ with $\|P_1\| < \delta < \|P_0\|/3$, so that each x_i is either equal to some y_j or is included in exactly one interval (y_{j-1}, y_j). Let M_i, m_i, M_j^*, m_j^* be the sups and infs for P_0 and P_1, respectively. If $y_0 < y_1 < \cdots < y_k \le x_1 < y_{k+1}$, then

$$\sum_{j=1}^{k} (M_j^* - m_j^*)(y_j - y_{j-1}) \le (M_1 - m_1)(x_1 - x_0).$$

Similar estimates hold for the terms $(M_j^* - m_j^*)\Delta y_j$ for y_j between x_1 and x_2, and so forth. This includes all terms of $U(f, P_1) - L(f, P_1)$ except those with some $x_i \in (y_{j-1}, y_j)$. There are at most $N - 1$ of these, and their sum is less than $(N - 1)(M - m)\delta$, where M and m are bounds for f. Thus,

$$U(f, P_1) - L(f, P_1) \le U(f, P_0) - L(f, P_0) + (N - 1)(M - m)\delta$$

whenever $\|P_1\| < \delta < \|P_0\|/3$. Continue, and show that $|R(f, P, c) - \int_a^b f| < 2\varepsilon$ whenever $\|P\| \le \|P_1\|$. $\bullet$

XVI

log x AND e^x

Our array of explicit functions is pretty thin so far because of the difficulty of defining nonarithmetic functions. Functions like x^2 and x^n are defined in any field, but most of our favorites from calculus require some sort of limit for their definition. Thus $\sin x$, e^x, and $\log x$ are meaningless in an arbitrary field. Recall that we do have a definition of a^x for any $a > 0$. The defining procedure was the following: First, x^n is strictly increasing on $[0, \infty)$ and continuous, and so there is an inverse function. This defines $x^{1/n}$ for $x \geq 0$ and $n \in \mathbb{N}$. That leads to definitions $x^{m/n} = \left(x^{1/n}\right)^m$, and $x^{-m/n} = \left(x^{m/n}\right)^{-1}$, so that x^r is defined for all $x \geq 0$ and all rational numbers r. Now we fix the x — call it a so it looks fixed — and consider the function a^r defined for rational numbers r. We showed that a^r is uniformly continuous on the set of rationals in any finite interval, and hence that a^r extends to a unique continuous function a^x for all x in any bounded interval. Since the interval is arbitrary, a^x is defined for $x \in \mathbb{R}$. Thus we know what e^x means whenever we decide what the number e is, but we still have no simple way of finding $\frac{d}{dx} e^x$ from this definition.

Instead of fighting through the old definition of a^x to calculate its derivative, we start with the definition of $\log x$ as the familiar integral, from which its properties are easy to verify. The properties of the inverse function, e^x, follow from those of $\log x$.

We define $\log x$ for $x > 0$ by

$$\log x = \int_1^x \frac{1}{t}\, dt. \tag{1}$$

It is immediate from (1) that $\frac{d}{dx} \log x = \frac{1}{x}$ for all $x > 0$, since $\frac{1}{t}$ is continuous. Therefore, $\log x$ is a strictly increasing function on $(0, \infty)$.

PROPOSITION 1: *For all $x > 0$ and $y > 0$,*

$$\log xy = \log x + \log y.$$

Proof: Let y be any positive constant, and define the functions $f(x) = \log xy$, $g(x) = \log x + \log y$. It is easy to check that $f'(x) = g'(x) = 1/x$ for all $x > 0$, and so f and g differ by a constant (Problem 10, Chapter 13). Since $f(1) = g(1) = \log y$, f and g are identical. ∎

PROBLEM 1: (i) Show that $\log x^n = n \log x$ for all $x > 0$ and all integers n. (N.B. *integers*, not natural numbers.)

(ii) Show that $\log x$ is one-to-one on $(0, \infty)$ onto $(-\infty, \infty)$. •

From Problem 1 we know there is a unique number x_0 such that $\log x_0 = 1$. This number is designated e, and numerical calculations with the integral will show that e is approximately 2.718.

To complete the connection between $\log x$ and e^x, we let $E(x)$ denote the inverse to $\log x$. (We now know what e^x is, but we do not yet know that $E(x) = e^x$.) Thus for all $x > 0$, $E(\log x) = x$, and for all $x \in \mathbb{R}$, $\log E(x) = x$. It follows that for all x, y,

$$\log (E(x)E(y)) = \log E(x) + \log E(y) = x + y.$$

Since we also know that $\log E(x + y) = x + y$ for all x, y, and $\log x$ is one-to-one, we have the identity

$$E(x + y) = E(x)E(y). \tag{2}$$

Since $\log e = 1$, $E(1) = e$, and from (2) we get

$$E(2) = e^2, \quad E(3) = e^3, \quad \ldots, \quad E(n) = e^n, \quad \ldots.$$

We also conclude from (2) that

$$E\left(\frac{1}{n} + \frac{1}{n} + \cdots + \frac{1}{n}\right) = E(1) = e = E\left(\frac{1}{n}\right)^n,$$

so that $E(1/n) = e^{1/n}$. Consequently $E(m/n) = e^{m/n}$ for all $m, n \in \mathbb{N}$. From (2) we also conclude (see Problem 2) that $E(x - y) = E(x)/E(y)$ for all x, y, and so

$$E\left(-m/n\right) = \left(e^{m/n}\right)^{-1},$$

which gives us $E(r) = e^r$ for all rationals r. We know that E is continuous, and that there is only one continuous function, e^x, that agrees with e^r on the rationals. Therefore, $E(x) = e^x$, and e^x is the inverse of $\log x$.

PROBLEM 2: Use (2) to show that $E(x - y) = E(x)/E(y)$ for all x, y. •

PROBLEM 3: Show that $\frac{d}{dx}e^x = e^x$ for all x, and conclude that e^x has derivatives of all orders. Does $\log x$ have derivatives of all orders? Hint: You know that e^x and $\log x$ are both differentiable, and $\log e^x = x$ for all x. Differentiate both sides using the chain rule. •

PROBLEM 4: Show that e^x is a convex function. •

PROBLEM 5: (i) Show that $\frac{1}{n}\log n \longrightarrow 0$ as $n \longrightarrow \infty$. Hint: If you show that $(\log n)/n$ is monotone, then it suffices to consider a subsequence n_k, for example, $n_k = 2^k$.

(ii) Show that $x/e^x \longrightarrow 0$ as $x \longrightarrow \infty$. •

PROBLEM 6: Extend the result of Problem 1 by showing that $\log a^{1/n} = \frac{1}{n}\log a$ and $\log a^{m/n} = \frac{m}{n}\log a$ for all $a > 0$ and all integers m and n. Then show that $\log a^r = r\log a$ for all $a > 0$ and all $r \in \mathbb{R}$. •

PROBLEM 7: Show that $\lim_{n\to\infty}\left(1 + \frac{1}{n}\right)^n = e$. Hint: Consider the logarithm and apply l'Hospital's rule to $\lim_{x\to 0}\frac{1}{x}\log(1 + x)$. •

PROBLEM 8: Show that $\frac{d}{dx}a^x = a^x\log a$. Hint: From Problem 6, $a^x = e^{x\log a}$. •

PROBLEM 9: Define $\sinh x = \frac{1}{2}(e^x - e^{-x})$, $\cosh x = \frac{1}{2}(e^x + e^{-x})$. Verify the following.

(i) $\frac{d}{dx}\cosh x = \sinh x$; $\frac{d}{dx}\sinh x = \cosh x$.
(ii) $\cosh^2 x - \sinh^2 x = 1$.
(iii) $\sinh x$ is strictly increasing on $\mathbb{R}$ onto $\mathbb{R}$, and $\sinh^{-1} x = \log\left(x + \sqrt{x^2 + 1}\right)$. •

PROBLEM 10: Show that $2.6 < e < 2.8$. Hint: Find partitions P_1 of $[1, 2.6]$ and P_2 of $[1, 2.8]$ such that $U(P_1, 1/x) < 1$ and $L(P_2, 1/x) > 1$. •

PROBLEM 11: (i) Show that if f is decreasing on $[a, b]$ and $P = \{x_0, \ldots, x_n\}$ is the uniform partition of $[a, b]$ such that $\Delta x = x_i - x_{i-1} = (b - a)/n$ for all i, then $U(f, P) - L(f, P) \le \Delta x(f(a) - f(b))$.

(ii) Let $P = \{1, 1.1, 1.2, \ldots, 2.7\}$. Estimate the error between $(U(1/x, P) + L(1/x, P))/2$ and $\log 2.7$. •

XVII

UNORDERED SUMS AND INFINITE SERIES

Recall that finite sums are defined inductively, starting with the axiomatic assumption that $x_1 + x_2$ makes sense. The inductive step, assuming that $x_1 + \cdots + x_n$ has been defined, is

$$x_1 + \cdots + x_n + x_{n+1} = (x_1 + \cdots + x_n) + x_{n+1},$$

which makes sense since the right side is the sum of *two* numbers. The fact that such finite sums can be rearranged or grouped arbitrarily is proved by tedious and uninstructive inductive arguments, which by now we feel free to omit.

Now we want to extend the idea of summation from finite sets of numbers to infinite sets of numbers. Our goal and pious hope is that the infinite sum we define will behave like a finite sum. In particular, we would like to be able to rearrange and group the summands at will. For example, a critical step in many arguments is the equality

$$\sum_{i=1}^{\infty}\sum_{j=1}^{\infty} x_{ij} = \sum_{j=1}^{\infty}\sum_{i=1}^{\infty} x_{ij}.$$

This is indeed an identity for the unordered sums we define first, but not, alas, for the familiar infinite series of the calculus.

We start with an indexed family of numbers — one number x_α for each α in some index set A — and we want to add up all the numbers x_α for $\alpha \in A$. The set A is not assumed to be finite or even countable. We define the sum, denoted $\sum_{\alpha \in A} x_\alpha$ or $\sum_A x_\alpha$ or $\sum_\alpha x_\alpha$, as the limit of finite sums. To make a net of the finite sums, we partially order the *finite* subsets of A by inclusion, so that $F < G$ means $F \subset G$. The finite subsets of A are thus a directed set. If S_F denotes the sum of the finite number of elements x_α for $\alpha \in F$, then $\{S_F\}$ is a net on the finite subsets of A, and we define

$$\sum_{\alpha \in A} x_\alpha = \lim_F S_F = \lim_F \sum_{\alpha \in F} x_\alpha.$$

Notice that if A is a finite set, then the new definition of $\sum_{\alpha \in A} x_\alpha$ coincides with our familiar fourth grade concept. The sum $\sum_{\alpha \in A} x_\alpha$ is called the **unordered sum**, and we say the sum **converges** when the net $\{S_F\}$ converges, and that the x_α are **summable over** A. The definition of limit in this case is the following: $\sum_{\alpha \in A} x_\alpha = s$ if and only if for each $\varepsilon > 0$ there is a finite subset F of A such that $\left| \sum_{\alpha \in G} x_\alpha - s \right| < \varepsilon$ whenever G is a finite set such that $F \subset G \subset A$.

PROBLEM 1: Let x and y be two functions on an index set A, and assume that $\sum_{\alpha \in A} x_\alpha = s$ and $\sum_{\alpha \in A} y_\alpha = t$. Show that

(i) $\sum_{\alpha \in A} k x_\alpha = ks$ for any number k;
(ii) $\sum_{\alpha \in A} (x_\alpha + y_\alpha) = s + t$;
(iii) $\sum_{\alpha \in A} (x_\alpha - y_\alpha) = s - t$.

Hint: There really is not much to prove here. The idea is for you to pause and see that the foregoing results are just special cases of previously proved facts about convergent nets, and elementary facts about finite sums.   ●

Recall that a *sequence* $\{x_n\}$ converges if and only if it is a Cauchy sequence; that is, if given $\varepsilon > 0$, there is N such that $|x_n - x_m| < \varepsilon$ for all n and m greater than N. We similarly define a net $\{x_\alpha\}$ to be a **Cauchy net** if given $\varepsilon > 0$, there is α_0 such that $|x_\alpha - x_\beta| < \varepsilon$ whenever $\alpha > \alpha_0$ and $\beta > \alpha_0$. A net $\{x_\alpha\}$ converges if and only if it is a Cauchy net (see Problem 2), and that provides a convenient criterion for the convergence of unordered sums (see Problem 3).

PROBLEM 2: Show that a net $\{x_\alpha\}$ converges if and only if it is Cauchy net. Hint: To show that the Cauchy condition is sufficient for convergence, pick α_n for each n such that $|x_\alpha - x_\beta| < 1/n$ if $\alpha > \alpha_n$ and $\beta > \alpha_n$; that is, let α_n correspond to $\varepsilon = 1/n$. The indices α_n can be chosen so that $\alpha_1 < \alpha_2 < \alpha_3 < \cdots$. Show that $\{x_{\alpha_n}\}$ is a Cauchy sequence. Let ℓ be the limit of the sequence $\{x_{\alpha_n}\}$, and show that ℓ is the limit of the net.   ●

PROBLEM 3: (i) The sum $\sum_{\alpha \in A} x_\alpha$ converges if and only if for each $\varepsilon > 0$ there is a finite set $F \subset A$ such that $\left| \sum_{\alpha \in G} x_\alpha \right| < \varepsilon$ for every finite set G that is disjoint from F. Hint: We of course know things like $\sum_{F \cup G} x_\alpha = \sum_F x_\alpha + \sum_G x_\alpha$ for disjoint *finite* sets F and G. Use Problem 2.
(ii) If $\sum_{\alpha \in A} x_\alpha$ converges, and $B \subset A$, then $\sum_{\alpha \in B} x_\alpha$ converges.   ●

The summation properties of the following proposition hold for unordered sums, and of course for finite sums, but they do not hold for infinite series.

PROPOSITION 1: *(i) If A and B are two disjoint sets, then $\sum_{A \cup B} x_\alpha$ converges if and only if $\sum_A x_\alpha$ and $\sum_B x_\alpha$ both converge, and then $\sum_{A \cup B} x_\alpha = \sum_A x_\alpha + \sum_B x_\alpha$.*

(ii) If $A^+ = \{\alpha \in A : x_\alpha \geq 0\}$ and $A^- = \{\alpha \in A : x_\alpha < 0\}$, and $\sum_A x_\alpha$ converges, then $\sum_A x_\alpha = \sum_{A^+} x_\alpha + \sum_{A^-} x_\alpha$ and $\sum_A |x_\alpha| = \sum_{A^+} x_\alpha - \sum_{A^-} x_\alpha$.

(iii) $\sum_A x_\alpha$ converges if and only if $\sum_A |x_\alpha|$ converges, and $\left| \sum_A x_\alpha \right| \leq \sum_A |x_\alpha|$.

Proof: (i) If $\sum_{A \cup B} x_\alpha$ converges, then both $\sum_A x_\alpha$ and $\sum_B x_\alpha$ converge by Problem 3 (ii). Assume, then, that $\sum_A x_\alpha = s_A$ and $\sum_B x_\alpha = s_B$. Let $\varepsilon > 0$. There is a finite set $F_A \subset A$ such that $\left| \sum_G x_\alpha - s_A \right| < \varepsilon/2$ if $F_A \subset G \subset A$, and there is a finite set $F_B \subset B$ such that $\left| \sum_G x_\alpha - s_B \right| < \varepsilon/2$ if $F_B \subset G \subset B$. Let $F = F_A \cup F_B$, and let G be a finite subset of $A \cup B$ containing F. Then G is the disjoint union of $G \cap A \supset F_A$ and $G \cap B \supset F_B$, so that

$$\left| \sum_G x_\alpha - (s_A + s_B) \right| \leq \left| \sum_{G \cap A} x_\alpha - s_A \right| + \left| \sum_{G \cap B} x_\alpha - s_B \right| < \varepsilon;$$

that is,

$$\sum_{A \cup B} x_\alpha = s_A + s_B = \sum_A x_\alpha + \sum_B x_\alpha.$$

Parts (ii) and (iii) follow readily from part (i). ∎

PROBLEM 4: Write out the details of the proofs of parts (ii) and (iii) of Proposition 1. •

The last parts of Proposition 1 are worthy of special notice. *Convergence of unordered sums is a form of absolute convergence.*

One of the apparent differences between the unordered sums just defined and the familiar infinite series of the calculus is that an unordered sum allows an uncountable number of summands, since no countability restriction was placed on the index sets. In fact, except for a welter of zeros, only countable sums make sense.

PROPOSITION 2: *If $\sum_{\alpha \in A} x_\alpha$ converges, then the set $\{\alpha \in A : x_\alpha \neq 0\}$ is countable.*

Proof: Assume that there are uncountably many α such that $x_\alpha \neq 0$. Then there is some $n \in \mathbb{N}$ such that $|x_\alpha| \geq 1/n$ for uncountably many α. If there are uncountably many α with $x_\alpha \geq 1/n$, then there are uncountably many such α outside any finite set F, and the Cauchy criterion of Problem 2 surely fails. ∎

Proposition 2 makes it clear that the difference between unordered sums and the more familiar infinite series is not how many terms you can sum, but whether you order the summands. We recall now some facts about series, and we will see how series differ from unordered sums.

If $\{x_n\}$ is a sequence of numbers, then the infinite **series** $\sum_{n=1}^{\infty} x_n$ is defined to be the limit of **partial sums** $s_n = x_1 + \cdots + x_n$; that is, $\sum_{n=1}^{\infty} x_n = \lim s_n$, and convergence of the series is equivalent to convergence of the sequence $\{s_n\}$. Notice that a specific order of the summands x_i is involved in the definition of $\{s_n\}$. A reordering of $\{x_n\}$ gives a different sequence of partial sums, which may or may not converge. We emphasize the difference between the notation $\sum_{\mathbb{N}} x_n$ for the unordered sum of a sequence of elements $\{x_n\}$, and $\sum_{n=1}^{\infty} x_n$ for the limit the sequence $\{s_n\}$ where $s_n = x_1 + \cdots + x_n$.

PROPOSITION 3: *If $\sum_{n \in \mathbb{N}} x_n = s$, then $\sum_{n=1}^{\infty} x_n = s$, but not conversely.*

Proof: Assume the unordered sum converges to s: $\sum_{n \in \mathbb{N}} x_n = s$. Let $\varepsilon > 0$. There is a finite set $F \subset \mathbb{N}$ such that $\left| \sum_{G} x_n - s \right| < \varepsilon$ for every finite set G with $F \subset G \subset \mathbb{N}$. Let N be the largest number in F and let $\mathbb{N}_n = \{1, \ldots, n\}$ for each n. If $n \geq N$, $\mathbb{N}_n \supset F$ and so $\left| \sum_{\mathbb{N}_n} x_i - s \right| < \varepsilon$; that is, $|s_n - s| < \varepsilon$ if $n \geq N$, where

$$s_n = \sum_{\mathbb{N}_n} x_i = \sum_{i=1}^{n} x_i.$$

This shows that $s_n \longrightarrow s$, or $\sum_{n=1}^{\infty} x_n = s$. We will show that the converse is false in the next chapter by showing that convergence of $\sum_{n=1}^{\infty} x_n$ does not imply the convergence of $\sum_{n=1}^{\infty} |x_n|$, whereas we know from Proposition 1 that $\sum_{\mathbb{N}} x_n$ converges if and only if $\sum_{\mathbb{N}} |x_n|$ converges. ∎

The series $\sum_{n=1}^{\infty} x_n$ is called **absolutely convergent** provided $\sum_{n=1}^{\infty} |x_n|$ converges. To see that an absolutely convergent series is convergent, let

$$s_n = x_1 + \cdots + x_n,$$
$$\bar{s}_n = |x_1| + \cdots + |x_n|.$$

If $\sum_{n=1}^{\infty} x_n$ is absolutely convergent, then $\{\bar{s}_n\}$ is a Cauchy sequence. If $m > n$, then

$$|s_m - s_n| = |x_{n+1} + \cdots + x_m|$$
$$\leq |x_{n+1}| + \cdots + |x_m|$$
$$= \bar{s}_m - \bar{s}_n.$$

Therefore, $\{s_n\}$ is a Cauchy sequence if $\{\bar{s}_n\}$ is a Cauchy sequence, and $\sum_{n=1}^{\infty} x_n$ converges if $\sum_{n=1}^{\infty} |x_n|$ converges.

PROPOSITION 4: *If the series $\sum_{n=1}^{\infty} x_n$ is absolutely convergent, then the unordered sum $\sum_{n \in \mathbb{N}} x_n$ converges, and consequently $\sum_{n=1}^{\infty} x_n = \sum_{n \in \mathbb{N}} x_n$ by Proposition 3.*

Proof: We will show that if $\sum_{n=1}^{\infty} |x_n|$ converges, then $\sum_{n \in \mathbb{N}} |x_n|$ converges, and both sums are the sup of all finite sums of the $|x_n|$. Let $\bar{s}_n = |x_1| + \cdots + |x_n|$ and let $\bar{s} = \sup \bar{s}_n = \sum_{n=1}^{\infty} |x_n|$. If F is any finite subset of $\mathbb{N}$, then $F \subset \mathbb{N}_n = \{1, \ldots, n\}$ where n is the largest element in F. Therefore, for each finite $F \subset \mathbb{N}$ there is n such that $S_F \leq \bar{s}_n$, where $S_F = \sum_{n \in F} |x_n|$. It follows that $\sum_{n \in \mathbb{N}} |x_n|$ converges since the increasing net $\{S_F\}$ is bounded, and $\lim_F S_F \leq \bar{s}$. However, each $\bar{s}_n$ is a finite sum $S_{\mathbb{N}_n}$, and so $\lim_F S_F \geq \lim \bar{s}_n = s$ and $\sum_{n \in \mathbb{N}} |x_n| = \sum_{n=1}^{\infty} |x_n|$. Since $\sum_{\mathbb{N}} |x_n|$ converges, $\sum_{\mathbb{N}} x_n$ converges and $\sum_{\mathbb{N}} x_n = \sum_{n=1}^{\infty} x_n$. ∎

PROBLEM 5: Let $\sum_{n=1}^{\infty} x_n$ be an absolutely convergent series, and let $\{y_k\}$ be a rearrangement of $\{x_n\}$; that is, let n_k be a one-to-one function on $\mathbb{N}$ onto $\mathbb{N}$ such that $y_k = x_{n_k}$ for each k. Show that $\sum_{n=1}^{\infty} x_n = \sum_{k=1}^{\infty} y_k$. That is, absolutely convergent *series* can be rearranged at will. Hint: This is simply an accounting exercise. If you put together the right facts from the preceding results in the right order, you do not have to do any additional work.   •

PROBLEM 6: Show that if $\sum_{n=1}^{\infty} x_n$ and $\sum_{n=1}^{\infty} y_n$ both converge absolutely, then $\sum_{n=1}^{\infty} (x_n + y_n)$ converges absolutely. Hint: It is easy to show that the partial sums $\sum_{n=1}^{N} |x_n + y_n|$ are bounded. However, it may be more instructive to notice that absolute convergence of a series is equivalent to convergence of the net of finite sums, and the sum of two convergent nets is convergent.   •

The reordering of terms is permissible in unordered sums, and in absolutely convergent series, but not in conditionally convergent series. The other property of finite sums that we need frequently is the grouping property. This fails badly for series. For example, the series

$$(1 - 1) + (1 - 1) + (1 - 1) + \cdots$$

converges to zero, but the series

$$1 - 1 + 1 - 1 + 1 + \cdots$$

diverges. A more convincing example involves the fact that changing the order of summation, which is a kind of grouping, is not always possible. Consider the series

$$1 - 1 + \frac{1}{2} - \frac{1}{2} + \frac{1}{3} - \frac{1}{3} + \cdots. \tag{1}$$

This series converges to zero. Now let $\{x_{mn}\}$ be a square array of numbers such that every row $n = $ constant is the series (1). Then $s_n = \sum_{m=1}^{\infty} x_{mn} = 0$ for all n, and so

$$\sum_{n=1}^{\infty} \sum_{m=1}^{\infty} x_{mn} = \sum_{n=1}^{\infty} s_n = 0.$$

However, $\sum_{n=1}^{\infty} x_{mn}$, the sum up the mth column, diverges for each m, since x_{mn} is constant for fixed m, and hence $\sum_{m=1}^{\infty} \sum_{n=1}^{\infty} x_{mn}$ is nonsense.

The foregoing iterated sums illustrate one kind — the most important kind — of grouping of the terms of a countable sum. The double series $\sum_{n=1}^{\infty} \sum_{m=1}^{\infty} x_{mn}$ groups the terms in each row n = constant to get row sums $\sum_{m=1}^{\infty} x_{mn} = s_n$, and then adds the sums s_n. The iterated series in the other order would group the terms in each column m = constant, but these sums are all $\pm\infty$.

We show in the next two propositions that the iterated sum grouping is legitimate if the unordered sum converges, and that the unordered sum converges if either iterated sum converges absolutely. The moral of this chapter is therefore that you can do any kind of manipulation you like with an unordered sum, or with an absolutely convergent series.

PROPOSITION 5: *If $\sum_{N\times N} x_{mn}$ converges, then both iterated series converge, and*

$$\sum_{N\times N} x_{mn} = \sum_{m=1}^{\infty} \sum_{n=1}^{\infty} x_{mn} = \sum_{n=1}^{\infty} \sum_{m=1}^{\infty} x_{mn}.$$

Proof: Assume the unordered sum $\sum_{N\times N} x_{mn}$ converges, with sum s. Then $\sum_{N\times N} |x_{mn}|$ converges, and we let $\bar{s}$ be its sum. Any finite sum of terms $|x_{mn}|$ is bounded by $\bar{s}$. Hence finite sums of terms $|x_{mn}|$ across any row are bounded, and so each row sum $\sum_{m=1}^{\infty} |x_{mn}|$ converges, and consequently $\sum_{m=1}^{\infty} x_{mn}$ converges for each n. Let $s_n = \sum_{m=1}^{\infty} x_{mn}$. Let $\varepsilon > 0$, and let F be a finite subset of $N \times N$ such that $|\sum_G x_{mn} - s| < \varepsilon$ if $G \supset F$. Let N be any number larger than all m and n with $(m, n) \in F$. Thus all the terms x_{mn} corresponding to the finite set F lie in the N by N square. In each row $n = 1, 2, \ldots, N$, take enough terms — say M terms — so that

$$|x_{1n} + x_{2n} + \cdots + x_{Mn} - s_n| < \frac{\varepsilon}{N};$$

that is,

$$\left| \sum_{m=1}^{M} x_{mn} - s_n \right| < \frac{\varepsilon}{N}.$$

We can find one M that works for all N rows, and we can take $M \geq N$. Then

$$\left| \sum_{n=1}^{N} \sum_{m=1}^{M} - \sum_{n=1}^{N} s_n \right| < \varepsilon.$$

The double sum contains all the terms x_{mn} for $(m, n) \in F$, and so this sum is within ε of s. It follows that

$$\left| \sum_{n=1}^{N} s_n - s \right| < 2\varepsilon.$$

for any N larger than all m, n, with $(m, n) \in F$; that is, $\sum_{n=1}^{\infty} s_n = s$, or

$$\sum_{n=1}^{\infty} \sum_{m=1}^{\infty} x_{mn} = s.$$ ∎

The next proposition shows what you check before you begin to manipulate.

PROPOSITION 6: *If $\sum_{m=1}^{\infty} \sum_{n=1}^{\infty} |x_{mn}|$ converges or if $\sum_{n=1}^{\infty} \sum_{m=1}^{\infty} |x_{mn}|$ converges, then $\sum_{\mathbb{N} \times \mathbb{N}} x_{mn}$ converges, and consequently*

$$\sum_{\mathbb{N} \times \mathbb{N}} x_{mn} = \sum_{m=1}^{\infty} \sum_{n=1}^{\infty} x_{mn} = \sum_{n=1}^{\infty} \sum_{m=1}^{\infty} x_{mn}.$$

Proof: Assume $\sum_{m=1}^{\infty} |x_{mn}| = \bar{s}_n$, and $\sum_{n=1}^{\infty} \bar{s}_n = \bar{s}$. We show that every finite sum of terms $|x_{mn}|$ is bounded by $\bar{s}$, and consequently $\sum_{\mathbb{N} \times \mathbb{N}} |x_{mn}|$ and $\sum_{\mathbb{N} \times \mathbb{N}} x_{mn}$ converge. Let F be any finite subset of $\mathbb{N} \times \mathbb{N}$, and let all (m, n) in F lie under the line $y = N$; that is, $n \leq N$ for all $(m, n) \in F$. Let F_n be the elements of F in the nth row. Then $\sum_{m \in F_n} |x_{mn}| \leq \bar{s}_n$. Hence

$$\sum_{(m,n) \in F} |x_{mn}| = \sum_{n=1}^{N} \sum_{m \in F_n} |x_{mn}| \leq \sum_{n=1}^{N} \bar{s}_n \leq \bar{s}.$$

All finite sums $\sum_F |x_{mn}|$ are bounded, so that $\sum_{\mathbb{N} \times \mathbb{N}} |x_{mn}|$ converges, and hence $\sum_{\mathbb{N} \times \mathbb{N}} x_{mn}$ converges.   ∎

PROBLEM 7: An ostensibly more general grouping theorem for unordered sums can be stated as follows. Let A be any index set, and let $A = \bigcup_i A_i$ where the sets A_i, for $i \in I$, are disjoint subsets of A. Here I is another index set. The unordered sum $\sum_A x_\alpha$ can be gotten by first adding the A_i groups, and then adding those sums; that is,

$$\sum_A x_\alpha = \sum_{i \in I} \sum_{\alpha \in A_i} x_\alpha.$$

Hint: The proof of Proposition 5 will work if you think of the elements x_α for $\alpha \in A_i$ as the elements of the ith row, with sum s_i, so that $\sum_{i \in I} \sum_{\alpha \in A_i} x_\alpha = \sum_{i \in I} s_i$.   •

PROBLEM 8: Changing the order of integration in an iterated integral $\int_0^{\infty} \int_0^{\infty} f(x, y) \, dx \, dy$ runs into the same kind of problem as changing the order in an iterated sum. Define a function $f(x, y)$ for $x \geq 0$, $y \geq 0$ such that $\int_0^{\infty} f(x, y) \, dx = 0$ for all y, and $\int_0^{\infty} f(x, y) \, dy$ diverges for all x. Hint: Compare the double sum of numbers x_{mn} where each row is given by the series (1).   •

PROBLEM 9: (i) Let D be a directed set with an element β such that there is no $\alpha \neq \beta$ with $\beta < \alpha$. If $\{x_\alpha\}$ is a net on D, then $\lim_\alpha x_\alpha = x_\beta$.

(ii) Show that if A is a finite index set, then the unordered sum of the numbers x_α for $\alpha \in A$ is the same as the usual sum. •

PROBLEM 10: (i) Show that every sequence $\{s_n\}$ is the sequence of partial sums of some series $\sum x_n$.

(ii) Is it true that every net $\{S_F\}$ is the net of finite sums of some unordered sum $\sum_{\alpha \in A} x_\alpha$? Hint: Show that if $\{S_F\}$, $F \in D$, is a net of finite sums, then there must be elements $F \in D$ (corresponding to singletons) such that there is no $G \in D$ with $G \neq F$ and $G < F$. Find a directed set D that does not have this property. •

PROBLEM 11: The condition $\lim x_n = 0$ is necessary for the convergence of the series $\sum x_n$. Find an analogous condition for the unordered sum $\sum_{\alpha \in A} x_\alpha$. •

XVIII

THE CALCULUS OF SERIES

n this chapter we review some of the useful facts about series. We are therefore concerned with sums $\sum_{n=1}^{\infty} x_n$, defined to be the limit of a specific sequence of finite sums $\{s_n\}$, with $s_n = x_1 + \cdots + x_n$. The series $\sum_{n=1}^{\infty} x_n$ **converges** if $\{s_n\}$ converges; otherwise, the series **diverges**. If $\sum_{n=1}^{\infty} |x_n|$ converges, then $\sum_{n=1}^{\infty} x_n$ **converges absolutely**. We have seen that every absolutely convergent series is convergent. If $\sum_{n=1}^{\infty} x_n$ converges but $\sum_{n=1}^{\infty} |x_n|$ diverges, then $\sum_{n=1}^{\infty} x_n$ **converges conditionally**. We will write $\sum x_n$ when there is no need to be specific about the indices. Although we are now concerned with series rather than unordered sums, it is useful to keep in mind the results of the last chapter. If a series converges absolutely, then the corresponding unordered sum converges, and the sums are of course the same. The unordered sum is indifferent to rearrangement and grouping, and it follows that the series can also be rearranged or grouped, *if it converges absolutely*.

We will need later the fact that in a conditionally convergent series, the series of positive terms and the series of negative terms both diverge. The next problem has the details.

PROBLEM 1: Let $\{x_n\}$ be a sequence and define

$$x_n^+ = \begin{cases} x_n & \text{if } x_n \geq 0, \\ 0 & \text{if } x_n < 0; \end{cases} \qquad x_n^- = \begin{cases} x_n & \text{if } x_n < 0, \\ 0 & \text{if } x_n \geq 0. \end{cases}$$

Thus $x_n = x_n^+ + x_n^-$ for each n. For each n, define

$$s_n = x_1 + \cdots + x_n,$$

$$\bar{s}_n = |x_1| + \cdots + |x_n|,$$

$$s_n^+ = x_1^+ + \cdots + x_n^+,$$

$$s_n^- = x_1^- + \cdots + x_n^-,$$

so that $s_n = s_n^+ + s_n^-$ and $\bar{s}_n = s_n^+ - s_n^-$. Show that if $\sum x_n$ converges, then both $\{s_n^+\}$ and $\{s_n^-\}$ converge, or both sequences diverge. In the first case $\sum x_n$ converges absolutely, and in the second case $\sum x_n$ converges conditionally. •

The series $\sum 1/n$ is called the **harmonic series**. Notice that $1/3 + 1/4 \geq 1/2$, $1/5 + \cdots + 1/8 \geq 1/2$, $1/9 + \cdots + 1/16 \geq 1/2$, and so on, so that if $N = 1 + 2 + 4 + \cdots + 2^n$, then $s_N \geq n/2$; hence the partial sums $\{s_n\}$ are not bounded. Therefore, the harmonic series diverges. The most celebrated conditionally convergent series is obtained by alternating the signs in the harmonic series. By grouping the terms in the partial sums of $\sum(-1)^{n+1}1/n$, it is easy to show that $\{s_{2n}\}$ is increasing, $\{s_{2n+1}\}$ is decreasing, and for all n,

$$0 \leq s_{2n} < s_{2n+1} \leq 1.$$

Therefore, both the monotone sequences $\{s_{2n}\}$ and $\{s_{2n+1}\}$ converge. Since $s_{2n+1} - s_{2n} = 1/(2n + 1)$, both sequences have the same limit, and so $\{s_n\}$ converges and $\sum(-1)^{n+1}1/n$ converges.

PROBLEM 2: Show that $\sum(-1)^{n+1}x_n$ converges for any sequence $\{x_n\}$ such that $x_n \geq 0$ for all n, $x_1 \geq x_2 \geq x_3 \geq \cdots$, and $x_n \longrightarrow 0$. •

Notice that $x_n \longrightarrow 0$ is a necessary condition for convergence of $\sum x_n$. If $\sum x_n$ converges and $\{s_n\}$ is the sequence of partial sums, then $\{s_n\}$ is a Cauchy sequence, and given $\varepsilon > 0$ there is N so $|s_n - s_{n-1}| = |x_n| < \varepsilon$ if $n \geq N$.

To prove that the harmonic series diverges, we grouped the terms as follows to show the partial sums are unbounded:

$$1 + \frac{1}{2} + \left(\frac{1}{3} + \frac{1}{4}\right) + \left(\frac{1}{5} + \cdots + \frac{1}{8}\right) + \left(\frac{1}{9} + \cdots + \frac{1}{16}\right) + \cdots.$$

The same idea can be used as a test for convergence of other positive series. Suppose $a_1 \geq a_2 \geq a_3 \geq \cdots$. Then, if $m = 2^{n+1} - 1$,

$$s_m = a_1 + (a_2 + a_3) + (a_4 + \cdots + a_7) + \cdots + (a_{2^n} + \cdots + a_m)$$

$$\leq a_1 + 2a_2 + 4a_4 + \cdots + 2^n a_{2^n}.$$

An alternative grouping is, with $k = 2^n - 2^{n-1} + 1$,

$$s_{2^n} = a_1 + a_2 + (a_3 + a_4) + \cdots + (a_k + \cdots + a_{2^n})$$

$$\geq a_1 + a_2 + 2a_4 + 4a_8 + \cdots + 2^{n-1}a_{2^n}.$$

A judicious interpretation of these inequalities will take care of the following problem.

PROBLEM 3: If $\{x_n\}$ is a decreasing nonnegative sequence, then $\sum x_n$ converges if and only if the following series converges:

$$\sum 2^n x_{2^n} = x_1 + 2x_2 + 4x_4 + 8x_8 + \cdots . \qquad \bullet$$

Perhaps the most versatile series is the **geometric series** $\sum x^n$. If we let

$$s_n = 1 + x + x^2 + \cdots + x^n,$$

then $s_n - xs_n = 1 - x^{n+1}$, and

$$s_n = \frac{1}{1-x} - \frac{x^{n+1}}{1-x}.$$

If $|x| < 1$, then $x^n/(1-x) \longrightarrow 0$, and so $s_n \longrightarrow 1/(1-x)$. If $|x| \geq 1$, then $x^n \not\longrightarrow 0$, and $\sum x^n$ diverges. Thus,

$$\sum_{n=0}^{\infty} x^n = \frac{1}{1-x} \qquad \text{if } |x| < 1.$$

If a geometric series converges, it converges absolutely.

We have seen that convergence of series is of two types: absolute convergence, which provides a genuine extension of finite summation with all its usual properties, and conditional convergence, which depends on positive and negative summands occurring in just the right order to cancel each other out. We will be concerned henceforth with positive series, that is, series $\sum x_n$ with all $x_n \geq 0$. The partial sums of a positive series form an increasing sequence, and so a positive series converges if and only if the partial sums are bounded. The next proposition is embarrassingly simple, but we list it for the sake of completeness.

PROPOSITION 1: *(i) A series $\sum_{n=1}^{\infty} x_n$ converges if and only if each series $\sum_{n=N}^{\infty} x_n$ converges.*
(ii) If $\sum x_n$ and $\sum y_n$ are positive series and $x_n \leq y_n$ for all $n \geq N$, then the convergence of $\sum y_n$ implies the convergence of $\sum x_n$, and the divergence of $\sum x_n$ implies the divergence of $\sum y_n$.

Two common tests for convergence are the **root test**, which is of great importance for power series, and the **ratio test**. Both tests effect a comparison with geometric series, and use Proposition 1 to determine convergence or divergence.

PROPOSITION 2: *(i) If $\lim \sqrt[n]{|x_n|} = r$, then $\sum x_n$ converges absolutely if $r < 1$ and diverges if $r > 1$.*
(ii) If $\lim |x_{n+1}/x_n| = r$, then $\sum x_n$ converges absolutely if $r < 1$ and diverges if $r > 1$.

Proof: (i) Let $\sqrt[n]{|x_n|} \longrightarrow r < 1$, and let $0 \le r < s < 1$. There is N so $\sqrt[n]{|x_n|} \le s$ if $n \ge N$, so that $|x_n| \le s^n$ if $n \ge N$. Since $0 < s < 1$, $\sum s^n$ is a convergent geometric series, and so $\sum |x_n|$ converges by Proposition 1. If $\sqrt[n]{|x_n|} \longrightarrow r > 1$, then there is N so $\sqrt[n]{|x_n|} \ge 1$ if $n \ge N$, and hence $|x_n| \ge 1$ if $n \ge N$. Since $x_n \nrightarrow 0$, $\sum x_n$ diverges. The proof of part (ii) is similar, and is given as the following problem. ∎

PROBLEM 4: Prove the ratio test, Proposition 2(ii). •

The root and ratio tests depend on the convergence of a specific sequence: in one case the sequence $\left\{\sqrt[n]{|x_n|}\right\}$, and in the other case the sequence $\left\{|x_{n+1}/x_n|\right\}$. Suppose $\{x_n\}$ is a sequence of nonnegative numbers such that $\sqrt[n]{x_n} \longrightarrow r$, with $0 < r < 1$, so that $\sum x_n$ converges. If we replace, say, every other term of the series by zero, the resulting series will obviously also converge, since its partial sums are smaller, and therefore bounded. However, if $\sum y_n$ is the new series, with every other x_n replaced by zero, then $\left\{\sqrt[n]{y_n}\right\}$ diverges, and the root test fails. Similar observations apply to the ratio test, which requires that all x_n be nonzero for the sequence of ratios even to be defined.

Now we check the proof of the root test to see what is really used, and whether it can be modified to cover the case $\sum y_n$ above. For the root test all we used was that fact that there is $s < 1$ and N such that $\sqrt[n]{|x_n|} < s$ for all $n \ge N$. This will be true if no *subsequence* of $\left\{\sqrt[n]{|x_n|}\right\}$ converges to a number greater than or equal to 1. We are led to the following proposition.

PROPOSITION 3: *If $\{x_n\}$ is a bounded sequence, then the largest limit of any subsequence of $\{x_n\}$ is $\lim_{n\to\infty} \sup_{k\ge n} x_k$ and the smallest limit of any subsequence is $\lim_{n\to\infty} \inf_{k\ge n} x_k$. These two limits exist, and there are subsequences that converge to them.*

Proof: Assume $\{x_n\}$ is bounded, and let

$$s_n = \inf_{k\ge n} x_k,$$

$$t_n = \sup_{k\ge n} x_k.$$

Both $\{s_n\}$ and $\{t_n\}$ are bounded sequences, with $\{s_n\}$ increasing and $\{t_n\}$ decreasing, and so both sequences converge. We let

$$r = \lim_{n\to\infty} t_n = \lim_{n\to\infty} \sup_{k\ge n} x_k$$

and show that some subsequence converges to r, and no subsequence converges to a number larger than r. The corresponding statements for $\lim_{n\to\infty} s_n = \lim_{n\to\infty} \inf_{k\ge n} x_k$ are left to Problem 5.

It is easy to see that no subsequence converges to a number larger than r for if $x_{n_k} \longrightarrow \ell > r$, then $t_n = \sup_{k \geq n} x_k \geq \ell$ for all n, contradicting $\lim t_n = r$. Now we show that some subsequence does converge to r. For each $\varepsilon > 0$, there is N such that $r \leq t_n < r + \varepsilon$ for all $n \geq N$. Since $t_n = \sup_{k \geq n} x_k$, there is x_k with $k \geq n$ such that $t_n - \varepsilon < x_k \leq t_n$. Hence for each ε there are arbitrarily large k such that $r - \varepsilon < x_k < r + \varepsilon$. It follows, as you will show in Problem 5, that there is a subsequence $\{x_{n_k}\}$ with $x_{n_k} \longrightarrow r$. ∎

We introduce the following natural notation:

$$\limsup_{n \to \infty} x_n = \lim_{n \to \infty} \sup_{k \geq n} x_k,$$

$$\liminf_{n \to \infty} x_n = \lim_{n \to \infty} \inf_{k \geq n} x_k.$$

With the natural conventions about $\pm\infty$, $\limsup x_n$ and $\liminf x_n$ make sense whether or not $\{x_n\}$ is bounded, and these extended "numbers" are still the largest and smallest "limits" of subsequences. For example, if $\lim x_n = +\infty$, then $\limsup x_n = +\infty$ and $\liminf x_n = +\infty$. The reader is asked to check these possibilities in Problem 5.

PROBLEM 5: (i) Show that if $\{x_n\}$ is bounded, some subsequence converges to $\liminf x_n$, and no subsequence converges to a smaller number.

(ii) Investigate the possibilities for unbounded sequences. Show, for example, that $\limsup x_n = +\infty$ if and only if $\{x_n\}$ is not bounded above, and $\liminf x_n = +\infty$ if and only if $\lim x_n = +\infty$.

(iii) Show that $\lim x_n$ exists if and only if $\limsup x_n$ and $\liminf x_n$ are equal (finite numbers). •

PROBLEM 6: Let $\{x_n\}$ be a bounded sequence, and let $r = \limsup x_n$. Show that for every $\varepsilon > 0$ there are infinitely many n such that $x_n > r - \varepsilon$, and at most a finite number of n such that $x_n > r + \varepsilon$. Show that these conditions, for every $\varepsilon > 0$, imply that $r = \limsup x_n$. •

Now consider again the root test for the positive series $\sum x_n$. If $\limsup \sqrt[n]{x_n} = r < 1$, and $r < s < 1$, then by Problem 6 there is some N such that $\sqrt[n]{x_n} < s$ for all $n \geq N$. This is what we used to compare $\sum x_n$ with the convergent geometric series $\sum s^n$, so that we have the following more general root test.

PROPOSITION 4: If $\sum x_n$ is a positive series, then $\sum x_n$ converges if $\limsup \sqrt[n]{x_n} < 1$, and the series diverges if $\limsup \sqrt[n]{x_n} > 1$.

Let f be a nonnegative function on $[1, \infty)$ such that f is integrable on every interval $[1, n]$. If $f(x)$ lies between $f(n)$ and $f(n + 1)$ for all $x \in [n, n + 1]$, then the series $\sum f(n)$ can be compared with the improper integral $\int_1^\infty f(x)\, dx$. If $f(n) \geq f(x) \geq f(n + 1)$ for all n and all $x \in [n, n + 1]$, then

$$f(1) + f(2) + \cdots + f(n) \geq \int_1^2 f + \int_2^3 f + \cdots + \int_n^{n+1} f = \int_1^{n+1} f.$$

Similarly,

$$f(2) + \cdots + f(n) \le \int_1^2 f + \int_2^3 f + \cdots + \int_{n-1}^n f = \int_1^n f.$$

It follows from the first inequality that if $\sum f(n)$ converges, the integral $\int_1^\infty f$ converges, and from the second that if $\int_1^\infty f$ converges, then $\sum f(n)$ converges. The usual applications are to monotone decreasing functions, but the inequalities $f(n) \ge f(x) \ge f(n+1)$ on $[n, n+1]$ are all that is used.

PROPOSITION 5 (The Integral Test): *If f is nonnegative and integrable on every interval $[1, n]$, and $f(n) \ge f(x) \ge f(n+1)$ for all n and all $x \in [n, n+1]$, then $\sum f(n)$ converges if and only if $\int_1^\infty f$ converges.*

PROBLEM 7: (i) A series of the form $\sum 1/n^p$ is called a *p-series*. Show that a *p*-series converges if and only if $p > 1$.

(ii) Show that $\sqrt[n]{x_n} \longrightarrow 1$ and $x_{n+1}/x_n \longrightarrow 1$ for any *p*-series, and therefore nothing can be determined from the root test or ratio test in the cases that the limiting root or ratio is one. Hint: See Problem 5, Chapter 16. •

PROBLEM 8: Use the integral test to show that $\sum \dfrac{1}{n \log n}$ diverges. How about $\sum \dfrac{1}{n(\log n) \log(\log n)}$? •

If we replace a sequence $\{x_n\}$ by the sequence $\{\sigma_n\}$ of initial averages,

$$\sigma_n = \frac{x_1 + \cdots + x_n}{n},$$

then the new sequence $\{\sigma_n\}$ is less skittish and more likely to converge. For example, if $\{x_n\}$ is the divergent sequence $1, 0, 1, 0, 1, \ldots$, then $\{\sigma_n\}$ is the sequence $1, \frac{1}{2}, \frac{2}{3}, \frac{1}{2}, \frac{3}{5}, \ldots$, which converges to $\frac{1}{2}$. The sequence $\{\sigma_n\}$ is called the **sequence of Cesàro means** of $\{x_n\}$. The principal importance of this Cesàro smoothing is its application to the sequence $\{s_n\}$ of partial sums of a series. For example, if f is a continuous function on $\mathbb{R}$ with period 2π, then the Fourier series of f may not converge to f, but the Cesàro means of the partial sums converge to f uniformly. The verification of this fact is appropriate to a course in Fourier series, but we mention the result here to show the importance of what is called **Cesàro summability**.

PROPOSITION 6 (Cesàro Summation): *If $\sum x_n$ has partial sums $\{s_n\}$, and $\sum x_n = S$, so that $s_n \longrightarrow S$, and $\sigma_n = (s_1 + \cdots + s_n)/n$, then $\sigma_n \longrightarrow S$. (The point is that $\{\sigma_n\}$ might converge usefully even when $\{s_n\}$ does not, or when we do not know if $\{s_n\}$ converges, but if $\{s_n\}$ does converge, then $\{\sigma_n\}$ converges to the same number.)*

Proof: Assume $s_n \longrightarrow S$ and consider the difference $\sigma_n - S$:

$$\sigma_n - S = \frac{1}{n}(x_1 + \cdots + x_n) - S$$

$$= \frac{1}{n}[(x_1 - S) + \cdots + (x_n - S)].$$

Let M be a bound for the terms $|s_n - S|$. For any $\varepsilon > 0$, pick N such that $|s_n - S| < \varepsilon$ if $n \geq N$. For $n \geq N$ we have

$$|\sigma_n - S| \leq \frac{1}{n}\left[|x_1 - S| + \cdots + |x_N - S|\right]$$

$$+ \frac{1}{n}\sum_{N+1}^{n}|x_k - S|$$

$$\leq \frac{1}{n} \cdot NM + \frac{n-N}{n}\varepsilon$$

$$< \frac{NM}{n} + \varepsilon.$$

Now pick $N_1 > N$ such that $NM/N_1 < \varepsilon$. Then if $n \geq N_1$, $|\sigma_n - S| < 2\varepsilon$.   ■

Cesàro summation is a way of assigning a generalized sum to some divergent series, with the Cesàro sum equaling the real sum if the series does converge. Abel summation is another such technique. If $\{a_n\}$ is a bounded sequence, then $\sum a_n r^n$ converges for all $r \in [0, 1)$ and possibly $\lim_{r \to 1^-} \sum a_n r^n$ exists even if $\sum a_n$ diverges. However, if $\sum a_n = s$, then $\lim_{r \to 1^-} \sum a_n r^n = s$. The details are in the following problem.

PROBLEM 9: If $\sum_{n=0}^{\infty} a_n = s$, then $\lim_{r \to 1^-} \sum_{n=0}^{\infty} a_n r^n = s$. Hint: Let $s_n = a_0 + \cdots + a_n$ and assume $|s_n - s| < \varepsilon$ if $n \geq N$. Let

$$f(r) = \sum_{n=0}^{\infty} a_n r^n$$

$$= s_0 + (s_1 - s_0)r + (s_2 - s_1)r^2 + \cdots$$

$$= \sum_{n=0}^{\infty} s_n r^n - r \sum_{n=0}^{\infty} s_n r^n$$

$$= (1 - r)\sum_{n=0}^{\infty} s_n r^n.$$

Since $\sum r^n = 1/(1-r)$,

$$s = s(1-r)\sum_{n=0}^{\infty} r^n = (1-r)\sum_{n=0}^{\infty} s r^n,$$

and

$$|f(r) - s| = \left|(1 - r)\sum_{n=0}^{\infty}(s_n - s)r^n\right|$$

$$\leq |1 - r|\sum_{n=0}^{N}|s_n - s|r^n + \varepsilon$$

$$< |1 - r|\sum_{n=0}^{N}|s_n - s| + \varepsilon. \qquad \bullet$$

PROBLEM 10: Show that $\sum_{n=0}^{\infty}(-1)^n$ is Abel summable to $1/2$; that is, show that $\lim_{r \to 1-}\sum_{n=0}^{\infty}(-1)^n r^n = 1/2$. $\qquad \bullet$

PROBLEM 11: Show that $\sum x_n^2$ converges if $\sum x_n$ converges absolutely, and that the absolute convergence is necessary. $\qquad \bullet$

PROBLEM 12: Let $x_n \geq 0$ for all n and $x_n \neq 0$ for some n. Let $y_n = (x_1 + \cdots + x_n)/n$, and show that $\sum y_n$ diverges. $\qquad \bullet$

PROBLEM 13: If $\sum x_n$ and $\sum y_n$ are positive series and $\lim x_n/y_n = p > 0$, then $\sum x_n$ converges if and only if $\sum y_n$ converges. Suppose $p = 0$? $p = \infty$? $\quad \bullet$

PROBLEM 14: State and prove a more general ratio test using $\limsup |x_{n+1}/x_n|$ instead of $\lim |x_{n+1}/x_n|$. Give an example where the new test applies and the old test does not. $\qquad \bullet$

PROBLEM 15: Let $p > 1$ and $a > 1$. Show the following series converge.

(i) $\sum \log n/n^p$
(ii) $\sum n^p/a^n$
(iii) $\sum a^n/n!$

Hint: For (i), let $p = 1 + 2\varepsilon$ and show $\log n/n^\varepsilon \longrightarrow 0$. $\qquad \bullet$

XIX

SEQUENCES AND SERIES
OF FUNCTIONS

Suppose $\{f_n(x)\}$ is a sequence of functions with common domain A. We can consider $\{f_n(x)\}$ as a family of sequences of numbers — one sequence for each $x \in A$ — or we can try to define a kind of convergence $f_n \longrightarrow f$ of functions that indicates that the graphs of the f_n everywhere get close to the graph of f. In the first case, if $f_n(x) \longrightarrow f(x)$ for each $x \in A$, we say $\{f_n\}$ converges to f **pointwise on** A. The convergence of $\{f_n(x_1)\}$ has nothing to do with the convergence of $\{f_n(x_2)\}$ if $x_1 \neq x_2$. To express the second idea — the convergence of the graphs — we say that $\{f\}$ **converges to** f **uniformly on** A if for any given $\varepsilon > 0$ there is N such that $|f_n(x) - f(x)| < \varepsilon$ for *all* $x \in A$ and all $n \geq N$. That is, all f_n for $n \geq N$ are uniformly close to the limit f.

To illustrate the difference, let f_n be the function on $[0, 1]$ whose triangular graph consists of the segments from $(0, 0)$ to $(1/n, n)$, from $(1/n, n)$ to $(2/n, 0)$, and from $(2/n, 0)$ to $(1, 0)$. For each $x \in [0, 1]$, there is N_x such that $f_n(x) = 0$ for all $n \geq N_x$. Therefore, $f_n \longrightarrow 0$ pointwise on $[0, 1]$, even though the maximum vertical distance between the graphs of f_n and the zero function increases with n. Notice that in this example there is for each x and each ε a number N_x, which depends on both x and ε, such that $|f_n(x)| < \varepsilon$ if $n \geq N_x$. For a sequence $\{g_n\}$ to converge uniformly to 0 on $[0, 1]$, there must be for each $\varepsilon > 0$ a single number N that works for all x. Uniform convergence on a set A obviously implies pointwise convergence on A.

PROBLEM 1: (i) Let $f_n(x) = x^n$ for $0 \le x \le 1$. Show that $\{f_n\}$ converges pointwise on $[0, 1]$ (to what function?), but does not converge uniformly.

(ii) Show that $x^n(1 - x)^n$ converges to 0 uniformly on $[0, 1]$. •

We would like the limits of sequences of functions to enjoy some of the properties of the individual functions f_n. For example, if all f_n are continuous, or differentiable, or integrable, or convex, does the limit function have the same property? The following problem shows that pointwise convergence is a miserable failure with regard to continuity and integrability. The situation with regard to differentiability is even worse, but harder to illustrate.

PROBLEM 2: (i) Let $f_n(x) = x^n$ on $[0, 1]$, and let $f(x) = \lim f_n(x)$. Show that f is not continuous even though all f_n are continuous.

(ii) Let $r_1, r_2, r_3, \ldots$ be an enumeration of all the rationals in $[0, 1]$, and let f_n be the function that is one on $r_1, \ldots, r_n$ and zero elsewhere. Show that $\int_0^1 f_n = 0$ for all n but $f(x) = \lim f_n(x)$ is not integrable.

(iii) Assume all f_n are convex on (a, b) and $f_n \longrightarrow f$ on (a, b). Is f convex? •

As we might expect, uniform convergence carries more genetic information than pointwise convergence, and we pursue this idea now.

PROPOSITION 1: *If* $\{f_n\}$ *is a sequence of continuous functions on a set A, and* $f_n \longrightarrow f$ *uniformly on A, then* f *is continuous on A.*

Proof: Let $x_0 \in A$, and let $\{x_k\}$ be any sequence in A that converges to x_0. We have to show that

$$\lim_{k \to \infty} f(x_k) = f(x_0). \tag{1}$$

Let $\varepsilon > 0$. For all n and x_k, the triangle inequality gives us

$$|f(x_k) - f(x_0)| \le |f(x_k) - f_n(x_k)| + |f_n(x_k) - f_n(x_0)| + |f_n(x_0) - f(x_0)|. \tag{2}$$

We pick N such that $|f_n(x) - f(x)| < \varepsilon$ for all $x \in A$ if $n \ge N$. In particular, for $n = N$, the first and last terms on the right side of (2) are less than ε whatever x_k is, so that

$$|f(x_k) - f(x_0)| < \varepsilon + |f_N(x_k) - f_N(x_0)| + \varepsilon.$$

Since f_N is continuous, there is K such that if $k \ge K$, $|f_N(x_k) - f_N(x_0)| < \varepsilon$; thus if $k \ge K$,

$$|f(x_k) - f(x_0)| < 3\varepsilon,$$

and f is continuous at the arbitrary point $x_0 \in A$. ∎

In general, if $\{f_n\}$ is a sequence of continuous functions converging pointwise to f on a set A, and $x_k \longrightarrow x_0 \in A$, we have

$$f(x_k) = \lim_{n \to \infty} f_n(x_k)$$

for each k because of convergence at each x, and

$$f_n(x_0) = \lim_{k \to \infty} f_n(x_k)$$

for each n, because each f_n is continuous. Therefore, the condition (1) that f be continuous can be written

$$\lim_{k \to \infty} \lim_{n \to \infty} f_n(x_k) = \lim_{n \to \infty} \lim_{k \to \infty} f_n(x_k). \tag{3}$$

The equality (3) of the two iterated limits is exactly the condition that the limit of continuous functions be continuous. Uniform convergence is sufficient to imply that (3) holds.

PROBLEM 3: Show that (3) need not hold if $f_n \longrightarrow f$ pointwise.   •

Uniform convergence also behaves much better with respect to integration than pointwise convergence, although to some extent this is also a reflection on our definition of the integral. The Lebesgue integral we will study later is more generous toward pointwise limits. For the Riemann integral, an identity like

$$\int_a^b \lim f_n = \lim \int_a^b f_n \tag{4}$$

is hopeless, as we saw in Problem 2. However, (4) does hold if the convergence is uniform.

PROPOSITION 2: *If $\{f_n\}$ is a sequence of integrable functions on $[a, b]$, and $f_n \longrightarrow f$ uniformly on $[a, b]$, then f is integrable and $\int_a^b f_n \longrightarrow \int_a^b f$.*

Proof: To show that f is integrable, let $\varepsilon > 0$ and pick a function f_N that is ε–close to f on $[a, b]$: $f_N - \varepsilon < f < f_N + \varepsilon$. It follows that f is bounded, since f_N is. For any partition P, let M_i and m_i be the ith sup and inf for f, and M_i^*, m_i^* the corresponding numbers for f_N. Then for all i,

$$m_i^* - \varepsilon \leq m_i \leq M_i \leq M_i^* + \varepsilon,$$

and so

$$L(f_N, P) - \varepsilon(b - a) \leq L(f, P) \leq U(f, P) \leq U(f_N, P) + \varepsilon(b - a).$$

Since there is a partition P_0 with $U(f_N, P_0) - L(f_N, P_0) < \varepsilon$, we have

$$U(f, P_0) - L(f, P_0) < \varepsilon\,[1 + 2(b - a)].$$

Therefore, f is integrable, and

$$\left| \int_a^b f_n - \int_a^b f \right| = \left| \int_a^b (f_n - f) \right| \leq \int_a^b |f_n - f|. \tag{5}$$

For any given $\varepsilon > 0$, there is N such that $|f_n - f| < \varepsilon$ on all of $[a, b]$ if $n \geq N$, so that if $n \geq N$

$$\int_a^b |f_n - f| < \varepsilon(b - a).$$

It follows then from (5) that $\int_a^b f_n \longrightarrow \int_a^b f$. ∎

There is a useful Cauchy condition that guarantees uniform convergence. We will say that $\{f_n\}$ is **uniformly Cauchy** on a set A provided that, for any given $\varepsilon > 0$, there is N so that $|f_n(x) - f_m(x)| < \varepsilon$ for all $x \in A$ and all n and m greater than N.

PROBLEM 4: $\{f_n\}$ converges uniformly on A to some function f if and only if $\{f_n\}$ is uniformly Cauchy on A. Hint: The uniform Cauchy condition implies that a pointwise Cauchy condition holds, and that gives you the limit function f. •

Now let us see what we can hope for in the way of a theorem like "the derivative of the limit is the limit of the derivatives." Suppose $\{f_n\}$ is a sequence of differentiable functions on an interval such that $\{f_n'\}$ converges uniformly. This says that for large n the functions f_n are all nearly parallel to each other. Of course the f_n need not converge just because they are parallel, since we could have a situation like $f_n(x) = x + n$ with $f_n'(x) \longrightarrow 1$ uniformly. However, if $\{f_n\}$ converges at one point, then the graphs must come together uniformly over any bounded interval.

PROPOSITION 3: *If $\{f_n\}$ is a sequence of differentiable functions on an open interval (a, b), $\{f_n'\}$ converges uniformly on (a, b), and $\{f_n(x_0)\}$ converges for some $x_0 \in (a, b)$, then $\{f_n\}$ converges uniformly on (a, b) to a function f, the function f is differentiable, and $f_n' \longrightarrow f'$ uniformly on (a, b).*

Proof: Let $\varepsilon > 0$. Choose N so that $\left|f_n'(x) - f_m'(x)\right| < \varepsilon$ for all $x \in (a, b)$ and all $n, m \geq N$. In addition, let N be so large that $|f_n(x_0) - f_m(x_0)| < \varepsilon$ if $n, m \geq N$. We show that $\{f_n\}$ satisfies a uniform Cauchy condition. Apply the Mean Value Theorem to $f_n - f_m$ on any interval $[x_0, x]$ (or $[x, x_0]$), and get, for n, $m \geq N$,

$$|(f_n(x) - f_m(x)) - (f_n(x_0) - f_m(x_0))|$$
$$= |f_n'(c) - f_m'(c)||x - x_0|$$
$$< \varepsilon(b - a). \tag{6}$$

Since $|f_n(x_0) - f_m(x_0)| < \varepsilon$ if $n, m \geq N$, we see from (6) that $|f_n(x) - f_m(x)| < \varepsilon(1 + b - a)$ uniformly for $x \in (a, b)$ if $n, m \geq N$. Let f be the uniform limit of the f_n, and observe that f is continuous since it is the uniform limit of differentiable, hence continuous, functions. To show that f is differentiable and that $f_n' \longrightarrow f'$, consider any fixed point $x_1 \in (a, b)$. Let

$$q_n(x) = \begin{cases} \dfrac{f_n(x) - f_n(x_1)}{x - x_1} & \text{if } x \neq x_1, \\ f_n'(x_1) & \text{if } x = x_1. \end{cases}$$

Notice that q_n is a continuous function of x on (a, b). At points $x \neq x_1$ this is because f_n is continuous, and at $x = x_1$ because f_n is differentiable at x_1. We also define, for $x \neq x_1$,

$$q(x) = \frac{f(x) - f(x_1)}{x - x_1},$$

and observe that $q_n(x) \longrightarrow q(x)$ for all $x \neq x_1$. We show next that the continuous functions $\{q_n(x)\}$, which depend on some arbitrary but fixed x_1, converge uniformly for $x \neq x_1$. Again apply the Mean Value Theorem, this time to $f_n(x) - f_m(x)$ on $[x_1, x]$ or $[x, x_1]$, to get

$$|(f_n(x) - f_m(x)) - (f_n(x_1) - f_m(x_1))| = |f_n'(c) - f_m'(c)||x - x_1|.$$

Dividing by $|x - x_1|$, for $x \neq x_1$, we have for $n, m \geq N$,

$$|q_n(x) - q_m(x)| = |f_n'(c) - f_m'(c)| < \varepsilon.$$

Thus $\{q_n(x)\}$ converges uniformly on $(a, b) - \{x_1\}$. Since $\{q_n(x_1)\} = \{f_n'(x_1)\}$ also converges, $\{q_n\}$ converges uniformly on (a, b) (see Problem 5). The q_n are continuous functions on all of (a, b), and converge uniformly on (a, b), so that we can reverse the limits as in (3), and conclude

$$\lim_{x \to x_1} \lim_{n \to \infty} q_n(x) = \lim_{n \to \infty} \lim_{x \to x_1} q_n(x);$$

that is,

$$\lim_{x \to x_1} q(x) = \lim_{n \to \infty} f_n'(x_1),$$

$$f'(x_1) = \lim_{n \to \infty} f_n'(x_1).$$

This shows that f is differentiable, and $f'(x) = \lim f_n'(x)$ for all x, since x_1 was arbitrary. The convergence of the $\{f_n'\}$ is uniform by hypothesis. ∎

PROBLEM 5: Let $\{q_n\}$ be any sequence of functions on (a, b) such that $\{q_n\}$ converges uniformly on $(a, b) - \{x_1\}$, and $\{q_n(x_1)\}$ converges. Show that $\{q_n\}$ converges uniformly on (a, b). •

PROBLEM 6: The result of Proposition 3 is much easier to prove if we assume all the f_n' are continuous on (a, b). Use this additional assumption, and the fact that

$$f_n(x) = f_n(x_0) + \int_{x_0}^x f_n'$$

for all n and x, to show that $\{f_n\}$ converges uniformly to some f, and $f' = \lim f'_n$. Where do you use the continuity of f'_n? Could you just assume that all f'_n are integrable instead? •

Now consider the problem of approximating a function f by a sequence of polynomials $P_n(x)$ in such a way that $P_n(x) \longrightarrow f(x)$ on some interval. For simplicity, we will initially consider the interval $[0, r)$, and assume that f has as many derivatives as we want on $[0, r)$. The tangent function $P_1(x) = f(0) + f'(0)x$ is the linear polynomial that best approximates f near zero. However, if f has a hairpin graph, then the difference between $f(x)$ and $P_1(x)$ can get large fast. Suppose that $|f''(x)|$ is small on $[0, r)$. Then f' cannot change much, and f should stay close to its tangent. To make this a quantitative estimate, let $m_2 \le f''(x) \le M_2$ on $[0, r)$, and let

$$f(x) - \left[f(0) + f'(0)x\right] = f(x) - P_1(x) = R_1(x).$$

Notice that $R_1(0) = R'_1(0) = 0$, and $R''_1(x) = f''(x)$. For $0 \le x < r$, since $R'_1(0) = 0$,

$$R'_1(x) = \int_0^x R''_1(x)\,dx = \int_0^x f''(x)\,dx.$$

Using the estimates $m_2 \le f''(x) \le M_2$ on $[0, r)$, we get

$$\int_0^x m_2\,dx \le \int_0^x f''(x)\,dx \le \int_0^x M_2\,dx,$$

$$m_2 x \le R'_1(x) \le M_2 x.$$

Now integrate again, using $R_1(0) = 0$, to get

$$\int_0^x m_2 x\,dx \le \int_0^x R'_1(x)\,dx \le \int_0^x M_2 x\,dx,$$

$$\frac{1}{2}m_2 x^2 \le R_1(x) \le \frac{1}{2}M_2 x^2. \tag{7}$$

Now let us read the foregoing calculations a little more carefully. Fix a value of x in $[0, r)$, and let $m_2 = \inf\{f''(t) : t \in [0, x]\}$, $M_2 = \sup\{f''(t) : t \in [0, x]\}$. All the preceding calculations work with these estimates m_2 and M_2 for the given fixed x. There is therefore, from (7), some number μ_x between m_2 and M_2 such that

$$R_1(x) = \frac{1}{2}\mu_x x^2.$$

If $f''(x)$ is continuous on $[0, x]$, then $\mu_x = f''(\xi)$ for some ξ_x between 0 and x. Thus, for some $\xi \in [0, x]$,

$$R_2(x) = \frac{1}{2}f''(\xi_x)x^2.$$

We have proved the following theorem.

PROPOSITION 4 (Taylor's Theorem for $n = 2$): *If f, f', f'' are continuous on $[0, r)$, then for each $x \in [0, r)$ there is ξ_x between 0 and x such that*

$$f(x) = f(0) + f'(0)x + \frac{1}{2}f''(\xi_x)x^2.$$

PROBLEM 7: Prove Taylor's Theorem for $n = 3$: if f, f', f'', f''' are continuous on $[0, r)$, then for each $x \in [0, r)$ there is $\xi_x \in [0, x]$ such that

$$f(x) = f(0) + f'(0)x + \frac{1}{2!}f''(0)x^2 + \frac{1}{3!}f'''(\xi_x)x^3. \qquad \bullet$$

The nth degree polynomial that agrees with f and its first n derivatives at 0 is

$$P_n(x) = f(0) + f'(0)x + \frac{1}{2!}f''(0)x^2 + \cdots + \frac{1}{n!}f^{(n)}(0)x^n.$$

The same computations that are used in the proof of Proposition 4 and Problem 7 yield the following theorem.

PROPOSITION 5 (Taylor's Theorem): *If f, f', $\ldots$, $f^{(n+1)}$ are continuous on $[0, r)$, then for each $x \in [0, r)$ there is $\xi_x \in [0, x]$ such that*

$$f(x) = P_n(x) + \frac{1}{(n + 1)!}f^{(n+1)}(\xi_x)x^{n+1}.$$

The result is not peculiar to intervals at zero, or intervals to the right of a given point (see Problem 8). Notice that for large n, the estimate for $f^{(n+1)}(x)$ need not be terribly small to yield a good estimate for $R_n(x) = f(x) - P_n(x)$. For example, if $|f^{(n+1)}(x)| \le M$ on $[0, \infty)$, then $|R_n(x)| \le \dfrac{M}{(n + 1)!}x^{n+1}$, and $R_n(x) \longrightarrow 0$ for all $x \in [0, \infty)$.

PROBLEM 8: (i) Assume that f has $n + 1$ continuous derivatives on $(-r, 0]$, and that $|f^{(n+1)}(x)| \le M$ on $(-r, 0]$. Show that $|f(x) - P_n(x)| \le \dfrac{1}{(n + 1)!}M|x|^{n+1}$ for $x \in (-r, 0]$. Hint: Let $g(x) = f(-x)$, and let $Q_n(x)$ be the Taylor polynomial for g. Check that $P_n(x) = Q_n(-x)$, so that, for $-r < x \le 0$,

$$|f(x) - P_n(x)| = |g(-x) - Q_n(-x)|$$

$$\le \frac{M}{(n + 1)!}(-x)^{n+1} = \frac{M}{(n + 1)!}|x|^{n+1}.$$

(ii) State Taylor's Theorem for $x \in (-r, r)$, if $|f^{(n+1)}(x)| \le M$ on $(-r, r)$. $\bullet$

PROBLEM 9: Let $P_n(x) = 1 + x + \dfrac{1}{2!}x^2 + \cdots + \dfrac{1}{n!}x^n$. Show that $P_n(x) \longrightarrow e^x$ uniformly on every bounded interval $(-r, r)$. $\bullet$

If $\{f_n(x)\}$ is a sequence of functions with common domain A, then we can form the series $\sum_{n=1}^{\infty} f_n(x)$. This series converges provided the sequence $\{s_n(x)\}$

of partial sums converges; that is, if for each $x \in A$,

$$s_n(x) = f_1(x) + \cdots + f_n(x),$$

then $\sum_{n=1}^{\infty} f_n(x) = f(x)$ if and only if $s_n(x) \longrightarrow f(x)$. If $s_n(x) \longrightarrow f(x)$ pointwise on A, then we say the series converges to $f(x)$ pointwise, and if $s_n(x) \longrightarrow f(x)$ uniformly on A, then the series converges uniformly. Many nice properties of functions are preserved by finite sums. Thus $s_n(x)$ is continuous or differentiable or integrable if each f_i is. We have seen that pointwise convergence is not likely to preserve these properties, but uniform convergence is. Therefore, the following simple test for uniform convergence of series is very useful.

PROPOSITION 6 (The Weierstrass M-Test): *If $\{f_n(x)\}$ is a sequence of functions on a set S, and $\{M_n\}$ is a sequence of numbers such that $|f_n(x)| \leq M_n$ for all n and all $x \in S$, and $\sum M_n$ converges, then $\sum f_n(x)$ converges uniformly on S.*

Proof: Let $s_n(x) = f_1(x) + \cdots + f_n(x)$, and let $t_n = M_1 + \cdots + M_n$. Since $\sum M_n$ converges, $\{t_n\}$ is a Cauchy sequence. Given $\varepsilon > 0$, there is N so that $t_n - t_m < \varepsilon$ if $n \geq m \geq N$. Therefore, for all x,

$$\begin{aligned} |s_n(x) - s_m(x)| &= |f_{m+1}(x) + \cdots + f_n(x)| \\ &\leq |f_{m+1}(x)| + \cdots + |f_n(x)| \\ &\leq M_{m+1} + \cdots + M_n \\ &= t_n - t_m < \varepsilon. \end{aligned}$$

Since $\{s_n(x)\}$ is uniformly Cauchy on S, the series $\sum f_n(x)$ converges uniformly on S. ∎

PROBLEM 10: Let $\{f_n(x)\}$ be a sequence of functions on $[a, b]$ such that $\sum f_n(x)$ converges uniformly on $[a, b]$ to $f(x)$.

   (i) If each f_n is continuous on $[a, b]$, then f is continuous on $[a, b]$.
   (ii) If each f_n is integrable on $[a, b]$, then f is integrable on $[a, b]$, and $\int_a^b f = \sum \int_a^b f_n$.
   (iii) If each f_n is differentiable on $[a, b]$ and $\sum f_n'(x)$ converges uniformly on $[a, b]$, then f is differentiable on $[a, b]$, and $f'(x) = \sum f_n'(x)$. •

PROBLEM 11: Show the sum of two convex functions is convex. State and prove a theorem for a series of convex functions. •

PROBLEM 12: (i) Show that for every n and every $x \neq -1$,

$$1 - x + x^2 - x^3 + \cdots + (-1)^n x^n = \frac{1}{1+x} + \frac{(-1)^n x^{n+1}}{1+x},$$

so that, for $x > 0$,

$$x - \frac{x^2}{2} + \frac{x^3}{3} - \cdots + \frac{(-1)^n x^{n+1}}{n+1} = \int_0^x \frac{dt}{1+t} + \int_0^x \frac{(-1)^n t^{n+1}}{1+t}\, dt.$$

(ii) The functions $x^{n+1}/(1 + x)$ converge pointwise but not uniformly on $[0, 1]$.

(iii) Nevertheless, $\int_0^1 [x^{n+1}/(1+x)]\,dx \longrightarrow 0$. Hint: The integrand converges to 0 uniformly on $[0, 1 - \varepsilon]$, and if $x^{n+1}/(1 + x)$ is less than 1 on $[1 - \varepsilon, 1]$, then $\int_{1-\varepsilon}^1 [x^{n+1}/(1 + x)]\,dx < \varepsilon$.

(iv) Conclude that $\sum_{n=1}^{\infty}(-1)^{n+1}x^n/n$ converges uniformly to $\log(1 + x)$ on $[0, 1]$ and, in particular,

$$\log 2 = 1 - \frac{1}{2} + \frac{1}{3} - \frac{1}{4} + \cdots .$$ ●

PROBLEM 13 (Taylor's Series): If f has derivatives of all orders on $\mathbb{R}$, and $|f^{(n)}(x)| \le M$ for all n and all x, then $f(x) = \sum_{n=0}^{\infty} \dfrac{f^{(n)}(0)}{n!} x^n$, and the series converges uniformly on every bounded interval $(-r, r)$. Show that $\sum_{n=1}^{\infty} \dfrac{f^{(n)}(0)}{(n - 1)!} x^{n-1}$ converges for all x, and converges uniformly on bounded intervals. Use Proposition 3 or Problem 6 to prove that $f'(x) = \sum_{n=1}^{\infty} \dfrac{f^{(n)}(0)}{(n - 1)!} x^{n-1}$. ●

PROBLEM 14: Let $C(x)$, $S(x)$ be two functions on $\mathbb{R}$ that have derivatives of all orders. Assume that $|C^{(n)}(x)| \le 1$, $|S^{(n)}(x)| \le 1$ for all x. Assume that the sequence $C(0)$, $C'(0)$, $C''(0), \ldots$ is the sequence $1, 0, -1, 0, 1, 0, -1, 0, \ldots$ and the sequence $S(0)$, $S'(0)$, $S''(0), \ldots$ is $0, 1, 0, -1, 0, 1, 0, -1, \ldots$. Show that $C'(x) = -S(x)$, $S'(x) = C(x)$ for all x. (We will use this result later when we define $\cos x$ and $\sin x$.) ●

PROBLEM 15: It is not true that $f(x)$ has a Taylor series expansion if f has derivatives of all orders. Let $f(x) = e^{-1/x^2}$ if $x \ne 0$ and $f(0) = 0$. Show that $f^{(n)}(x)$ exists for all x and all n, and that $f^{(n)}(0) = 0$ for all n. The Taylor series for $f(x)$ in powers of x is identically zero, but $f(x)$ is not identically zero. Hint: There are several inductions involved here. You will need to show that $e^{-1/x^2}x^{-k} \longrightarrow 0$ as $x \longrightarrow 0$ for every $k \in \mathbb{N}$. Notice that $e^{1/x^2} \ge \dfrac{1}{k!}\left(\dfrac{1}{x^2}\right)^k = \dfrac{1}{k!}\dfrac{1}{x^{2k}}$ for all $x \ne 0$. ●

PROBLEM 16: Show that e is not rational. Hint: From Problem 9, $e^{-1} = \lim s_n$, where

$$s_n = 1 - 1 + \frac{1}{2!} - \frac{1}{3!} + \cdots \pm \frac{1}{n!}.$$

Show that $n!s_n$ is an integer, and $n!(e^{-1} - s_n)$ is an integer if e is rational and n is large enough. The inequality $n!|e^{-1} - s_n| < 1/(n+1)$ leads to a contradiction. ●

PROBLEM 17: If $\{f_n\}$ is a uniformly bounded sequence on S, and A is a countable subset of S, then there is a subsequence of $\{f_n\}$ that converges at every point of A. Hint: Let $A = \{a_1, a_2, \ldots\}$. Let $f_{11}, f_{12}, \ldots$ be a subsequence of $\{f_n\}$

that converges at a_1, and let $f_{21}, f_{22}, \ldots$ be a subsequence of $\{f_{1k}\}$ that converges (also) at a_2, and so forth. Show that $\{f_{kk}\}$ is a subsequence of $\{f_n\}$, and converges at each a_i.   •

PROBLEM 18: (i) Use Taylor's Theorem to show that there is at most one differentiable function $h(x)$ on $\mathbb{R}$ such that $h'(x) = h(x)$ for all x and $h(0) = a$. (We know there is one such function, $h(x) = ae^x$, and so this shows there is exactly one.)

(ii) Use part (i) to show that $e^x e^y = e^{x+y}$ for all $x, y \in \mathbb{R}$. Hint: Fix y and let $f(x) = e^x e^y$, $g(x) = e^{x+y}$, and show $f'(x) = f(x)$, $g'(x) = g(x)$.   •

XX

TOPOLOGY IN $\mathbb{R}^2$

n this chapter we look at the topology of subsets of $\mathbb{R}^2$, and at properties of functions of two variables. Almost all our definitions and theorems for $\mathbb{R}^2$ have obvious extensions to $\mathbb{R}^n$. We will stick to $\mathbb{R}^2$, since the notation necessary to describe things in $\mathbb{R}^n$ is frequently clumsy enough to obliterate the content. We will make vigorous use of the usual picture of $\mathbb{R}^2$ as a plane furnished with two perpendicular lines, each of which has the elements of $\mathbb{R}$ etched thereon. The point (x, y) of "the plane" (i.e., of $\mathbb{R}^2$) is construed as the point whose perpendiculars to the axes land at the numbers x and y. Although the geometric imagery we have just introduced has absolutely no mathematical legitimacy for us, most of our intuition will come from the geometric picture. A *proof* of any statement about $\mathbb{R}^2$ or its subsets will depend on the axioms and theorems for $\mathbb{R}$. It is a curious fact that proofs and intuition come out of different ballparks. Mathematical proofs are deductive, but mathematical discovery is more often inductive.

The plane $\mathbb{R}^2$ becomes a linear space with the usual definition of vector sum and scalar multiplication:

$$(x_1, y_1) + (x_2, y_2) = (x_1 + x_2, y_1 + y_2),$$

$$c(x, y) = (cx, cy).$$

The distance from a point (x, y) to the origin $(0, 0)$ is given by the **norm**

$$\|(x, y)\| = \sqrt{x^2 + y^2}.$$

Now $\mathbb{R}^2$ is a **normed space**. The **inner product**, or **dot product**, of two points (or vectors) is defined by

$$(x_1, y_1) \bullet (x_2, y_2) = x_1 x_2 + y_1 y_2.$$

Now $\mathbb{R}^2$ is an **inner product space**. It is easy to check that the inner product satisfies all the requisite properties, and that

$$\|(x, y)\|^2 = (x, y) \bullet (x, y).$$

The following proposition lays out the basic inequalities for the norm and inner product.

PROPOSITION 1: *For all* $x_1, y_1, x_2, y_2, x_3, y_3$:

(i) $|(x_1, y_1) \bullet (x_2, y_2)| \leq \|(x_1, y_1)\| \|(x_2, y_2)\|$;

(ii) $\|(x_1, y_1) + (x_2, y_2)\| \leq \|(x_1, y_1)\| + \|(x_2, y_2)\|$;

(iii) $\|(x_1, y_1) - (x_2, y_1)\| = |x_1 - x_2|$,
$\|(x_1, y_1) - (x_1, y_2)\| = |y_1 - y_2|$,
$\|(x_1, y_1) - (x_2, y_2)\| \leq |x_1 - x_2| + |y_1 - y_2|$;

(iv) $\|(x_1, y_1) - (x_3, y_3)\| \leq \|(x_1, y_1) - (x_2, y_2)\| + \|(x_2, y_2) - (x_3, y_3)\|$.

PROBLEM 1: Verify the inequalities of Proposition 1. Hints: (i) Square both sides and observe that $(x_1 y_2 - x_2 y_1)^2 \geq 0$. (ii) Square both sides and use part (i). (iii) For any x, $\sqrt{x^2} = |x|$. (iv) This is a special case of (ii). •

The distance between points $P_1 = (x_1, y_1)$ and $P_2 = (x_2, y_2)$ is defined to be $\|P_1 - P_2\|$. The critical triangle inequality that a distance function must satisfy is part (iv) of Proposition 1: $\|P_1 - P_3\| \leq \|P_1 - P_2\| + \|P_2 - P_3\|$.

The **open disc of radius** r about P is

$$D(P, r) = \{Q : \|Q - P\| < r\}.$$

The **closed disc** is

$$\overline{D}(P, r) = \{Q : \|Q - P\| \leq r\}.$$

In Chapter 10 our definition of continuity for functions of two variables suggested that convergence in $\mathbb{R}^2$ would be defined as follows: $P_n = (x_n, y_n) \longrightarrow P_0 = (x_0, y_0)$ if and only if $x_n \longrightarrow x_0$ and $y_n \longrightarrow y_0$. Our actual definition is the following equivalent (see Problem 2) condition: $P_n = (x_n, y_n) \longrightarrow P_0 = (x_0, y_0)$ if and only if $\|P_n - P_0\| \longrightarrow 0$.

PROBLEM 2: Show that $\|P_n - P_0\| \longrightarrow 0$ if and only if $x_n \longrightarrow x_0$ and $y_n \longrightarrow y_0$, where $P_n = (x_n, y_n)$, $P_0 = (x_0, y_0)$. Hint: Cf. Proposition 1(iii). •

PROBLEM 3: Show that if f is continuous at (x_0, y_0) and $f(x_0, y_0) > 0$, then there is some $\delta > 0$ such that $f(x, y) > 0$ for all $(x, y) \in \mathcal{D}(f) \cap D((x_0, y_0), \delta)$. •

A set $U \subset \mathbb{R}^2$ is **open** if and only if for each $P \in U$ there is $r > 0$ so that $D(P, r) \subset U$. A set F is **closed** if and only if $\mathbb{R}^2 - F$ is open.

PROBLEM 4: (i) $D(P, r)$ is an open set.

(ii) The intersection of two open sets is open, and the union of any family of open sets is open.

(iii) The union of two closed sets is closed, and the intersection of any family of closed sets is closed. Hint: Part (iii) follows from part (ii) by set theoretic identities such as $(F_1 \cup F_2)' = F_1' \cap F_2'$, where prime (') denotes the complement in $\mathbb{R}^2$.   •

PROPOSITION 2: *A set F is closed if and only if F contains the limit of every convergent sequence in F.*

Proof: Suppose F is closed, so that $U = \mathbb{R}^2 - F$ is open, and let $\{P_n\}$ be a sequence in F such that $P_n \longrightarrow P_0$. We must show $P_0 \in F$. If $P_0 \notin F$, then $P_0 \in U$, and for some $\varepsilon > 0$ the disc $D(P_0, \varepsilon) \subset U$. The fact that $P_n \notin U$ for all n contradicts the assumption that $P_n \longrightarrow P_0$.

Now assume that F contains all limits of sequences in F, and let $U = \mathbb{R}^2 - F$. If U is not open, there is some $P_0 \in U$ such that every disc $D\left(P_0, 1/n\right)$ contains a point P_n of F. Thus P_0 is the limit of a sequence $\{P_n\}$ in F, contradicting the assumption.   ∎

PROBLEM 5: Let A be any subset of $\mathbb{R}^2$, and let $\overline{A}$ be the union of A and the set of all limits of sequences in A. Show that $\overline{A}$ is closed, and that $\overline{A}$ is the smallest closed set containing A. Hint: $\overline{A}$ certainly contains all limits of sequences in A, but does it contain all limits of sequences in $\overline{A}$? That is, you must show that if $P_n \longrightarrow P_0$ and each P_n is the limit of a sequence in A (possibly just the sequence $P_n, P_n, P_n, \ldots$), then $P_0 \in \overline{A}$. The second part follows, as you will explain, from the fact that if F is a closed set containing A, then F contains all limits of sequences in A.   •

PROBLEM 6: (i) Show that $(A \cup B)^- = \overline{A} \cup \overline{B}$ for any subsets A and B of $\mathbb{R}^2$, where $\overline{A}$ is defined in Problem 5.

(ii) How about $(A \cap B)^-$?   •

A set $S \subset \mathbb{R}^2$ is **bounded** if and only if there is a number M such that $\|P\| \leq M$ for all $P \in S$. Equivalently, S is bounded if and only if there is some number M' such that $|x| \leq M'$ and $|y| \leq M'$ for every point $(x, y) \in S$. This follows from the inequalities of Proposition 1, which imply that

$$\left\{(x, y) : |x| \leq M \text{ and } |y| \leq M\right\} \subset \left\{(x, y) : \|(x, y)\| \leq \sqrt{2}M\right\}$$

$$\subset \left\{(x, y) : |x| \leq \sqrt{2}M \text{ and } |y| \leq \sqrt{2}M\right\}.$$

A **sequence** $\{P_n\}$ is **bounded** if and only if there is M such that $\|P_n\| \leq M$ for all n — that is, if and only if the set $\{P_n : n \in \mathbb{N}\}$ is bounded.

A set K is **compact** if and only if K is both closed and bounded.

PROPOSITION 3 (The Bolzano–Weierstrass Theorem): *Every bounded sequence in $\mathbb{R}^2$ has a convergent subsequence.*

Proof: If $\{P_n\} = \{(x_n, y_n)\}$ is a bounded sequence, then $\{x_n\}$ and $\{y_n\}$ are bounded sequences in $\mathbb{R}$. Therefore, $\{x_n\}$ has a convergent subsequence: $x_{n_k} \longrightarrow x_0$. The sequence $\{y_{n_k}\}$ is still bounded, and so it has a convergent subsequence $y_{n_{k_\ell}} \longrightarrow y_0$. The corresponding subsequence $\{x_{n_{k_\ell}}\}$ converges to x_0, and so $P_{n_{k_\ell}} \longrightarrow (x_0, y_0)$. ∎

PROBLEM 7: If K is a subset of $\mathbb{R}^2$, then K is compact if and only if every sequence in K has a subsequence that converges to a point of K. •

PROBLEM 8: If $f(x, y)$ is continuous on a compact set $K \subset \mathbb{R}^2$, then f assumes a maximum value on K and a minimum value on K. •

PROBLEM 9: If $f(x, y)$ is continuous on a compact set $K \subset \mathbb{R}^2$, then $f[K]$ is a compact set in $\mathbb{R}$. •

PROPOSITION 4 (Heine–Borel Theorem): *If K is a compact subset of $\mathbb{R}^2$ and $\mathcal{U}$ is a family of open sets such that $K \subset \bigcup \mathcal{U}$, then there is a finite subset $\{U_1, \ldots, U_n\}$ of $\mathcal{U}$ such that $K \subset U_1 \cup \cdots \cup U_n$. (The usual chant is: Every open covering of a compact set has a finite subcovering.)*

Proof: Assume that K is compact (closed and bounded) and that $\mathcal{U}$ is an open covering of K that has no finite subcover. Since K is bounded, K is contained in some rectangle $R = [a, b] \times [c, d]$. Divide R in half both horizontally and vertically to get four congruent closed rectangles. If the intersection of K with each of these rectangles had a finite subcovering, then K would have a finite subcovering. Let $R_1 = [a_1, b_1] \times [c_1, d_1]$ be one of the rectangles such that $K \cap R_1$ has no finite subcovering. Here $[a_1, b_1]$ is one half of $[a, b]$, and $[c_1, d_1]$ is one half of $[c, d]$. Now divide R_1 into four quarters the same way and pick one, R_2, such that $K \cap R_2$ is not covered by any finite subset of $\mathcal{U}$. Continuing the quartering process, we define a sequence of rectangles $R_n = [a_n, b_n] \times [c_n, d_n]$ such that $K \cap R_n$ has no finite subcover. The sequences $\{a_n\}, \{b_n\}, \{c_n\}, \{d_n\}$ are monotonic and bounded, and $b_n - a_n = (b-a)/2^n$, $d_n - c_n = (d-c)/2^n$. Therefore $\lim a_n = \lim b_n = x_0$ and $\lim c_n = \lim d_n = y_0$. Each R_n contains a point (x_n, y_n) of K, and so $a_n \leq x_n \leq b_n$, $c_n \leq y_n \leq d_n$, and $(x_n, y_n) \longrightarrow (x_0, y_0)$. K is closed, so that $(x_0, y_0) \in K$, and therefore $(x_0, y_0) \in U_0$ for some $U_0 \in \mathcal{U}$. Since U_0 is open, there is some $r > 0$ such that that $D((x_0, y_0), r) \subset U_0$. Since $a_n \longrightarrow x_0$, $b_n \longrightarrow x_0$, and $c_n \longrightarrow y_0$, $d_n \longrightarrow y_0$, there is N so that $[a_N, b_N] \times [c_N, d_N] \subset D((x_0, y_0), r)$ — namely, take any N such that $|a_N - x_0| < r/2$, $|b_N - x_0| < r/2$, $|c_N - y_0| < r/2$, $|d_N - y_0| < r/2$. Since $[a_N, b_N] \times [c_N, d_N] = R_N \subset U_0$, the set $K \cap R_N$ obviously has a finite covering by sets of $\mathcal{U}$, viz. the single set U_0. ∎

PROBLEM 10: Show that a set $K \subset \mathbb{R}^2$ is compact if and only if every open covering of K has a finite subcovering. Hint: The proof is similar to the proof for subsets of $\mathbb{R}$, and similar proofs work for any $\mathbb{R}^n$. •

The Heine–Borel condition of Problem 10 is equivalent to the closed–bounded set definition of compactness in Euclidean spaces. In general topological spaces, the Heine–Borel condition (every open cover has a finite subcover) is the *definition* of a compact set. Closed bounded sets in metric spaces other than $\mathbb{R}^n$ need not be compact.

Recall that a function f is continuous at a point $(x_0, y_0) \in \mathcal{D}(f)$ provided $f(x_n, y_n) \longrightarrow f(x_0, y_0)$ for every sequence $\{(x_n, y_n)\}$ in $\mathcal{D}(f)$ such that $(x_n, y_n) \longrightarrow (x_0, y_0)$. This is equivalent to the following ε–δ condition: f is continuous at $(x_0, y_0) \in \mathcal{D}(f)$ if and only if for each $\varepsilon > 0$ there is $\delta > 0$ such that $|f(P) - f(P_0)| < \varepsilon$ for all $P \in \mathcal{D}(f)$ satisfying $\|P - P_0\| < \delta$. A function f is **uniformly continuous on a set** $S \subset \mathbb{R}^2$ if for every $\varepsilon > 0$, there is $\delta > 0$ such that $|f(P_1) - f(P_2)| < \varepsilon$ for all $P_1, P_2 \in \mathcal{D}(f)$ with $\|P_1 - P_2\| < \delta$. For uniform continuity, each $\varepsilon > 0$ must find a δ that works uniformly over the set.

PROPOSITION 5: *If f is continuous on a compact set K, then f is uniformly continuous on K.*

Proof: Assume that f is continuous on K but not uniformly continuous. That means that there is some $\varepsilon > 0$ for which no δ works uniformly over K; that is, for each $\delta = 1/n$ there are points P_n and Q_n in K with $\|P_n - Q_n\| < 1/n$ but $|f(P_n) - f(Q_n)| \geq \varepsilon$. Let $\{P_{n_k}\}$ be a convergent subsequence, with $P_{n_k} \longrightarrow P_0 \in K$. Since $\|P_n - Q_n\| < 1/n$ for all n, $Q_{n_k} \longrightarrow P_0$. Since f is continuous at P_0, there is N such that if $k \geq N$, $|f(P_{n_k}) - f(P_0)| < \varepsilon/2$ and $|f(Q_{n_k}) - f(P_0)| < \varepsilon/2$. Consequently, $|f(P_{n_k}) - f(Q_{n_k})| < \varepsilon$ if $k \geq N$, which is a contradiction. ∎

Here is another proof of the preceding proposition, using the Heine–Borel characterization of compactness. Both of these proofs will work in any metric space, and hence do not depend on the Euclidean structure. Let f be continuous on the compact set K, and let $\varepsilon > 0$. For each $P_0 \in K$, there is $\delta_0 > 0$ such that if $P \in D(P_0, \delta_0)$, $|f(P) - f(P_0)| < \varepsilon/2$. The open sets $D(P_0, \delta_0/2)$ form an open covering of K, and so some finite family covers K:

$$ K \subset D\left(P_1, \frac{\delta_1}{2}\right) \cup \cdots \cup D\left(P_n, \frac{\delta_n}{2}\right), $$

where $P_1, \ldots, P_n \in K$ and each δ_i corresponds to $\varepsilon/2$ at P_i. Let Q_1, Q_2 be any two points of K with $\|Q_1 - Q_2\| < \delta = \min_i(\delta_i/2)$, and let $Q_1 \in D(P_i, \delta_i/2)$. Then $Q_2 \in D(P_i, \delta_i)$, and we have the inequalities

$$ |f(Q_1) - f(P_i)| < \frac{\varepsilon}{2}, \qquad |f(Q_2) - f(P_i)| < \frac{\varepsilon}{2}. $$

It follows that $|f(Q_1 - f(Q_2)| < \varepsilon$. Hence, whenever $\|Q_1 - Q_2\| < \delta$, $|f(Q_1) - f(Q_2)| < \varepsilon$, and f is uniformly continuous on K.

PROBLEM 11: Any finite union of compact sets is compact. Hint: Surely you will want to consider just two compact sets. •

PROBLEM 12: Every closed disc $\overline{D}(P, r)$ is compact. •

PROBLEM 13: Let R_n be the set of all squares of the form $[(i - 1)/2^n, i/2^n] \times [(j - 1)/2^n, j/2^n] = R(n; i, j)$ for $i, j = 0, \pm 1, \pm 2, \ldots$.

(i) Show that any open set U is the union of squares $R(n; i, j)$.

(ii) Any finite union of squares $R(n; i, j)$ is compact.

(iii) Any open set U is a union $U = \bigcup K_n$, where each K_n is compact and $K_1 \subset K_2 \subset \cdots$. Hint: Let K_n be the union of all squares $R(n; i, j)$ that are subsets of $U \cap [-n, n] \times [-n, n]$.   •

The condition of the following theorem becomes the definition of continuity for functions in general topological spaces.

PROPOSITION 6: *If f is defined on an open set $G \subset \mathbb{R}^2$, then f is continuous on G if and only if $f^{-1}[U]$ is an open set for every open set $U \subset \mathbb{R}$.*

Proof: First suppose that f is continuous on G, and G is open. Let U be an open subset of $\mathbb{R}$, and let $V = f^{-1}[U]$. Let $P_0 \in V$, and let $f(P_0) = y_0 \in U$. There is some $\varepsilon > 0$ such that $(y_0 - \varepsilon, y_0 + \varepsilon) \subset U$. Since f is continuous at P_0, for this ε there is $\delta_0 > 0$ such that $|f(P) - f(P_0)| < \varepsilon$ if $\|P - P_0\| < \delta_0$ and $P \in G$. Since G is open, there is $\delta > 0$ such that $P \in G$ if $\|P - P_0\| < \delta$, and we choose such a δ with $\delta \leq \delta_0$. Then $P \in G$ and $|f(P) - f(P_0)| < \varepsilon$ if $\|P - P_0\| < \delta$; that is, $f(P) \in (f(P_0) - \varepsilon, f(P_0) + \varepsilon) \subset U$ if $P \in D(P_0, \delta)$, or $D(P_0, \delta) \subset f^{-1}[U]$, and $f^{-1}[U]$ is open.

Now assume that $f^{-1}[U]$ is open for every open set U, and in particular for every open interval. Let $P_0 \in G$, let $f(P_0) = y_0$, and let $\varepsilon > 0$. Since $(y_0 - \varepsilon, y_0 + \varepsilon)$ is open, $f^{-1}\left[(y_0 - \varepsilon, y_0 + \varepsilon)\right]$ is an open set containing P_0. Therefore, there is $\delta > 0$ such that $D(P_0, \delta) \subset f^{-1}\left[(y_0 - \varepsilon, y_0 + \varepsilon)\right]$. This implies that if $\|P - P_0\| < \delta$, then $|f(P) - y_0| < \varepsilon$. Since $f(P_0) = y_0$, we have the ε–δ condition for f to be continuous at P_0.   ∎

PROBLEM 14 (Dini's Theorem): If $\{f_n\}$ is a decreasing sequence of continuous functions that converges pointwise to a continuous function f on a compact set K, then $f_n \longrightarrow f$ uniformly on K. Hint: You can replace f_n by $f_n - f$ and so assume that $f_n \geq 0$ and $f_n \longrightarrow 0$ on K. For each $P \in K$, there is N_P such that $0 \leq f_{N_P}(P) < \varepsilon$. Since f_{N_P} is continuous, $f_{N_P}(Q) < \varepsilon$ for all Q in some disc $D(P, r_P)$. A finite number of the discs cover K: $K \subset D(P_1, r_1) \cup \cdots \cup D(P_n, r_n)$. If $M = \max\{N_{P_1}, \ldots, N_{P_n}\}$, then $0 \leq f_M(P) < \varepsilon$ for all $P \in K$. Where do you use the continuity of f? Is the statement true if f is not continuous?   •

PROBLEM 15: Define the **diameter** of a bounded set $S \subset \mathbb{R}^2$ to be diam $S = \sup\{\|P - Q\| : P, Q \in S\}$. Let $\{K_n\}$ be a sequence of compact sets in $\mathbb{R}^2$ such that $K_1 \supset K_2 \supset K_3 \supset \cdots$, and diam $K_n \longrightarrow 0$. Show that $\bigcap K_n$ is a singleton. •

PROBLEM 16: If diam $S = r$, are there necessarily points P and Q in S with $\|P - Q\| = r$? How about if S is closed?   •

PROBLEM 17: A set S is **dense** in a set T if $\overline{S} \supset T$ (see Problem 5). Show that S is dense in T if and only if for each $P \in T$ and each open set U containing P, there is $Q \in S \cap U$.   •

PROBLEM 18: Let U and V be open subsets of T such that both U and V are dense in T. Show that $U \cap V$ is dense in T. Hint: Let $t \in T$ and let $u_n \in U$, $u_n \longrightarrow t$. Each open set $U \cap D(u_n, 1/n)$ contains $v_n \in V$. (Why?) Show $v_n \in U \cap V$ and $v_n \longrightarrow t$.   •

XXI

CALCULUS OF TWO VARIABLES

he results of the last chapter show that the topological and continuity ideas in $\mathbb{R}^2$ are much like those in $\mathbb{R}$, and indeed much like those of any $\mathbb{R}^n$. To generalize the basic topological facts from $\mathbb{R}^2$ to $\mathbb{R}^n$ requires little more than the patience necessary to write down the extra subscripts. The differential properties we now investigate do, however, change significantly from $\mathbb{R}$ to $\mathbb{R}^2$ and $\mathbb{R}^n$.

We now want our (real-valued) functions to be defined on open sets, so that the derivatives make sense; hence, let $f(x, y)$ be defined on some open set in $\mathbb{R}^2$. If we fix $y(= y_0)$ and let $g(x) = f(x, y_0)$, then g is defined on some interval around x_0. If $g'(x_0)$ exists, we say f has a **partial derivative with respect to** x at (x_0, y_0). The notation is

$$g'(x_0) = f_x(x_0, y_0) = f_1(x_0, y_0)$$

$$= \lim_{h \to 0} \frac{f(x_0 + h, y_0) - f(x_0, y_0)}{h}.$$

The subscript x or 1 denotes the derivative with respect to the first variable, x. This is fine unless it turns out to be convenient to let the first variable be something other than x. Then, for example, it is not clear what $f_x(t, x)$ or $f_x(y, x)$ means. The subscript 1 for the derivative with respect to the first variable is unambiguous, but the notation f_x is useful and frequently more intuitive, so that we will usually use it and retain the notation f_1 for cases of dire necessity. The partial derivative of f

137

with respect to y (the second variable) is

$$f_2(x_0, y_0) = f_y(x_0, y_0) = \lim_{k \to 0} \frac{f(x_0, y_0 + k) - f(x_0, y_0)}{k}.$$

If f is a function of one variable and $f'(x_0)$ exists, then f is continuous at x_0 and the tangent line closely approximates f near x_0. The analogous statement for a function of two variables with partial derivatives at (x_0, y_0) is false. For example, if $f(x, y) = 0$ on both axes, then $f_x(0, 0) = f_y(0, 0) = 0$, no matter what f does off the axes, so that f need not even be continuous at $(0, 0)$. The example

$$f(x, y) = \begin{cases} \dfrac{xy}{x^2 + y^2} & \text{if } x^2 + y^2 \neq 0, \\ 0 & \text{if } x = y = 0, \end{cases} \tag{1}$$

has partial derivatives everywhere in $\mathbb{R}^2$ but is not continuous at $(0, 0)$.

PROBLEM 1: Check that the function (1) has both partial derivatives at every point (x_0, y_0), and hence is continuous in each variable separately, and that f is discontinuous at $(0, 0)$.   •

A function $L(x, y)$ is **affine** provided $L(x, y) = A + Bx + Cy$ for some constants A, B, C. The graph in $\mathbb{R}^3$ of an affine function will be called a plane. We will say that $f(x, y)$ is **differentiable** at (x_0, y_0) provided there is an affine function that closely approximates $f(x, y)$ near (x_0, y_0) in the following sense:

$$f(x, y) - f(x_0, y_0) = A(x - x_0) + B(y - y_0)$$
$$+ E(x, y)(x - x_0)$$
$$+ F(x, y)(y - y_0), \tag{2}$$

where $E(x, y) \longrightarrow 0$ and $F(x, y) \longrightarrow 0$ as $(x, y) \longrightarrow (x_0, y_0)$. The graph of the affine function $f(x_0, y_0) + A(x - x_0) + B(y - y_0)$ is called the **tangent plane** of f at (x_0, y_0).

PROBLEM 2: If f is differentiable at (x_0, y_0) and $A, B, E,$ and F are as in (2), then f is continuous at (x_0, y_0), f has both partial derivatives at (x_0, y_0), and $f_x(x_0, y_0) = A$, $f_y(x_0, y_0) = B$.   •

We have seen that the existence of both partial derivatives does not imply even the continuity of f, let alone differentiability. The useful sufficient condition for differentiability at (x_0, y_0) is the continuity of both partial derivatives on an open set around (x_0, y_0).

PROPOSITION 1: *If f, f_x, and f_y are continuous in some disc around (x_0, y_0), then f is differentiable at (x_0, y_0). It then follows from Problem 2 that the tangent plane for f at (x_0, y_0) is the graph of the function*

$$f(x_0, y_0) + f_x(x_0, y_0)(x - x_0) + f_y(x_0, y_0)(y - y_0).$$

Proof: Let (x_0, y_0) be any point in the disc where f, f_x, and f_y are continuous Consider the right-angle path that goes horizontally and then vertically from (x_0, y_0) to (x, y_0) to (x, y), and write

$$f(x, y) - f(x_0, y_0) = [f(x, y_0) - f(x_0, y_0)] + [f(x, y) - f(x, y_0)]. \qquad (3)$$

The function and its derivatives are continuous along the horizontal and vertical segments, and so the Mean Value Theorem can be applied to each of the brackets in (3):

$$f(x, y) - f(x_0, y_0) = f_x(c_x, y_0)(x - x_0) + f_y(x, d_y)(y - y_0), \qquad (4)$$

where c_x lies between x_0 and x, and d_y lies between y_0 and y. Let

$$E(x, y) = f_x(c_x, y_0) - f_x(x_0, y_0),$$
$$F(x, y) = f_y(x, d_y) - f_y(x_0, y_0). \qquad (5)$$

Since f_x and f_y are continuous at (x_0, y_0), and $(c_x, y_0) \longrightarrow (x_0, y_0)$, $(x, d_y) \longrightarrow (x_0, y_0)$ as $(x, y) \longrightarrow (x_0, y_0)$, it follows that $E(x, y) \longrightarrow 0$, $F(x, y) \longrightarrow 0$ as $(x, y) \longrightarrow (x_0, y_0)$. Substituting the expressions for $f_x(c_x, y_0)$ and $f_y(x, d_y)$ from (5) into (4), we get

$$f(x, y) - f(x_0, y_0) = [f_x(x_0, y_0) + E(x, y)] (x - x_0)$$
$$+ [f_y(x_0, y_0) + F(x, y)] (y - y_0),$$

with $E(x, y) \longrightarrow 0$, $F(x, y) \longrightarrow 0$ as $(x, y) \longrightarrow (x_0, y_0)$, and so f is differentiable at (x_0, y_0). ∎

If f is differentiable at (x_0, y_0), then the tangent plane at (x_0, y_0) hugs the graph of f near (x_0, y_0). It seems plausible then that the graph of f over every line in $\mathbb{R}^2$ through (x_0, y_0) has a tangent line, and hence a one-dimensional derivative along the line. Since lines in the plane other than the axes do not come with positive and negative directions, we have to specify a positive direction along each line through (x_0, y_0). Let $\mathbf{u} = (u_1, u_2)$ be a unit vector in $\mathbb{R}^2$, and define the **directional derivative** of f at $P_0 = (x_0, y_0)$ in the direction $\mathbf{u}$ by

$$D_{\mathbf{u}} f(x_0, y_0) = \lim_{h \to 0} \frac{f(P_0 + h\mathbf{u}) - f(P_0)}{h}$$

$$= \lim_{h \to 0} \frac{f(x_0 + hu_1, y_0 + hu_2) - f(x_0, y_0)}{h}. \qquad (6)$$

We show next that if f is differentiable at (x_0, y_0), then all the directional derivatives (6) exist. Let $A = f_x(x_0, y_0)$, $B = f_y(x_0, y_0)$, and assume that

$$f(x, y) - f(x_0, y_0) = A(x - x_0) + B(y - y_0) + E(x, y)(x - x_0) + F(x, y)(y - y_0), \qquad (7)$$

where $E(x, y)$ and $F(x, y)$ are functions that approach zero as $(x, y) \longrightarrow (x_0, y_0)$. Using (7), with $x - x_0 = hu_1$, $y - y_0 = hu_2$, we rewrite the difference quotient in

(6) as

$$\frac{Ahu_1 + Bhu_2 + Ehu_1 + Fhu_2}{h}.$$

As $h \longrightarrow 0$, $E \longrightarrow 0$ and $F \longrightarrow 0$, so that

$$D_{\mathbf{u}} f(x_0, y_0) = Au_1 + Bu_2$$

$$= f_x(x_0, y_0)u_1 + f_y(x_0, y_0)u_2.$$

The **gradient** of f at (x_0, y_0) is the vector (i.e., the point in $\mathbb{R}^2$)

$$\nabla f(x_0, y_0) = \mathbf{grad} f(x_0, y_0) = \bigl(f_x(x_0, y_0), f_y(x_0, y_0) \bigr).$$

If $\mathbf{u}$ is any unit vector, and f is differentiable at (x_0, y_0), then we have shown that

$$D_{\mathbf{u}} f(x_0, y_0) = \nabla f(x_0, y_0) \bullet \mathbf{u}.$$

PROBLEM 3: If f is differentiable at (x_0, y_0), then the maximum directional derivative of f at (x_0, y_0) is $\|\nabla f\|$, and this maximum is the directional derivative in the direction of $\nabla f(x_0, y_0)$.   •

Differentiability at (x_0, y_0) is a sufficient condition for all directional derivatives $D_{\mathbf{u}} f(x_0, y_0)$ to exist, but differentiability is not necessary, and even continuity is not necessary.

PROBLEM 4: Let $f(x, y) = 0$ if $x = 0$, and $f(x, y) = xy^2/(x^2 + y^4)$ if $x \neq 0$.

(i) Show that f is not continuous at $(0, 0)$.
(ii) Show that $D_{\mathbf{u}} f(0, 0)$ exists in every direction $\mathbf{u}$. Thus directional derivatives can exist without f being differentiable, or even continuous.   •

Now we will consider the problem of identifying local maxima and minima of functions of two variables, and we start by considering second order derivatives. If $f(x, y)$ has both first partials in some open set, then these functions $f_x(x, y)$ and $f_y(x, y)$ are again functions of two variables, which might themselves have partial derivatives. There are four possible second partials for f: f_{xx}, f_{xy}, f_{yx}, and f_{yy}. Experience with simple functions indicates that usually $f_{xy}(x, y) = f_{yx}(x, y)$, and we give next sufficient conditions for this identity.

PROPOSITION 2: *If f, f_x, f_y, f_{xy}, and f_{yx} are continuous in an open set U, then $f_{xy}(x, y) = f_{yx}(x, y)$ for all $(x, y) \in U$.*

Proof: Assume that f and its first partials and mixed second partials are continuous in U. Let $P_0 = (a, b) \in U$, and let $D(P_0, r) \subset U$. Let h and k be small positive numbers such that $P_2 = (a + h, b + k) \in D(P_0, r)$. Let $P_1 = (a + h, b)$ and $P_3 = (a, b + k)$, so that $P_0 P_1 P_2 P_3$ are the corners, listed counterclockwise, of the rectangle with lower left corner $P_0 = (a, b)$ and upper right corner $P_2 = (a + h, b + k)$. All the points of the rectangle are in $D(P_0, r)$ and therefore in U, and so the continuity assumptions hold. Let

$$F(h, k) = f(P_2) - f(P_1) - f(P_3) + f(P_0).$$

We will show that there are points (ξ, η) and (ξ_1, η_1) inside $P_0P_1P_2P_3$ such that

$$f_{yx}(\xi, \eta) = f_{xy}(\xi_1, \eta_1). \tag{8}$$

The points (ξ, η) and (ξ_1, η_1) depend on h and k, and as $(h, k) \longrightarrow (0, 0)$, (ξ, η) and (ξ_1, η_1) both approach (a, b). Thus we can conclude from (8) that $f_{yx}(a, b) = f_{xy}(a, b)$, where (a, b) is any point of U.

To show that (8) holds, let

$$\varphi(y) = f(a + h, y) - f(a, y),$$

so that

$$\varphi(b + k) - \varphi(b) = f(P_2) - f(P_3) - \big(f(P_1) - f(P_0)\big)$$
$$= F(h, k).$$

The Mean Value Theorem applies to φ on $[b, b + k]$, and so there is $\eta \in (b, b + k)$ such that

$$F(h, k) = \varphi(b + k) - \varphi(b)$$
$$= \varphi'(\eta)k$$
$$= \big[f_y(a + h, \eta) - f_y(a, \eta)\big] k.$$

Now apply the Mean Value Theorem to f_y on $[a, a + h]$; there is $\xi \in [a, a + h]$ such that

$$F(h, k) = \big[f_y(a + h, \eta) - f_y(a, \eta)\big] k$$
$$= f_{yx}(\xi, \eta)hk.$$

If

$$\psi(x) = f(x, b + k) - f(x, b),$$

then

$$\psi(a + h) - \psi(a) = f(P_2) - f(P_1) - \big[f(P_3) - f(P_0)\big]$$
$$= F(h, k).$$

Applying the Mean Value Theorem to ψ on $[a, a + h]$ and then to $f_x(\xi, b + k) - f_x(\xi, b)$ on $[b, b+k]$, we get a number $\xi_1 \in [a, a+h]$ and a number $\eta_1 \in [b, b+k]$ such that

$$F(h, k) = f_{xy}(\xi_1, \eta_1)kh.$$

Thus $f_{xy}(\xi_1, \eta_1) = f_{yx}(\xi, \eta)$ holds for points (ξ, η) and (ξ_1, η_1) arbitrarily close to (a, b), so that both $f_{xy}(a, b)$ and $f_{xy}(a, b)$ are limits of the same sequence, and they are therefore equal.   ∎

With the simplification that the mixed second partials are equal, we can give a formula for the second directional derivative that will be useful in determining local extreme values. We calculate the second derivative of $f(x, y)$ in the direction $\mathbf{u} = (u_1, u_2)$ as follows: for (x, y) near (a, b),

$$D_{\mathbf{u}} f(x, y) = f_x(x, y)u_1 + f_y(x, y)u_2.$$

Therefore, if $f_{xy} = f_{yx}$,

$$D_{\mathbf{u}}^2 f(x, y) = D_{\mathbf{u}}\left[f_x(x, y)u_1 + f_y(x, y)u_2 \right]$$
$$= f_{xx}(x, y)u_1^2 + f_{xy}(x, y)u_1 u_2$$
$$+ f_{yx}(x, y)u_2 u_1 + f_{yy}(x, y)u_2^2$$
$$= f_{xx}(x, y)u_1^2 + 2f_{xy}(x, y)u_1 u_2 + f_{yy}(x, y)u_2^2.$$

Recall that for a function g of one variable, if $g'(x_0) = 0$ and $g''(x_0) > 0$, then g has a local minimum at x_0. For two variables we might suppose that if $D_{\mathbf{u}} f(x_0, y_0) = 0$ and $D_{\mathbf{u}}^2 f(x_0, y_0) > 0$ for every $\mathbf{u}$, then f would have a local minimum at (x_0, y_0). It is certainly true that under these conditions, f would have a local minimum along every line through (x_0, y_0), but that is not enough to show that f has a local minimum at (x_0, y_0).

PROBLEM 5: Let $f(x, y) = (y - 2x^2)(y - x^2)$.

(1) Show that $D_{\mathbf{u}} f(0, 0) = 0$ for all $\mathbf{u}$, and $D_{\mathbf{u}}^2 f(0, 0) > 0$ for all $\mathbf{u}$ except $(1, 0)$. Hence, except for the x axis, f has a local minimum at $(0, 0)$ along every line through $(0, 0)$ by the second derivative test. Clearly, $f(x, 0) = 2x^4$ also has a local minimum at $(0, 0)$.

(2) Show that $f(0, 0)$ is not a local minimum. Hint: $f(x, y)$ is positive above the parabola $y = 2x^2$ and below the parabola $y = x^2$, and $f(x, y)$ is negative between the parabolas. (Notice that every line through $(0, 0)$ other than the x axis goes from the region $y < 0$, where $f(x, y) > 0$; through $(0, 0)$, where $f = 0$; and then initially into the region $y > 2x^2$, where $f(x, y) > 0$. Since f goes from positive to zero to positive along each line through $(0, 0)$, this is another proof that $f(0, 0)$ is a local minimum on each line.)   •

The sufficient condition for a function of two variables to have a local maximum or minimum involves the Taylor's Theorem representation along each line. Let f and its first and second partial derivatives be continuous on an open set containing (x_0, y_0). If $\mathbf{u} = (u_1, u_2)$ is any unit vector, then the values of f along the line through (x_0, y_0) in the direction $\mathbf{u}$ are

$$g(t) = f(x_0 + tu_1, y_0 + tu_2).$$

The derivatives of g, which are the directional derivatives of f in the direction $\mathbf{u}$, are

$$g'(t) = f_x(x_0 + tu_1, y_0 + tu_2)u_1 + f_y(x_0 + tu_1, y_0 + tu_2)u_2,$$
$$g''(t) = f_{xx}u_1^2 + 2f_{xy}u_1u_2 + f_{yy}u_2^2,$$

where the second partials of f are evaluated at points along the line as in the formula for $g'(t)$. Taylor's formula for g and $n = 2$ is

$$g(t) = g(0) + g'(0)t + \frac{1}{2!}g''(\tau)t^2 \tag{9}$$

where τ is some number between 0 and t. If f_x and f_y are zero at (x_0, y_0), which is obviously a necessary condition for f to have a local max or min at (x_0, y_0), then from (9) we get

$$f(x_0 + tu_1, y_0 + tu_2) = f(x_0, y_0) + \frac{1}{2!}\left[f_{xx}u_1^2 + 2f_{xy}u_1u_2 + f_{yy}u_2^2\right], \tag{10}$$

where the second partials are evaluated at some point

$$(x_0 + \tau u_1, y_0 + \tau u_2) \tag{11}$$

with τ between 0 and t. For f to have a local minimum at (x_0, y_0), the bracketed term in (10) must be positive for all (u_1, u_2) and all sufficiently small τ. This leads us to consider a general **quadratic form**:

$$Q = Au_1^2 + 2Bu_1u_2 + Cu_2^2$$
$$= A\left[u_1^2 + 2\frac{B}{A}u_1u_2 + \frac{B^2}{A^2}u_2^2\right] + \left(C - \frac{B^2}{A}\right)u_2^2$$
$$= A\left(u_1 + \frac{B}{A}u_2\right)^2 + \frac{CA - B^2}{A}u_2^2$$
$$= A\left[\left(u_1 + \frac{B}{A}u_2\right)^2 + \frac{CA - B^2}{A^2}u_2^2\right].$$

If $B^2 - AC < 0$, then both terms inside the bracket are positive, and Q has the same sign as A for all values of u_1 and u_2. The quantity $B^2 - AC$ is called the **discriminant** of the quadratic form. If $B^2 - AC < 0$, then $A \neq 0$, and the division is legitimate.

Now return to (10) and let $A = f_{xx}(x_0, y_0)$, $B = f_{xy}(x_0, y_0)$, $C = f_{yy}(x_0, y_0)$. If $B^2 - AC < 0$, then the discriminant $f_{xy}^2 - f_{xx}f_{yy}$ of the quadratic form in (10) will be negative near (x_0, y_0) by continuity of the partials, and the quadratic form will have the same sign as $A = f_{xx}(x_0, y_0)$. Therefore, the bracket in (10) will be positive near (x_0, y_0) if $f_{xx}(x_0, y_0) > 0$ and negative near (x_0, y_0) if $f_{xx}(x_0, y_0) < 0$. This proves the following proposition.

PROPOSITION 3: *If f and its first and second partial derivatives are continuous in an open set containing (x_0, y_0), $f_x(x_0, y_0) = f_y(x_0, y_0) = 0$, and $f_{xy}^2(x_0, y_0) - f_{xx}(x_0, y_0)f_{yy}(x_0, y_0) < 0$, then f has a local minimum at (x_0, y_0) if $f_{xx}(x_0, y_0) > 0$, and f has a local maximum at (x_0, y_0) if $f_{xx}(x_0, y_0) < 0$.*

PROBLEM 6: Let $Q(u_1, u_2) = Au_1^2 + 2Bu_1u_2 + Cu_2^2$ where A, B, C are numbers. Show that if $B^2 - AC > 0$, then Q will have both positive and negative values. What can you say about the case $B^2 - AC = 0$?   •

Now we turn to the question of implicit functions — when does an equation in x and y define y as a function of x? The equations we consider are $F(x, y) = 0$ for some function F defined on an open set. If $F(x_0, y_0) = 0$ and there is a function f of one variable defined on some interval I around x_0 such that $f(x_0) = y_0$ and $F(x, f(x)) = 0$ for all $x \in I$, then we say that f is **defined implicitly by the equation $F(x, y) = 0$.** The next proposition gives some sufficient conditions on F that ensure that the equation $F(x, y) = 0$ does define y as some function $f(x)$ locally near a point (x_0, y_0) where $F(x_0, y_0) = 0$.

PROPOSITION 4 (Implicit Function Theorem): *If F and F_y are continuous on some open set U containing (x_0, y_0), and $F(x_0, y_0) = 0$, $F_y(x_0, y_0) \neq 0$, then there is a continuous function f defined on some interval $I = (x_0 - \delta, x_0 + \delta)$ such that $F(x, f(x)) = 0$ for all $x \in I$.*

Proof: Assume that F and F_y are continuous near (x_0, y_0), with $F(x_0, y_0) = 0$, and to be specific assume that $F_y(x_0, y_0) > 0$. Then there is some disc $D((x_0, y_0), r)$ such that the functions are continuous and $F_y > 0$ in this disc. There is a small square $S = [x_0 - s, x_0 + s] \times [y_0 - s, y_0 + s]$ centered at (x_0, y_0) and contained in the disc where F and F_y are continuous and $F_y > 0$. Since $F_y > 0$ in S, F is strictly increasing along each vertical segment in S. In particular, $F(x_0, y_0 - s) < 0$ and $F(x_0, y_0 + s) > 0$. It follows that $F(x, y_0 - s) < 0$ and $F(x, y + s) > 0$ for all x in some interval $I = [x_0 - \delta, x_0 + \delta]$, with $\delta < s$. For each $x \in I$, F increases from a negative value at $(x, y_0 - s)$ to a positive value at $(x, y_0 + s)$. By the Intermediate Value Theorem, and the fact that F is strictly increasing on each vertical segment, there is for each $x \in I$ a unique y between $y_0 - s$ and $y_0 + s$ such that $F(x, y) = 0$. This unique y, for $x \in I$, will be $f(x)$; thus, $F(x, f(x)) = 0$ for each $x \in I$, and f is the implicitly defined function. The proof that f is continuous is the following problem.   ∎

PROBLEM 7: Show that the function f of Proposition 4 is continuous on $[x_0 - \delta, x_0 + \delta]$. Hint: Let $x_n \in I = [x_0 - \delta, x_0 + \delta]$, and let $x_n \longrightarrow a \in I$. The points $(x_n, f(x_n))$ all lie in the compact set $I \times [y_0 - s, y_0 + s]$. The sequence $\{f(x_n)\}$ lies in the compact set $[y_0 - s, y_0 + s]$ and so has a convergent subsequence. Let $\{f(x_{n_k})\}$ be any convergent subsequence, with $f(x_{n_k}) \longrightarrow b$. Then $(x_{n_k}, f(x_{n_k})) \longrightarrow (a, b)$. Since $F(x_{n_k}, f(x_{n_k})) = 0$ for all n_k, $F(a, b) = 0$, and $b = f(a)$. That is, every convergent subsequence of $\{f(x_n)\}$, for $x_n \longrightarrow a$, satisfies $f(x_{n_k}) \longrightarrow f(a)$. Therefore, $f(x_n) \longrightarrow f(a)$, because otherwise ....   •

PROBLEM 8: Say something about the uniqueness (or lack thereof) of the function(s) f determined under the hypotheses of Proposition 4. Notice that $F(x, y) = 0$ does not always determine a unique f — for example, $F(x, y) = (y - x^2)(y + x^2)$ at $(0, 0)$.   •

PROBLEM 9: Find all local maxima and minima of the following functions.

(i) $f(x, y) = x^2 + y^2 - xy - 4x + 5y + 10$
(ii) $f(x, y) = 6y - 2x - x^2 - y^2$ •

PROBLEM 10: Find the point (x, y) such that the sum of the squares of the distances from (x, y) to the vertices (x_1, y_1), (x_2, y_2), (x_3, y_3) of a triangle is a minimum.   •

XXII

COMPLEX NUMBERS

In the last chapter we dressed up the plane $\mathbb{R}^2$ and made it into a normed linear space, or vector space. Cauchy sequences $\{(x_n, y_n)\}$ were defined in the usual way in terms of the norm $\|(x, y)\| = \sqrt{x^2 + y^2}$. All Cauchy sequences in $\mathbb{R}^2$ converge, and so $\mathbb{R}^2$ is a **complete** normed space. Now we give $\mathbb{R}^2$ a multiplication that, together with the linear space addition, makes $\mathbb{R}^2$ a field. The two field operations $+$ and $\cdot$ on $\mathbb{R}^2$ are defined as follows:

$$(x_1, y_1) + (x_2, y_2) = (x_1 + x_2, y_1 + y_2),$$
$$(x_1, y_1) \cdot (x_2, y_2) = (x_1 x_2 - y_1 y_2, x_1 y_2 + x_2 y_1). \tag{1}$$

We will refer to the pairs (x, y) as **complex numbers** when we have multiplication in mind as a possibility, and then refer to $\mathbb{R}^2$ as the **complex plane**, and use the symbol $\mathbb{C}$ instead of $\mathbb{R}^2$. So a pair (x, y) is a *point* in $\mathbb{R}^2$, a *two-vector*, or *complex number* depending on our mood at the moment. We retain the usual norm and distance of $\mathbb{R}^2$, but now the norm is called the **absolute value** and denoted $|(x, y)|$ instead of $\|(x, y)\|$. Thus $|(x, y)| = \sqrt{x^2 + y^2}$, and the distance between $z_1 = (x_1, y_1)$ and $z_2 = (x_2, y_2)$ is

$$|z_1 - z_2| = |(x_1, y_1) - (x_2, y_2)| = \sqrt{(x_1 - x_2)^2 + (y_1 - y_2)^2}.$$

PROBLEM 1: Show that $\mathbb{C}$ with the operations (1) satisfies all the Axioms I through V. Hint: The identities for addition and multiplication are of course no longer called 0 and 1, but you can show that there are complex numbers (x_0, y_0) and (x_1, y_1) that have the appropriate properties: $(x_0, y_0) + (x, y) = (x, y)$ and $(x_1, y_1) \cdot (x, y) = (x, y)$, for all (x, y). •

The complex numbers enjoy all the *algebraic* properties of the real numbers, since these properties are simply the consequences of the Axioms I through V. There is no order relation on $\mathbb{C}$ that satisfies Axioms VI, VII, and VIII, and by "algebraic properties" we explicitly exclude any reference to order.

There is an obvious one-to-one correspondence between the complex numbers $(x, 0)$ and the real numbers x. This correspondence preserves the algebraic operations, which is why we can use the same symbols, $+$ and $\cdot$, for the operations in the two fields $\mathbb{R}$ and $\mathbb{C}$. Thus

$$(x, 0) + (y, 0) = (x + y, 0),$$
$$(x, 0) \cdot (y, 0) = (x \cdot y, 0).$$

The complex number $(0, 1)$ is denoted i, and the definition (1) is designed so that $i^2 = (-1, 0)$. Since $(1, 0)$ is the multiplicative unit in $\mathbb{C}$, and $(-1, 0)$ is its additive inverse, i^2 is the negative of the multiplicative unit. With the order axioms in $\mathbb{R}$, we proved that $-1 < 0$ and that no square is negative. Hence no square can be the additive inverse of the multiplicative unit if the order axioms hold. Since $i^2 = (-1, 0)$, it is clear that there can be no order relation in $\mathbb{C}$ that satisfies Axioms VI, VII, and VIII. A statement such as "$r < s$," therefore, always implies that r and s are real numbers.

It follows from the definition of multiplication that for every $y \in \mathbb{R}$,

$$(y, 0) \cdot (0, 1) = (y, 0)i = (0, y).$$

Therefore, any complex number (x, y) can be written

$$(x, y) = (x, 0) + (y, 0)i.$$

We now agree that a complex number of the form $(x, 0)$ will be denoted simply by x, and called **real**. The complex number (x, y) can now be written

$$(x, y) = x + yi = x + iy. \tag{2}$$

With this convention, addition and multiplication in $\mathbb{C}$ now look like this:

$$(x_1 + iy_1) + (x_2 + iy_2) = (x_1 + x_2) + i(y_1 + y_2),$$
$$(x_1 + iy_1) \cdot (x_2 + iy_2) = (x_1x_2 - y_1y_2) + i(x_1y_2 + x_2y_1).$$

The point of all the verbiage preceding (2) is to make it clear that we have not somehow stumbled on a new number, i, that does most of the things a real number does, but occasionally exhibits aberrant behavior like $i^2 = -1$. The expression $x + iy$ is simply shorthand for $(x, 0) + (y, 0)(0, 1)$, and we can perform algebraic operations on expressions $x + iy$ just as though x and y were real numbers, because the complex numbers $(x, 0)$ obey exactly the same algebraic rules as the real numbers x. We henceforth consider $\mathbb{R}$, in the guise of the complex numbers $x = x + 0i$, as a subset of $\mathbb{C}$ without further expostulation.

Since distance is the same in $\mathbb{C}$ and $\mathbb{R}^2$, and the topology of either is defined in terms of the distance, the topology in $\mathbb{C}$ is the same as the topology of $\mathbb{R}^2$. Open and closed sets are the same. Sequences have the same limits. Cauchy sequences converge. The triangle inequality holds for absolute value because it holds for the norm; that is, for all $z_1, z_2 \in \mathbb{C}$,

$$|z_1 + z_2| \le |z_1| + |z_2|.$$

PROBLEM 2: Show that for all z_1, z_2,

$$\big||z_1| - |z_2|\big| \le |z_1 - z_2|.$$

Hint: $|z_1| = |z_1 + z_2 - z_2| \le |z_1 + z_2| + |z_2|$.   •

PROBLEM 3: Show that for all $z = x + iy$,

$$\frac{1}{\sqrt{2}} \left(|x| + |y|\right) \le |z| \le |x| + |y|.$$ •

PROBLEM 4: (i) Show that every nonzero complex number $A + iB$ has two square roots; that is, there are two numbers $x + iy$ such that $x^2 - y^2 = A$ and $2xy = B$. Hint: Take care of the two cases $B = 0$, $A > 0$, and $B = 0$, $A < 0$ first; then assume $B \ne 0$. There is a temptation when dealing with high school type algebraic equations to fall into high school type mental sloth. Do not succumb; take care to state whether the conditions (equations) you impose on x and y are necessary or sufficient for roothood.

(ii) Show that every quadratic equation $az^2 + bz + c = 0$, where a, b, c are complex numbers and $a \ne 0$, has one or two roots depending on whether $b^2 - 4ac = 0$ or not. Hint: Complete the square and use part (i).   •

Open and closed discs in $\mathbb{C}$ are denoted

$$D(z_0, r) = \{z : |z - z_0| < r\},$$
$$\overline{D}(z_0, r) = \{z : |z - z_0| \le r\}.$$

Recall that a set $F \subset \mathbb{C}$ is closed if F contains the limit of every convergent sequence in F, and a set G in $\mathbb{C}$ is open if $\mathbb{C} - G$ is closed.

PROBLEM 5: (i) A set $G \subset \mathbb{C}$ is open if and only if for each $z \in G$ there is $r > 0$ such that $D(z, r) \subset G$.

(ii) Each open disc $D(z_0, r)$ is an open set.

(iii) Each closed disc $\overline{D}(z_0, r)$ is a closed set.   •

The **conjugate** of the complex number $z = x + iy$ is $\bar{z} = x - iy$. The mapping $f(z) = \bar{z}$ sends every point of $\mathbb{C}$ to its reflection in the x axis. Notice that

$$z\bar{z} = (x + iy)(x - iy) = x^2 - (iy)^2 = x^2 + y^2 = |z|^2. \tag{3}$$

From (3) it follows that if $z \neq 0$,

$$z^{-1} = \frac{\bar{z}}{|z|^2} = \frac{x}{x^2 + y^2} - i\frac{y}{x^2 + y^2}.$$

If $z = x + iy$, then x is called the **real part** of z, denoted $\operatorname{Re} z$, and y is the **imaginary part** of z, denoted $\operatorname{Im} z$. Both $\operatorname{Re} z$ and $\operatorname{Im} z$ are real numbers. The following identities are useful:

$$\operatorname{Re} z = \frac{1}{2}(z + \bar{z}),$$

$$\operatorname{Im} z = \frac{1}{2i}(z - \bar{z}).$$

PROBLEM 6: Verify the following identities.

  (i) $\overline{z \pm w} = \bar{z} \pm \bar{w}$
 (ii) $\overline{zw} = \bar{z}\,\bar{w}$
(iii) $\overline{z/w} = \bar{z}/\bar{w}$
(iv) $|-z| = |z| = |\bar{z}|$
 (v) $|zw| = |z||w|$
(vi) $|z/w| = |z|/|w|$ •

PROBLEM 7: Show that the equation $2z^4 - z + 4 = 0$ has no roots in $D(0, 1)$.   •

PROBLEM 8: Show that for all z_1, z_2, $|z_1 + z_2|^2 + |z_1 - z_2|^2 = 2\left(|z_1|^2 + |z_2|^2\right)$. (The geometric interpretation is that the sum of the squares of the diagonals of any parallelogram equals the sum of squares of the sides. If the parallelogram (with vertices $0, z_1, z_1 + z_2, z_2$) is a rectangle, this is the Pythagorean Theorem.)   •

Continuity for complex-valued functions of a complex variable (and we include the possibility that $\mathcal{D}(f) \subset \mathbb{R} \subset \mathbb{C}$) is defined in the usual way: f is continuous at a point z_0 if $z_0 \in \mathcal{D}(f)$ and $f(z_n) \longrightarrow f(z_0)$ whenever $\{z_n\}$ is a sequence in $\mathcal{D}(f)$ such that $z_n \longrightarrow z_0$. The sum, product, and quotient of continuous functions are again continuous, and the proofs are the same as for real functions (see Problem 9).

PROBLEM 9: (i) Addition and multiplication of complex numbers are continuous functions of two complex variables; that is, if $z_n \longrightarrow z_0$ and $w_n \longrightarrow w_0$, then $z_n + w_n \longrightarrow z_0 + w_0$ and $z_n w_n \longrightarrow z_0 w_0$.
  (ii) If f and g are complex-valued functions that are continuous at z_0, then $f + g$ and $f \cdot g$ are continuous at z_0, and f/g is continuous at z_0 if $g(z_0) \neq 0$.
(iii) If g is a continuous complex-valued function on $S \subset \mathbb{C}$, and f is continuous on $g[S]$, then $f \circ g$ is continuous on S.   •

PROBLEM 10: (i) If $f(z) = \bar{z}$ and $g(z) = |z|$, then f and g are continuous on $\mathbb{C}$.

(ii) If $h(z) = z^{-1}$ for $z \in \mathbb{C} - \{0\}$, then $h(z) = \bar{z}/|z|^2 = f(z)/g(z)^2$ is continuous on $\mathbb{C} - \{0\}$.   •

Complex-valued functions of a complex variable can be pictured as transformations of plane figures into other plane figures. For example, one important group of functions maps all lines and circles into lines or circles. The word "group" is used here in its mathematical sense, and we shall study these functions in detail a little later. We start with a specific example.

PROPOSITION 1: *Let* a *be a complex number,* $|a| < 1$, *and let* $F(z) = (z - a)/(1 - \bar{a}z)$. F *is a continuous one-to-one function on* $\mathbb{C} - \{1/\bar{a}\}$ *onto* $\mathbb{C} - \{-1/\bar{a}\}$. F *maps* $D(0, 1)$ *onto itself, maps the unit circle onto itself, and maps* $\mathbb{C} - (\overline{D}(0, 1) \cup \{1/\bar{a}\})$ *onto* $\mathbb{C} - (\overline{D}(0, 1) \cup \{-1/\bar{a}\})$.

Proof: F is obviously continuous except at the one point $z = 1/\bar{a}$ where it is not defined. To see that F is one-to-one, we let $w = (z - a)/(1 - \bar{a}z)$, and show that if $w \neq -1/\bar{a}$ there is exactly one z satisfying the equation. The following are equivalent if $w \neq -1/\bar{a}$:

$$w(1 - \bar{a}z) = z - a,$$
$$z(1 + \bar{a}w) = w + a, \tag{4}$$
$$z = \frac{w + a}{1 + \bar{a}w}.$$

Now let $|z| = 1$, so that $z^{-1} = \bar{z}$. Then

$$|F(z)| = \frac{|z - a|}{|1 - \bar{a}z|} = \frac{|z||1 - a\bar{z}|}{|1 - \bar{a}z|} = 1.$$

Therefore, F maps $\Gamma = \{z : |z| = 1\}$ into itself. The inverse function, as we see from (4), is

$$F^{-1}(w) = \frac{w + a}{1 + \bar{a}w} = \frac{w - (-a)}{1 - (-\bar{a})w}.$$

Clearly $F^{-1}(w)$ is the same sort of function as F, and so F^{-1} maps Γ *onto* Γ. Hence F maps points z not on Γ either into the inside or into the outside of the circle. Now we show that F maps $D(0, 1)$ onto $D(0, 1)$. Let $|z_0| < 1$. Notice that $|F(0)| = |-a| < 1$, so that $F(0) \in D(0, 1)$. Let $g(t) = F(tz_0)$ for $0 \leq t \leq 1$, so that g maps $[0, 1]$ onto the F-image of the segment from 0 to z_0. If $|F(z_0)| > 1$, then $|g(0)| < 1$, $|g(1)| > 1$, and $|g(t)|$ is continuous. By the Intermediate Value Theorem, $|g(t_0)| = |F(t_0z_0)| = 1$ for some t_0. This is a contradiction, since $|t_0z_0| < 1$ and F maps only points of Γ into Γ. Therefore, $|F(z)| < 1$ for all $z \in D(0, 1)$. If $|a| < 1$, then $|-a| < 1$ and the same argument shows that F^{-1} maps every point of $D(0, 1)$ into $D(0, 1)$. That leaves only one possibility for points outside $\overline{D}(0, 1)$. F maps all such points, except $1/\bar{a}$, into the same set except for $-1/\bar{a}$. Since F^{-1} does the same thing, F and F^{-1} are *onto* maps.   ∎

PROBLEM 11: (i) If $F(z) = (z-a)/(1-\bar{a}z)$ and $|a| = 1$, then F is a constant: $F(z) = -a$ for all $z \neq a$.

(ii) If $|a| > 1$, then F maps Γ onto Γ and, except for $\pm 1/\bar{a}$, F maps $D(0, 1)$ onto $\mathbb{C} - \bar{D}(0, 1)$ and $\mathbb{C} - \bar{D}(0, 1)$ onto $D(0, 1)$. •

Before we look at a general class of functions that behave more or less like the F of Proposition 1 and Problem 11, we discuss the complex equations for lines and circles. A **line** is a set of the form $\gamma = \{\alpha + \beta t : t \in \mathbb{R}\}$, where α and β are complex numbers and $\beta \neq 0$. Think of α and β as vectors, and γ is seen to be all points that can be gotten to by starting at α and going in the direction β (or the direction $-\beta$).

PROBLEM 12: If z_1 and z_2 are distinct points on the line $\gamma = \{\alpha + t\beta : t \in \mathbb{R}\}$, then $\gamma = \{(1 - t)z_1 + tz_2 : t \in \mathbb{R}\}$. (Notice that these two representations of γ are just complex forms for the point–slope and two-point equations of a line.) •

A **circle** is a set of the form $\{z : |z - z_0| = r > 0\}$. It is easy to see that the mapping $f(z) = \bar{z}$ sends circles onto circles, since $|z - z_0| = r$ if and only if

$$r = |\overline{(z - z_0)}| = |\bar{z} - \bar{z}_0| = |f(z) - \bar{z}_0|.$$

It is also true, but not obvious, that $f(z) = z^{-1}$ sends circles not through 0 onto circles.

PROPOSITION 2: If γ is a circle, $0 \notin \gamma$, and $f(z) = z^{-1}$, then $f[\gamma]$ is a circle. If $0 \in \gamma$, then $f[\gamma - \{0\}]$ is a line.

Proof: The Cartesian equation of any circle γ has the form

$$A(x^2 + y^2) + Bx + Cy + D = 0, \tag{5}$$

where A, B, C, D are real numbers. The point $z = x + iy$ lies on γ if and only if

$$Az\bar{z} + \frac{1}{2}B(z + \bar{z}) + \frac{1}{2i}C(z - \bar{z}) + D = 0. \tag{6}$$

If $w = f(z) = 1/z$ for some z on (6), then $z = 1/w$ and

$$A\left(\frac{1}{w\bar{w}}\right) + \frac{B}{2}\left(\frac{1}{w} + \frac{1}{\bar{w}}\right) + \frac{C}{2i}\left(\frac{1}{w} - \frac{1}{\bar{w}}\right) + D = 0,$$
$$A + \frac{B}{2}(\bar{w} + w) + \frac{C}{2i}(\bar{w} - w) + Dw\bar{w} = 0. \tag{7}$$

If the circle γ does not pass through 0, then $D \neq 0$. Hence $f[\gamma]$, which is the set of w satisfying (7), is a circle. If γ passes through 0, then $D = 0$ and $f[\gamma]$ is the line

$$A + \frac{1}{2}B(\bar{w} + w) + \frac{1}{2i}C(\bar{w} - w) = 0,$$

or

$$A + Bu - Cv = 0,$$

where $w = u + iv = f(z)$ for $z \in \gamma$.   ∎

PROBLEM 13: Show that if γ is a line (Equation (5) with $A = 0$) and $f(z) = z^{-1}$, then $f[\gamma]$ is a line if $0 \in \gamma$ ($D = 0$), and $f[\gamma]$ is a circle if $0 \notin \gamma$ ($D \neq 0$).   •

Proposition 2 and Problem 13 can be summed up as follows: the mapping $f(z) = z^{-1}$ sends circles and lines into circles and lines. A circle is necessarily a bounded set, and a line is necessarily an unbounded set. The function $f(z) = z^{-1}$ will be bounded on any set that is bounded away from 0, and unbounded on any line or circle containing 0. That tells you which circles and lines map onto circles, and which circles and lines map onto lines. Moreover, 0 is in the closure of $f[F]$ for any unbounded set F, which says that f maps every line onto a line or circle through 0.

PROBLEM 14: Show that every line, $\{\alpha + t\beta : t \in \mathbb{R}\}$, is a closed set, and every circle, $\{z : |z - z_0| = r\}$, is a compact set.   •

Define the **distance between a point z_0 and a set F** as follows:

$$\text{dist}(z_0, F) = \inf \left\{ |z_0 - w| : w \in F \right\},$$

and the **distance between two sets E and F** by

$$\text{dist}(E, F) = \inf \left\{ |z - w| : z \in E, w \in F \right\}.$$

PROBLEM 15: (i) Show that if F is a closed set and $z_0 \notin F$, then $\text{dist}(z_0, F) > 0$, and there is $w_0 \in F$ such that $\text{dist}(z_0, F) = |z_0 - w_0|$.

(ii) If E is compact and F is closed, and $E \cap F = \varnothing$, then $\text{dist}(E, F) > 0$ and there are points $z_0 \in E$, $w_0 \in F$ such that $\text{dist}(E, F) = |z_0 - w_0|$. Hint: If $|z_n - w_n| \longrightarrow \text{dist}(E, F)$, then both $\{z_n\}$ and $\{w_n\}$ are bounded sequences if E is compact.

(iii) Show there are disjoint closed sets E and F, with $E \subset \mathbb{C}$, $F \subset \mathbb{C}$, such that $\text{dist}(E, F) = 0$.   •

A mapping of the form

$$F(z) = \frac{az + b}{cz + d} \tag{8}$$

where a, b, c, d are complex numbers and

$$ad - bc \neq 0, \tag{9}$$

is called a **Moebius transformation**, or **linear fractional transformation**. The world is neatly divided into two classes with respect to the word "Moebius": most people mispronounce the word badly, and the rest sound ridiculous in pronouncing it correctly. We will therefore stick to "linear fractional transformation," and usually abbreviate to "LF map" or "LF transformation."

The LF maps (8) play an important role in geometric function theory. The condition (9) is simply the stipulation that $F(z)$ is not a constant function.

PROPOSITION 3: *Each LF transformation (8) is a continuous one-to-one function on $\mathbb{C} - \{-d/c\}$ onto $\mathbb{C} - \{a/c\}$. If $F(z) = (az + b)/(cz + d)$, then $F^{-1}(w) = (-dw + b)/(cw - a)$.*

PROBLEM 16: Prove Proposition 3. Notice that the critical quantity "$ad - bc$" that shows that F is not constant is the same for both F and F^{-1}. •

The LF maps form a **group** under composition. That is, there is an identity map I such that $I \circ F(z) = F \circ I(z) = F(z)$ for all LF maps F, and each LF map F has an inverse map $F^{-1} = G$ such that $F \circ G(z) = G \circ F(z) = I(z)$. Composition of functions is necessarily a distributive operation, since $(F \circ G) \circ H$ and $F \circ (G \circ H)$ mean the same thing:

$$(F \circ G) \circ H(z) = F\big(G\big(H(z)\big)\big) = (F \circ (G \circ H))(z).$$

The group identity is

$$I(z) = z = \frac{1 \cdot z + 0}{0 \cdot z + 1}.$$

Proposition 3 shows that each LF map has an inverse map, which is again an LF function. Therefore, the LF maps form a group provided $F \circ G$ is again an LF map whenever F and G are.

PROBLEM 17: Let $F(z) = (az + b)/(cz + d)$ and $G(z) = (\alpha z + \beta)/(\gamma z + \delta)$ with $ad - bc \neq 0$, $\alpha\beta - \delta\gamma \neq 0$. Compute $F \circ G(z)$, and show that $F \circ G$ is an LF map. Hint: Do not forget to show that $F \circ G$ is not constant. •

Let $F(z) = (az + b)/(cz + d)$ be any LF transformation. The following are identities for $c \neq 0$ and $z \neq -d/c$:

$$\frac{az + b}{cz + d} = \frac{\frac{a}{c}(cz + d) + b - \frac{ad}{c}}{cz + d}$$

$$= \frac{a}{c} + \frac{bc - ad}{c} \frac{1}{cz + d}. \tag{10}$$

The identity (10) shows that each LF transformation is a composition of the following simple LF maps:

$F_1(z) = cz;$

$F_2(z) = z + d; \qquad F_2(F_1(z)) = cz + d;$

$F_3(z) = z^{-1}; \qquad F_3(F_2(F_1(z))) = \dfrac{1}{cz + d};$

$$F_4(z) = \frac{bc - ad}{c} z; \qquad F_4\left(F_3\left(F_2\left(F_1(z)\right)\right)\right) = \frac{bc - ad}{c} \cdot \frac{1}{cz + d};$$

$$F_5(z) = z + \frac{a}{c}; \qquad F_5\left(F_4\left(F_3\left(F_2\left(F_1(z)\right)\right)\right)\right) = F(z).$$

Each of the simple LF maps F_1, F_2, F_3, F_4, F_5 maps lines and circles onto lines or circles. The map $F_3(z) = 1/z$ was treated in Proposition 2. You are asked to check the easy verifications for F_1 and F_4 (multiplication) and F_2 and F_5 (translation) in Problem 18. It follows that every LF map sends circles and lines onto circles or lines.

PROBLEM 18: Let $F(z) = az + b$ with $a \neq 0$ (so F comprises all possibilities for F_1 and F_2 above). Show that F maps every line $\{\alpha + t\beta : t \in \mathbb{R}\}$ onto a line, and maps every circle $\{z : |z - z_0| = r\}$ onto a circle. N.B. "onto" not "into."   •

PROBLEM 19: Let $F(z) = (1 + z)/(1 - z)$. Show that $\operatorname{Re} F(z) = \operatorname{Re}(1 + z)/(1 - z) = (1 - |z|^2)/|1 - z|^2$. Hence F maps $D(0, 1)$ into the right half-plane $\operatorname{Re} w > 0$ and maps $\mathbb{C} - \overline{D}(0, 1)$ into the left half-plane. Show that both these maps are onto, and conclude that F maps the unit circle onto the y axis. (The function $(1 - |z|^2)/|1 - z|^2$ is called the **Poisson kernel**, and it plays a critical role in the proof that every continuous function on $|z| = 1$ can be extended continuously onto $\overline{D}(0, 1)$ so as to be harmonic (i.e., the real part of an analytic function) in $D(0, 1)$.

PROBLEM 20: Show that no nonconstant polynomial $P(z) = a_0 + a_1 z + \cdots + a_n z^n$, with $n \geq 1$, is bounded. Hint: Prove the stronger statement: $\lim_{\|z\| \to \infty} |P(z)| = \infty$, using

$$P(z) = z^n \left[a_n + \frac{a_{n-1}}{z} + \cdots + \frac{a_0}{z^n} \right].$$

What is the net which is involved in the preceding limit?   •

XXIII

CURVES IN THE PLANE

Integrals along curves, sometimes called line integrals, play an important role in many parts of analysis, and a critical role in complex analysis. To discuss integrals over curves we first need to know something about curves, and about their lengths. Here is an outline of how the length of a curve γ is defined. Consider a partition of γ, consisting of points $z_0, z_1, z_2, \ldots, z_n$ on the curve, starting at one end, z_0, with the points z_i picked in order along the curve until we get to the other end z_n. The line segments joining each z_i to z_{i+1} form a polygonal path p, whose length is

$$\ell(p) = \sum_{i=1}^{n} |z_i - z_{i-1}|.$$

The lengths $\ell(p)$ are approximations to the length of γ, and we define the **length of** γ to be $\sup \ell(p)$, where the sup is taken over all polygonal paths.

For the preceding definition to make sense, we have to clean up a few details. For instance: What is a curve? What is an end of a curve? How many ends does a curve have? What does it mean to pick points z_i *in order* along a curve? Clearly you cannot just pick some points on a curve and join them any which way to form a polygonal path, and expect the length of that path to have anything to do with the curve. We proceed to tidy up these details.

A **curve** is the one-to-one continuous image of an interval. That is, γ is a curve in $\mathbb{C}$ provided there is a continuous one-to-one function $\zeta(t)$ on some interval $[a, b]$ such that $\gamma = \{\zeta(t) : a \leq t \leq b\}$.

PROBLEM 1: (i) A curve is a compact set.

(ii) If $\zeta(t)$ is a continuous one-to-one function on $[a, b]$ onto γ, then $\zeta^{-1}(z)$ is a continuous function on γ onto $[a, b]$. Hint: Let $\{z_n\}$ be a sequence in γ such that $z_n \longrightarrow z_0$. Let $t_n = \zeta^{-1}(z_n)$. If $\{t_n\}$ does not converge, then $\limsup t_n \neq \liminf t_n$ and you can show that there are two different subsequences that converge to two different points — say, $t_{n_k} \longrightarrow t_0$ and $t_{m_k} \longrightarrow t_0' \neq t_0$. Since $z_{n_k} \longrightarrow z_0 = \zeta(t_0)$ and $z_{m_k} \longrightarrow z_0 = \zeta(t_0')$, you have a contradiction.   •

Let γ be a curve, and let $\gamma = \zeta[a, b]$, where ζ is continuous and one-to-one. There are of course many other functions that map other intervals onto γ. If ζ_1 on $[a_1, b_1]$ and ζ_2 on $[a_2, b_2]$ are two such functions, then by Problem 1, $\zeta_2^{-1} \circ \zeta_1$ is a continuous one-to-one function on $[a_1, b_1]$ onto $[a_2, b_2]$. As we have shown earlier, this implies that the real function $\zeta_2^{-1} \circ \zeta_1$ is either strictly increasing or strictly decreasing. If $\zeta_2^{-1} \circ \zeta_1$ is strictly increasing, then ζ_1 and ζ_2 map the left endpoints a_1 and a_2 of their intervals onto the same point $\alpha \in \gamma$, and ζ_1 and ζ_2 trace out the points of γ in the same order at t runs from a_1 or a_2 to b_1 or b_2. If $\zeta_2^{-1} \circ \zeta_1$ is decreasing, then ζ_1 and ζ_2 trace out the curve in opposite directions. In any case, ζ_1 and ζ_2 map their intervals' endpoints onto the same two points of γ, and these are the **endpoints of** γ. If we specify that one of them, α, is the **initial point**, then all one-to-one functions ζ that map a real interval $[a, b]$ onto γ with $\zeta(a) = \alpha$ will trace out the points of γ in the same order. From now on all curves will have a specified initial point and terminal point, and so there is a well-defined order of points along the curve. A partition $p = \{z_0, z_1, \ldots, z_n\}$ of points in order along γ now makes sense, and each partition p of γ corresponds to a partition $\{t_0, t_1, \ldots, t_n\}$ in the domain $[a, b]$ of any one-to-one function ζ that maps onto γ with $\zeta(a) = \alpha$.

Now we can define the length of γ:

$$\ell(\gamma) = \sup \{\ell(p) : p \text{ is a partition of } \gamma\}$$

$$= \sup \left\{ \sum_{j=1}^{n} |z_j - z_{j-1}| : \{z_0, \ldots, z_n\} \text{ is a partition of } \gamma \right\}.$$

If $\ell(\gamma)$ is finite, then we say γ is **rectifiable**, and $\ell(\gamma)$ is its **length**.

If $q = \{w_0, w_1, \ldots, w_N\}$ is a partition of γ, and every point of $p = \{z_0, \ldots, z_n\}$ is a point of q, then we say that q is a **refinement** of p. If q is a refinement of p, then it follows from the triangle inequality that $\ell(q) \geq \ell(p)$. If ζ maps the interval $[a, b]$ onto γ, q and p are partitions of γ, and q, p correspond to partitions Q, P of $[a, b]$, then q is a refinement of p if and only if Q is a refinement of P.

PROBLEM 2: Let γ be a curve, and let a, b be any two complex numbers with $a \neq 0$. Let $a\gamma + b = \{az + b : z \in \gamma\}$. Show that $a\gamma + b$ is a curve, that $a\gamma + b$ is rectifiable if and only if γ is rectifiable, and that $\ell(a\gamma + b) = |a|\ell(\gamma)$. In particular, if $|a| = 1$, then the two curves have the same length.   •

The derivative of a complex-valued function of a real variable is defined in the usual way. Let $f(t) = u(t) + iv(t)$ for $a < t < b$. Then

$$f'(t_0) = \lim_{h \to 0} \frac{f(t_0 + h) - f(t_0)}{h}$$

$$= \lim_{h \to 0} \frac{u(t_0 + h) - u(t_0)}{h} + i \frac{v(t_0 + h) - v(t_0)}{h}.$$

It is clear that $f'(t_0)$ exists if and only if both $u'(t_0)$ and $v'(t_0)$ exist, and then $f'(t_0) = u'(t_0) + iv'(t_0)$.

A curve γ is **smooth** provided γ is the image of a one-to-one function ζ such that $\zeta'(t) = x'(t) + iy'(t)$ is continuous. We will call such a function ζ a **smooth function**, and say that ζ maps its domain $[a, b]$ smoothly onto γ.

PROPOSITION 1: *If ζ is a smooth function and $\gamma = \{\zeta(t) : a \leq t \leq b\}$, so that γ is a smooth curve, then γ is rectifiable and*

$$\ell(\gamma) = \int_a^b \left\{ x'(t)^2 + y'(t)^2 \right\}^{1/2} dt = \int_a^b |\zeta'(t)| \, dt.$$

Proof: Let $\zeta(t) = x(t) + iy(t)$ be a smooth function on $[a, b]$, and consider the function

$$F(s, t) = \sqrt{x'(s)^2 + y'(t)^2}.$$

The function F is continuous on the compact set $[a, b] \times [a, b]$, and hence it is uniformly continuous. For each $\varepsilon > 0$, there is $\delta > 0$ such that

$$|F(s_1, t_1) - F(s_2, t_2)| < \frac{\varepsilon}{b - a} \tag{1}$$

whenever $|s_1 - s_2| < \delta$ and $|t_1 - t_2| < \delta$. Let $\{t_0, t_1, \ldots, t_n\}$ be a partition of $[a, b]$ such that $|t_i - t_{i-1}| < \delta$ for all i, and in addition

$$\left| \int_a^b \sqrt{x'(t)^2 + y'(t)^2} \, dt - \sum_{i=1}^n \sqrt{x'(c_i)^2 + y'(c_i)^2} \, \Delta t_i \right| < \varepsilon$$

for any choices $c_i \in (t_{i-1}, t_i)$. From (1) we see that

$$\left| \sum_{i=1}^n \sqrt{x'(c_i)^2 + y'(c_i)^2} \, \Delta t_i - \sum_{i=1}^n \sqrt{x'(c_i)^2 + y'(d_i)^2} \, \Delta t_i \right| < \varepsilon$$

whenever c_i and d_i are any points in (t_{i-1}, t_i). Thus

$$\left| \int_a^b \sqrt{x'(t)^2 + y'(t)^2} \, dt - \sum_{i=1}^n \sqrt{x'(c_i)^2 + y'(d_i)^2} \, \Delta t_i \right| < 2\varepsilon \tag{2}$$

for any choices $c_i, d_i \in (t_{i-1}, t_i)$. The estimate (2) will hold *a fortiori* for any refinement of $\{t_0, \ldots, t_n\}$, with any choices c_i, d_i in each new subinterval.

Now let $p = \{z_0, \ldots, z_n\}$ be any polygonal approximation to γ that contains all the points $\zeta(t_i)$. Thus $z_i = \zeta(s_i)$ for some partition $\{s_0, \ldots, s_N\}$ that is a refinement of $\{t_0, \ldots, t_n\}$. The length of the polygonal path p is

$$
\begin{aligned}
\ell(p) &= \sum_{i=1}^{N} |\zeta(s_i) - \zeta(s_{i-1})| \\
&= \sum_{i=1}^{N} \sqrt{(x(s_i) - x(s_{i-1}))^2 + (y(s_i) - y(s_{i-1}))^2} \\
&= \sum_{i=1}^{N} \sqrt{(x'(c_i)\,\Delta s_i)^2 + (y'(d_i)\,\Delta s_i)^2} \\
&= \sum_{i=1}^{N} \sqrt{x'(c_i)^2 + y'(d_i)^2}\,\Delta s_i,
\end{aligned}
\tag{3}
$$

where each pair c_i, d_i is chosen in accordance with the Mean Value Theorem on $[s_{i-1}, s_i]$. From (2) and (3) we see that

$$
\left| \ell(p) - \int_a^b \sqrt{x'(t)^2 + y'(t)^2}\,dt \right| < 2\varepsilon
$$

for any polygonal path $p = \{z_0, \ldots, z_N\}$ that corresponds to a partition $\{s_0, \ldots, s_N\}$ that refines $\{t_0, \ldots, t_n\}$. It follows that $\sup \ell(p) < \infty$, so that γ is rectifiable, and

$$
\ell(\gamma) = \int_a^b \sqrt{x'(t)^2 + y'(t)^2}\,dt = \int_a^b |\zeta'(t)|\,dt.
\tag{4}
$$

Notice that in (4) ζ is *any* function that maps smoothly onto γ. ∎

If γ_1 is a curve with initial point α and final point β, and γ_2 is a curve with initial point β and final point η, then $\gamma_1 + \gamma_2$ will denote the union of γ_1 and γ_2, with the points ordered from α to β to η. We now include as curves all such finite sums $\gamma_1 + \gamma_2 + \cdots + \gamma_n$, where the order along the sum is determined by the order on each γ_i. The length of such a curve is the sum of the lengths of the γ_i. If each γ_i is a smooth curve, the sum is a **piecewise smooth curve**. Notice that curves are now allowed to intersect themselves. If the initial point of γ_1 and the terminal point of γ_n coincide, then γ is a **closed curve**. If the γ_j have no other points in common, then γ is a **simple closed curve**. For example, the sides of a square or triangle, joined in order, form a piecewise smooth simple closed curve. A simple closed curve can also be parameterized with a single function: $\gamma = \{\zeta(t) : a \le t \le b\}$, where $\zeta(a) = \zeta(b)$ but ζ is one-to-one on every proper subinterval $[a, c] \subset [a, b]$.

PROBLEM 3: (i) Let $\zeta(t) = \dfrac{1}{\sqrt{1 + t^2}} + i\dfrac{t}{\sqrt{1 + t^2}}$ for $t \in \mathbb{R}$. Show that ζ is a smooth function on $\mathbb{R}$ and that ζ maps $\mathbb{R}$ one-to-one onto the part of the unit circle $\Gamma = \{z : |z| = 1\}$ in the right half-plane, $\{z : \operatorname{Re} z > 0\}$.

(ii) Show that if $\Gamma(t)$ is the part of the unit circle that is the image of $[0, t]$, then $\ell\,(\Gamma(t)) = \int_0^t \dfrac{ds}{1 + s^2}$.

(iii) Show that $\int_0^\infty \dfrac{dt}{1 + t^2}$ converges. Hint: We have yet to define the trigonometric functions, and so mention of the arctangent function here would be gauche. Compare the integral with $\int_1^\infty \dfrac{dt}{t^2}$. •

As we saw in Problem 3, the length of the arc of the unit circle from 1 to $\dfrac{1}{\sqrt{2}} + \dfrac{1}{\sqrt{2}}i$ is $\int_0^1 \dfrac{dt}{1 + t^2}$. We now define the number π by

$$\pi = 4\int_0^1 \frac{dt}{1 + t^2} = 2\int_{-1}^1 \frac{dt}{1 + t^2}. \tag{5}$$

Numerical approximations to the integral show that π is approximately 3.1416.

PROBLEM 4: Show that the length of the unit circle $\Gamma = \{(x, y) : x^2 + y^2 = 1\}$ is 2π. Hint: We know from (5) that if $\gamma = \zeta[0, 1]$ for the function ζ of Problem 3 (so that γ is the arc from 1 to $\dfrac{1}{\sqrt{2}} + \dfrac{1}{\sqrt{2}}i$), then $\ell(\gamma) = \dfrac{\pi}{4}$. Show that the arc γ_2 from $\dfrac{1}{\sqrt{2}} + \dfrac{1}{\sqrt{2}}i$ to i has the same length by Problem 2, since

$$\gamma_2 = \left(\frac{1}{\sqrt{2}} + \frac{1}{\sqrt{2}}i\right)\gamma.$$

Therefore, the arc of Γ in the first quadrant has length $\dfrac{\pi}{2}$. Et cetera. •

A set $E \subset \mathbb{C}$ is **connected** if and only if there do not exist two nonempty sets A and B such that $E = A \cup B$ and $\overline{A} \cap B = \varnothing$, $A \cap \overline{B} = \varnothing$.

PROPOSITION 2: *Any union of connected sets all containing a given point z_0 is connected.*

Proof: Let $\{E_\alpha\}$ be a family of connected sets, all of which contain z_0, and let $E = \bigcup E_\alpha$. Assume E is not connected, so that there are nonempty sets A and B with $A \cup B = E$, $\overline{A} \cap B = A \cap \overline{B} = \varnothing$. Let $z_0 \in A$ and let z_1 be any point of B. If E_α is any one of the sets containing z_1, then both z_0 and z_1 are in E_α. Let $A_\alpha = E_\alpha \cap A$, $B_\alpha = E_\alpha \cap B$. Then A_α and B_α are nonempty sets (since $z_0 \in A_\alpha$ and $z_1 \in B_\alpha$) the union of which is E_α. Moreover,

$$\overline{A_\alpha} \cap B_\alpha \subset \overline{A} \cap B = \varnothing,$$
$$A_\alpha \cap \overline{B_\alpha} \subset A \cap \overline{B} = \varnothing,$$

which contradicts the fact that E_α is connected. ∎

If E is any set and $z \in E$, the **component of E containing** z is the largest connected subset of E that contains z. Since $\{z\}$ is connected and the union of all connected sets containing z is connected, the z component of E is the union of all connected subsets of E that contain z. Two components of E are either disjoint or identical. If E_1 and E_2 are two components of E and $z_0 \in E_1 \cap E_2$, then $E_1 \cup E_2$ is a connected set containing z_0. Since both E_1 and E_2 are maximal connected sets, $E_1 \cup E_2 = E_1 = E_2$. We have proved the following:

PROPOSITION 3: *Every subset $E \subset \mathbb{C}$ is the disjoint union of its components, which are its maximal connected subsets.*

PROBLEM 5: If G is a connected open set, then there do not exist nonempty open sets A and B such that $A \cup B = G$ and $A \cap B = \varnothing$. Hint: Show that $A \cap B = \varnothing$ implies $\overline{A} \cap B = \varnothing$ if B is open.   •

A set E is **arcwise connected** if for every two points $\alpha, \beta \in E$, there is a curve γ in E from α to β. Open and closed discs are both arcwise connected, since any α and β can be connected to the center by a segment, and thus connected to each other. Not all connected sets are arcwise connected. For example, let E consist of all points $(x, 0)$ for $0 \le x \le 1$, together with all points $(1/n, y)$ for $n \in \mathbb{N}, 0 \le y \le 1$, together with the point $(0, 1)$. The set E is connected, but there is no curve in E joining $(0, 0)$ and $(0, 1)$.

PROPOSITION 4: *Every arcwise connected set is connected.*

Proof: Assume E is arcwise connected but not connected, so that there are nonempty sets A and B with $A \cup B = E$, $\overline{A} \cap B = \varnothing$, $A \cap \overline{B} = \varnothing$. Let $\alpha \in A, \beta \in B$, and let $\gamma = \{\zeta(t) : a \le t \le b\}$ be a curve in E from α to β. Let $t_0 = \sup\{t : \zeta(t) \in A\}$. There is then a sequence $\{t_n\}$ in $[a, b]$ with $t_n \longrightarrow t_0$ and $\zeta(t_n) \in A$. Moreover, $\zeta(t) \in B$ for all $t > t_0$. Therefore, $\zeta(t_0) = \lim \zeta(t_n) \in \overline{A}$, and $\zeta(t_0) = \lim_{t \to t_0^+} \zeta(t) \in \overline{B}$. This is a contradiction, since $\zeta(t_0)$ is in either A or B, and in both of $\overline{A}$ and $\overline{B}$.   ■

For arbitrary sets in the plane, arcwise connectedness is a more restrictive condition. For open sets, connected and arcwise connected mean the same thing.

PROPOSITION 5: *A connected open set is arcwise connected.*

Proof: Let G be a connected open set, and let $\alpha \in G$. Let G_α be the set of all $\beta \in G$ such that α can be connected to β by a curve in G. Let β be any point of G_α. Since $\beta \in G$ and G is open, there is $r > 0$ such that $D(\beta, r) \subset G$. Any point z of $D(\beta, r)$ can be connected to α by a curve in G, namely, a curve from α to β in G followed by a radial segment from β to z. Therefore, $D(\beta, r) \subset G_\alpha$, and so G_α is an open connected (because arcwise connected) subset of G. The set $G - G_\alpha$ of points that cannot be connected to α by a curve in G is also open (see Problem 6), and so G is the union of two disjoint open sets G_α and $G - G_\alpha$, which is a contradiction by Problem 5 if $G - G_\alpha \ne \varnothing$.   ■

PROBLEM 6: Show that if G is an open set and $\alpha \in G$, then the set of points of G that cannot be connected to α by a curve in G is open.   ●

A connected open set G in $\mathbb{C}$ is called a **domain**. These sets, domains, are the natural domains of the analytic complex-valued functions we will study in Chapter 28.

PROBLEM 7: (i) Show that any two points of a domain G can be connected by a curve in G consisting of a finite number of horizontal and vertical segments (and thus a piecewise smooth curve).

(ii) Show that any two points of a domain can be joined by a smooth curve in G. Hint: Suppose $\overline{D}(1, 1) \subset G$, and $[0, 1]$, $[1, 1 + i]$ are two adjacent segments of a polygonal curve as in (i). Show there is a smooth curve in G from $\frac{1}{2}$ to $1 + \frac{1}{2}i$ that makes the path from 0 to $1 + i$ a smooth curve in G. Then change the scale if necessary.   ●

XXIV

TRIGONOMETRIC FUNCTIONS

T he functions $\cos \theta$ and $\sin \theta$ will be defined as the coordinates of the point P on the unit circle Γ such that the arc of Γ from 1 to P in the counterclockwise direction has length θ. Stated somewhat differently: if you start at 1 and go counterclockwise around the unit circle a distance θ, then the point you arrive at has coordinates $(\cos \theta, \sin \theta)$. If θ is negative, you go a distance $|\theta|$ around Γ in the clockwise direction to get to the point $(\cos \theta, \sin \theta)$. Now we must tighten up the foregoing description.

Let $\zeta(t)$ be the following smooth mapping of $\mathbb{R}$ onto $\Gamma^+ = \Gamma \cap \{z : \operatorname{Re} z > 0\}$:

$$\zeta(t) = \frac{1}{\sqrt{1 + t^2}} + \frac{t}{\sqrt{1 + t^2}} i. \qquad (1)$$

If

$$X(t) = \frac{1}{\sqrt{1 + t^2}} = \operatorname{Re} \zeta(t), \qquad Y(t) = \frac{t}{\sqrt{1 + t^2}} = \operatorname{Im} \zeta(t),$$

then $X(t)$ goes from 1 to 0 as t goes from 0 to ∞, and $Y(t)$ goes from 0 to 1 as t goes from 0 to ∞. Thus ζ maps $[0, \infty)$ one-to-one onto the part of Γ^+ in the first quadrant and maps $(-\infty, 0]$ one-to-one onto the part of Γ^+ in the fourth quadrant.

The counterclockwise direction along Γ is the direction from $\zeta(0) = 1$ to any $\zeta(t)$ for $t > 0$. This makes sense, as we saw in the last chapter.

We calculate $|\zeta'(t)|$ to get a formula for length along Γ:

$$X'(t) = -\frac{1}{2}(1 + t^2)^{-3/2}(2t) = \frac{-t}{(1 + t^2)^{-3/2}}, \tag{2}$$

$$Y'(t) = \frac{\sqrt{1 + t^2} - t\frac{1}{2}(1 + t^2)^{-1/2} \cdot 2t}{1 + t^2}$$

$$= \frac{1 + t^2 - t^2}{(1 + t^2)^{3/2}} = \frac{1}{(1 + t^2)^{3/2}}. \tag{3}$$

Therefore,

$$|\zeta'(t)| = \sqrt{X'(t)^2 + Y'(t)^2}$$

$$= \sqrt{\frac{t^2}{(1 + t^2)^3} + \frac{1}{(1 + t^2)^3}} = \frac{1}{1 + t^2}. \tag{4}$$

Let $\Gamma(t)$ be the curve from 1 to $\zeta(t)$ for $t \in \mathbb{R}$, so that $\Gamma(t)$ is an arc of Γ in the right half-plane. Let $\theta(t)$ be the length of $\Gamma(t)$ if $t > 0$, so that

$$\theta(t) = \int_0^t |\zeta'(s)|\, ds = \int_0^t \frac{ds}{1 + s^2}. \tag{5}$$

If $t < 0$, we let $\theta(t)$ be the negative of the length of $\Gamma(t)$, and so $\theta(t)$ is also given by (5) for $t < 0$. Now $\theta(t)$ is a smooth one-to-one function on $\mathbb{R}$ onto the signed lengths of arcs of Γ^+ that start at 1. We saw in Problem 4 of the last chapter that the length of Γ^+ is π, and indeed we defined π so this would be so. Therefore, θ is a one-to-one function on $\mathbb{R}$ onto $\left(-\frac{\pi}{2}, \frac{\pi}{2}\right)$. Let $T = \theta^{-1}$, so that T is defined on $\left(-\frac{\pi}{2}, \frac{\pi}{2}\right)$, and $T(\theta(t)) = t$ for all $t \in \mathbb{R}$. Since $\theta'(t) > 0$ for all t, T is differentiable, and for all t,

$$T'(\theta(t)) = \frac{1}{\theta'(t)} = 1 + t^2. \tag{6}$$

Now we define, for $\theta \in \left(-\frac{\pi}{2}, \frac{\pi}{2}\right)$,

$$\cos\theta = X(T(\theta)) = \operatorname{Re}\zeta(T(\theta)),$$
$$\sin\theta = Y(T(\theta)) = \operatorname{Im}\zeta(T(\theta)). \tag{7}$$

In other words, $(\cos\theta, \sin\theta)$ is the point on Γ^+ the distance of which from 1 along Γ (up if $\theta > 0$, down if $\theta < 0$) is θ. Now we calculate derivatives, using (2) and

(3), and hope that the results are not a surprise:

$$\cos' \theta = \frac{d}{d\theta} X(T(\theta))$$

$$= \frac{d}{d\theta} \frac{T(\theta)}{\sqrt{1 + T(\theta)^2}}$$

$$= \frac{-T(\theta)}{\left(1 + T(\theta)^2\right)^{3/2}} T'(\theta).$$

Now write $T(\theta)$ and $T'(\theta)$ in terms of t:

$$\cos' \theta(t) = -\frac{t}{(1 + t^2)^{3/2}} \cdot (1 + t^2)$$

$$= \frac{-t}{\sqrt{1 + t^2}}$$

$$= -Y(t) = -\sin \theta(t).$$

The same sort of computation leads to $\sin' \theta = \cos \theta$ (see Problem 1).

PROBLEM 1: Check that $\sin' \theta = \cos \theta$ for $-\frac{\pi}{2} < \theta < \frac{\pi}{2}$. •

Our definitions (7) give $\cos \theta$ and $\sin \theta$ only for $-\frac{\pi}{2} < \theta < \frac{\pi}{2}$. We now extend the definitions to include all real values of θ; the extension to complex values of θ will come later using power series. As $\theta \longrightarrow \frac{\pi}{2}-$, $t \longrightarrow \infty$, and Re $\zeta(t) \longrightarrow$ 0. Therefore, we define $\cos \frac{\pi}{2} = 0$, which makes $\cos \theta$ left-continuous at $\frac{\pi}{2}$. Similarly, define $\cos \left(-\frac{\pi}{2}\right) = 0$, $\sin \frac{\pi}{2} = 1$, $\sin \left(-\frac{\pi}{2}\right) = -1$, so that both functions are now continuous on $\left[-\frac{\pi}{2}, \frac{\pi}{2}\right]$. The definitions are now extended to all θ by agreeing that for all $\theta \in \mathbb{R}$,

$$\cos(\theta + \pi) = -\cos \theta, \qquad \sin(\theta + \pi) = -\sin \theta. \tag{8}$$

The values of $\cos \theta$ and $\sin \theta$ for $\frac{\pi}{2} \le \theta \le \pi$ are determined from (8) by their values on $\left[-\frac{\pi}{2}, 0\right]$, and the values for $-\pi \le \theta \le -\frac{\pi}{2}$ are determined by the values on $\left[0, \frac{\pi}{2}\right]$. Repeated application of (8) will determine $\cos \theta$ and $\sin \theta$ for all $\theta \in \mathbb{R}$, and give the periodicity relations

$$\cos(\theta + 2\pi) = \cos \theta, \qquad \sin(\theta + 2\pi) = \sin \theta.$$

Now we have $\cos\theta$ and $\sin\theta$ defined for all θ, with $\cos'\theta = -\sin\theta$ and $\sin'\theta = \cos\theta$ for $-\dfrac{\pi}{2} < \theta < \dfrac{\pi}{2}$, and therefore for all θ except possibly for odd multiples of $\dfrac{\pi}{2}$.

PROBLEM 2: (i) Show that $\cos'\theta = -\sin\theta$ and $\sin'\theta = \cos\theta$ for $\pi < \theta < \dfrac{3}{2}\pi$.

(ii) Show that $\cos\theta$ and $\sin\theta$ are continuous at $\dfrac{\pi}{2}$ and, therefore, continuous on $\mathbb{R}$. Hint: We know that $\cos\theta$ and $\sin\theta$ are left–continuous at $\dfrac{\pi}{2}$; that is, $\lim_{\theta\to\pi/2-}\cos\theta = \cos\frac{\pi}{2} = 0$, and $\lim_{\theta\to\pi/2-}\sin\theta = 1 = \sin\dfrac{\pi}{2}$. Show that $\lim_{\theta\to\pi/2+}\cos\theta = \lim_{\theta\to\pi/2+}\cos(\theta - \pi) = 0$, and so forth.   •

On $\left(-\dfrac{\pi}{2}, \dfrac{\pi}{2}\right)$, both $\cos\theta$ and $\sin\theta$ are differentiable, with $\cos'\theta = -\sin\theta$, $\sin'\theta = \cos\theta$. It follows from these identities that both $\cos\theta$ and $\sin\theta$ have second derivatives, third derivatives, and so forth. The Taylor series for $\cos\theta$ is

$$C(\theta) = \sum_{n=0}^{\infty}(-1)^n\frac{\theta^{2n}}{(2n)!}. \tag{9}$$

Since $\cos\theta$ has derivatives of all orders on $\left(-\dfrac{\pi}{2}, \dfrac{\pi}{2}\right)$ and all these derivatives are bounded by 1, the Taylor series (9) converges to $\cos\theta$ on $\left(-\dfrac{\pi}{2}, \dfrac{\pi}{2}\right)$. However, we saw earlier (Problem 14, Chapter 19) that the series (9) converges for all θ, that the function $C(\theta)$ defined by (9) has derivatives of all orders for all θ, and that

$$C'(\theta) = \sum_{n=1}^{\infty}(-1)^n\frac{\theta^{2n-1}}{(2n - 1)!}. \tag{10}$$

The derivatives of $\sin\theta$ at zero are the numbers $0, 1, 0, -1, 0, 1, 0, -1, \ldots$, and so the series (10) is the Taylor series for $-\sin\theta$, and the series (10) converges to $-\sin\theta$ on $\left(-\dfrac{\pi}{2}, \dfrac{\pi}{2}\right)$. Let $S(\theta)$ be the function on $\mathbb{R}$ defined by (10):

$$S(\theta) = -C'(\theta) = \sum_{n=1}^{\infty}(-1)^{n+1}\frac{\theta^{2n-1}}{(2n - 1)!}. \tag{11}$$

We know that $S(\theta) = \sin\theta$ for $-\dfrac{\pi}{2} < \theta < \dfrac{\pi}{2}$, but only for θ in this interval, because we only know that $\sin\theta$ has all derivatives in $\left(-\dfrac{\pi}{2}, \dfrac{\pi}{2}\right)$. Similarly we know that $\cos\theta = C(\theta)$ in $\left(-\dfrac{\pi}{2}, \dfrac{\pi}{2}\right)$, and only in this interval. However, the functions $C(\theta)$ and $S(\theta)$ defined by the series have derivatives everywhere.

By continuity, $\cos\theta = C(\theta)$ on $\left[-\frac{\pi}{2}, \frac{\pi}{2}\right]$. Therefore, $\cos\theta$ has a left derivative at $\frac{\pi}{2}$ and a right derivative at $-\frac{\pi}{2}$. To show that $\cos\theta$ has a right derivative at $\frac{\pi}{2}$, we use (8):

$$
\begin{aligned}
\cos'\left(\frac{\pi}{2}+\right) &= \lim_{h\to 0+} \frac{\cos\left(\frac{\pi}{2}+h\right) - \cos\frac{\pi}{2}}{h} \\
&= \lim_{h\to 0+} \frac{-\cos\left(-\frac{\pi}{2}+h\right) + \cos\left(-\frac{\pi}{2}\right)}{h} \\
&= \lim_{h\to 0} \frac{-C\left(-\frac{\pi}{2}+h\right) + C\left(-\frac{\pi}{2}\right)}{h} \\
&= -C'\left(-\frac{\pi}{2}\right) = S\left(-\frac{\pi}{2}\right) \\
&= \sin\left(-\frac{\pi}{2}\right) = -\sin\frac{\pi}{2}.
\end{aligned}
$$

Thus $\cos'\left(\frac{\pi}{2}-\right) = -\sin\frac{\pi}{2}$ and $\cos'\left(\frac{\pi}{2}+\right) = -\sin\frac{\pi}{2}$, so that $\cos'\left(\frac{\pi}{2}\right) = -\sin\frac{\pi}{2}$. The same sort of argument shows that $\sin'\left(\frac{\pi}{2}\right) = \cos\frac{\pi}{2}$. Hence both $\cos\theta$ and $\sin\theta$ have first derivatives at $\frac{\pi}{2}$ and, therefore, have derivatives of all orders at $\frac{\pi}{2}$, and therefore at all θ.

PROBLEM 3: (i) Show that $\sin'\left(\frac{\pi}{2}\right) = \cos\frac{\pi}{2}$.

(ii) Show that the existence of both first derivatives at $\frac{\pi}{2}$ implies the existence of derivatives of all orders at $\frac{\pi}{2}$ for both functions.

(iii) Show that if $\cos\theta$ and $\sin\theta$ have derivatives of all orders at $\frac{\pi}{2}$, then they have derivatives of all orders everywhere $\left(\text{i.e., at all integer multiples of } \frac{\pi}{2}\right)$.   •

Now that we know that $\cos\theta$ and $\sin\theta$ have derivatives of all orders for all θ, and that these derivatives are everywhere bounded by 1, Taylor's Theorem says that the series (9) and (11) converge to $\cos\theta$ and $\sin\theta$, respectively, on the whole line.

The trigonometric functions could simply be *defined* by their power series (9) and (11), but then their connection with points on the circle and with angles would be lost, or at least obscured. To reclaim some of the familiar trigonometry, let us define $\arg z$, for $z \neq 0$, to be the number θ in $(-\pi, \pi]$ such that $z/|z| = \cos\theta + i\sin\theta$. We define the *angle* between the positive x axis and the ray $\{tz : t \geq 0\}$ to be $\arg z$.

We also want to define the angle determined by any three distinct points z_1, z_0, z_2. We define this angle, denoted $\text{Ang}(z_1, z_0, z_2)$, in such a way that it is invariant under certain transformations. If $f_1(z) = z + c$ for some complex constant c, then f_1 is called a **translation**. If $f_2(z) = \alpha z$ for some α with $|\alpha| = 1$, then f_2 is called a **rotation**, for reasons we will explore later. We agree that $\text{Ang}(z_1, z_0, z_2)$ is invariant under translation and rotation; that is, for any $c \in \mathbb{C}$ and any α with $|\alpha| = 1$,

$$\text{Ang}(z_1, z_0, z_2) = \text{Ang}(z_1 - c, z_0 - c, z_2 - c),$$

$$\text{Ang}(z_1, z_0, z_2) = \text{Ang}(\alpha z_1, \alpha z_0, \alpha z_2).$$

Any three distinct numbers z_1, z_0, z_2 can be transformed by translation and rotation to a triple z_1^*, z_0^*, z_2^* with $z_0^* = 0$ and z_1^* real. We then define $\text{Ang}(z_1, z_0, z_2)$ to be $\arg z_2^*$. For two rays from the same point z_0, say

$$R_1 = \{z_0 + t(z_1 - z_0) : t \geq 0\},$$
$$R_2 = \{z_0 + t(z_2 - z_0) : t \geq 0\}, \tag{12}$$

we define the angle from R_1 to R_2 to be $\text{Ang}(z_2, z_0, z_1)$.

PROBLEM 4: Find a translation $f_1(z) = z + c$ and a rotation $f_2(z) = \alpha z$ such that $f_2 \circ f_1$ transforms z_1, z_0, z_2 to z_1^*, z_0^*, z_2^* with $z_0^* = 0$, $z_1^* > 0$ (which of course means real and positive).   •

PROBLEM 5: For the rays R_1 and R_2 of (12), let w_1 be any point of R_1 different from z_0 and let w_2 be any point of R_2 different from z_0. Show that $\text{Ang}(w_1, z_0, w_2) = \text{Ang}(z_1, z_0, z_2)$. In other words, the angle between two rays does not depend on the particular points z_1 and z_2 used to define the rays.   •

We agree that z_1, z_0, z_2 are the vertices of a right triangle with right angle at z_1 and hypotenuse the segment from z_0 to z_2 (the whole phrase is being defined) if the three points are distinct, and z_1, z_0, z_2 can be rigidly transformed (i.e., by translation and rotation) to points z_1^*, 0, z_2^* with $z_1^* > 0$, $\text{Re}\, z_2^* = z_1^*$. Thus we assume that z_1, z_0, z_2 can be rigidly transformed to $x > 0$, 0, $x + iy$. If $\theta = \text{Ang}(z_1, z_0, z_2) = \text{Ang}(x, 0, x + iy)$, then we have the familiar relations

$$\frac{\text{opposite side}}{\text{hypotenuse}} = \frac{y}{\sqrt{x^2 + y^2}} = \sin \theta,$$

$$\frac{\text{adjacent side}}{\text{hypotenuse}} = \frac{x}{\sqrt{x^2 + y^2}} = \cos \theta,$$

$$\frac{\text{opposite side}}{\text{adjacent side}} = \frac{y}{x} = \frac{\sin \theta}{\cos \theta}.$$

We now define $\tan \theta = \sin \theta / \cos \theta$, and the other trigonometric functions in the usual way.

Since $\sin \theta$ is one-to-one on $\left[-\dfrac{\pi}{2}, \dfrac{\pi}{2}\right]$ onto $[-1, 1]$, we can define an inverse function, $\arcsin x$, so that $\sin(\arcsin x) = x$ for all $x \in [-1, 1]$ and $\arcsin(\sin \theta) = \theta$ for all $\theta \in \left[-\dfrac{\pi}{2}, \dfrac{\pi}{2}\right]$.

PROBLEM 6: Show that $\dfrac{d}{dx} \arcsin x = \dfrac{1}{\sqrt{1-x^2}}$ for $-1 < x < 1$. Hint: You know that $\arcsin x$ is differentiable, because $\sin \theta$ is and $\sin' \theta = \cos \theta \neq 0$ on $\left(-\dfrac{\pi}{2}, \dfrac{\pi}{2}\right)$. Show that $\cos(\arcsin x) = \sqrt{1 - x^2}$, using only what we know. Be careful not to recrudesce here to the thoughtless computations of your calculus classes.　　　　　●

PROBLEM 7: Show that $\zeta(t) = \sqrt{1 - t^2} + it$ is a smooth mapping of $(-1, 1)$ onto Γ^+. Show that the length of the arc of Γ from 1 to $\dfrac{1}{\sqrt{2}} + \dfrac{1}{\sqrt{2}}i$, which we know is $\dfrac{\pi}{4}$, is given by $\int_0^{1/\sqrt{2}} \dfrac{dt}{\sqrt{1-t^2}}$. Thus (from Problem 6), $\arcsin^{-1}\left(\dfrac{1}{\sqrt{2}}\right) = \dfrac{\pi}{4}$.　　●

PROBLEM 8: Show that $\zeta(\theta) = \cos \theta + i \sin \theta$ is a smooth one-to-one mapping of $(-\pi, \pi]$ onto Γ.　　　　　●

To prove the sum formulas for $\sin \theta$ and $\cos \theta$, we will use the following consequence of Taylor's Theorem.

PROPOSITION 1: *There is at most one real-valued function $f(x)$ such that $f'(x)$ and $f''(x)$ exist on $\mathbb{R}$, $f(x) = -f''(x)$ for all $x \in \mathbb{R}$, and $f(x_0) = a$, $f'(x_0) = b$.*

Proof: Since f' and f'' exist everywhere, f and f' are continuous everywhere. It follows that f and f' are bounded on every bounded interval $[-N, N]$. Let $|f(x)| \leq M_N$ and $\left|f'(x)\right| \leq M_N$ on $[-N, N]$. Since $f''(x) = -f(x)$ for all x, $f'''(x)$ exists for all x, and $f'''(x) = -f'(x)$. Since f' is everywhere differentiable, f''' is too, and $f^{(4)}(x) = -f''(x)$, and so on. By induction, one shows that $f^{(k)}(x)$ exists for all x and all k, and $\left|f^{(k)}(x)\right| \leq M_N$ on $[-N, N]$. Moreover, the sequence $\{f^{(k)}(x_0)\}$ is $\{a, b, -a, -b, a, b, -a, -b, \ldots\}$. Taylor's Theorem says that for each $x \in [-N, N]$,

$$f(x) = \sum_{k=0}^{n} \frac{1}{k!} f^{(k)}(x_0)(x - x_0)^k + R_n(x),$$

where, for some ξ between x_0 and x,

$$R_n(x) = \frac{1}{(n + 1)!} f^{(k+1)}(\xi)(x - x_0)^{n+1}.$$

Since $\left|f^{(k+1)}(\xi)\right| \leq M_N$, for each $x \in [-N, N]$,

$$|R_n(x)| \leq M_N \frac{N^{n+1}}{(n+1)!}.$$

Since $N^n/n! \longrightarrow 0$ for any N, as $n \longrightarrow \infty$, $R_n(x) \longrightarrow 0$ uniformly on $[-N, N]$, for *any* N. That is, the Taylor series for f,

$$\sum_{k=0}^{\infty} \frac{f^{(k)}(x_0)}{k!}(x - x_0)^k, \tag{13}$$

converges to $f(x)$ for all $x \in \mathbb{R}$, and converges uniformly on each bounded interval. Since the coefficients $f^{(k)}(x_0)$ are determined by the conditions $f(x_0) = a$, $f'(x_0) = b$, $f''(x) = -f(x)$, there is only one such function, namely the function (13), that satisfies these conditions.   ∎

PROBLEM 9: Verify the identities

$$\cos(x \pm a) = \cos x \cos a \mp \sin x \sin a,$$

$$\sin(x \pm a) = \sin x \cos a \pm \cos x \sin a.$$

Hint: Let $f(x) = \cos(x + a)$ and let $g(x) = \cos x \cos a - \sin x \sin a$. Verify that f and f', g and g' are defined for all x, that $f''(x) = -f(x)$, $g''(x) = -g(x)$ for all x, and that $f(0) = \cos a = g(0)$, $f'(0) = -\sin a = g'(0)$. The identities for $\cos(x - a)$, $\sin(x - a)$ follow from $\cos(-x) = \cos x$, $\sin(-x) = -\sin x$.   •

From now on we will treat the trigonometric functions familiarly, and assume that we could prove whatever we use from the developments of this chapter. The "could" in that sentence is subjunctive, indicating politely that we certainly do not want to fool with elementary trigonometry anymore.

PROBLEM 10: Verify the following identities.

(i) $\cos^2 x = (1 + \cos 2x)/2$
(ii) $\sin^2 x = (1 - \cos 2x)/2$ •

PROBLEM 11: Assume that f is defined on $\mathbb{R}$, $f''(x)$ exists for all x, and $f''(x) + f(x) = 0$ for all x. Show that $f(x) = a \cos x + b \sin x$ for some numbers a and b. Hint: Let $g(x) = f(0) \cos x + f'(0) \sin x$, so that $g(0) = f(0), g'(0) = f'(0)$, and $g''(x) = -g(x)$ for all x.   •

PROBLEM 12: Use series to approximate $\cos 1$ and $\sin 1$ to three decimal places (i.e., error less than .0005).   •

XXV

LINE INTEGRALS

I n this chapter we will discuss the integral of a complex-valued function, first for a complex function $f(t) = u(t) + iv(t)$ defined on some real interval $a \leq t \leq b$, and then for a complex-valued function $f(z) = u(z) + iv(z)$ defined for z in some smooth curve γ in $\mathbb{C}$. The first situation appears to be a special case of the second, since real intervals $[a, b]$ are certainly smooth curves in $\mathbb{C}$. However, we will reduce all integrals over plane curves to integrals over real intervals, using a parameterizing function, so that in fact the second situation becomes a special case of the first.

Now let $f(t) = u(t) + iv(t)$ be defined for $t \in [a, b]$. We want to define $\int_a^b f(t)\, dt$. The Darboux definition of the integral is not a possibility, since it strongly involves the order relation, and there is no order among complex numbers. We therefore use the Riemann definition. If $P = \{t_0, t_1, \dots, t_n\}$ is a partition of $[a, b]$, and $c_j \in (t_{j-1}, t_j)$ for each j, then the Riemann sum for f, P, and c is

$$R(f, P, c) = \sum_{j=1}^{n} f(c_j)(t_j - t_{j-1})$$

$$= \sum_{j=1}^{n} u(c_j)(t_j - t_{j-1}) + i \sum v(c_j)(t_j - t_{j-1}).$$

The pairs (P, c) are ordered by refinement of the partition as usual, and so $R(f, P, c)$ is a complex-valued net on the directed set of all pairs (P, c). The net $R(f, P, c)$ will converge if and only if both real nets $R(u, P, c)$, $R(v, P, c)$ converge. In this case, f is integrable over $[a, b]$, and

$$\int_a^b f(t)\, dt = \lim_P R(f, P, c)$$

$$= \int_a^b u(t)\, dt + i \int_a^b v(t)\, dt. \tag{1}$$

In particular, if f is continuous on $[a, b]$, so that u and v are continuous, then u and v and f are integrable, and (1) holds.

The Fundamental Theorem of Calculus looks much the same for integrals of complex-valued functions of a real variable. Recall that if $F(t) = U(t) + iV(t)$, then $F'(t) = U'(t) + iV'(t)$. Now let $f(t) = u(t) + iv(t)$ be continuous on $[a, b]$ and let $F(t)$ be an antiderivative of $f(t)$ on $[a, b]$. (The existence of F is Problem 1.) If $F(t) = U(t) + iV(t)$, then

$$F'(t) = U'(t) + iV'(t) = u(t) + iv(t),$$

and so $U'(t) = u(t)$, $V'(t) = v(t)$. Therefore, applying the Fundamental Theorem to u and v separately, we have

$$\int_a^b f(t)\, dt = \int_a^b u(t) + i \int_a^b v(t)\, dt$$

$$= U(b) - U(a) + i[V(b) - V(a)]$$

$$= F(b) - F(a). \tag{2}$$

The usual differentiation formulas for sums, products, and quotients hold for complex-valued functions, and the proofs are the same (see Problem 2).

PROBLEM 1: Show that if $f(t) = u(t) + iv(t)$ is continuous on $[a, b]$, then f has an antiderivative on $[a, b]$.   •

PROBLEM 2: Let $f(t)$ and $g(t)$ be complex-valued functions on (a, b) such that $f'(t_0)$, $g'(t_0)$ exist for some $t_0 \in (a, b)$. Show the following:

(i) If $h(t) = f(t) + g(t)$, then $h'(t_0) = f'(t_0) + g'(t_0)$.
(ii) If $h(t) = f(t)g(t)$, then $h'(t_0) = f(t_0)g'(t_0) + g(t_0)f'(t_0)$.
(iii) If $h(t) = f(t)/g(t)$ and $g(t_0) \neq 0$, then
$$h'(t_0) = \left(g(t_0)f'(t_0) - f(t_0)g'(t_0)\right)/g(t_0)^2.$$ •

PROBLEM 3: (i) If $f(t)$ is a differentiable complex-valued function on $[a, b]$, and $g(t) = (f(t))^n$, then $g'(t) = nf(t)^{n-1}f'(t)$. Here $n = 0, \pm 1, \pm 2, \ldots$.
Hint: Start with $n = 2$ and use Problem 2(ii), (iii).

(ii) Show that $\int_a^b (f(t))^n f'(t)\, dt = \dfrac{1}{n+1}\left[f(b)^{n+1} - f(a)^{n+1}\right]$ for every $n \neq -1$.   •

Now we consider a complex-valued function $f(z) = u(z) + iv(z)$, defined for z on some smooth curve γ. Let $p = \{z_0, \ldots, z_n\}$ be a partition of γ, and let c_j be a point of γ between z_{j-1} and z_j for each j. This makes sense since we saw in Chapter 23 that the points of a curve are ordered once an initial point is determined. We let $z_j = x_j + iy_j$, $\Delta z_j = z_j - z_{j-1}$, $\Delta x_j = x_j - x_{j-1}$, $\Delta y_j = y_j - y_{j-1}$. The Riemann sum for f, p, and c is

$$R(f, p, c) = \sum_{j=1}^{n} f(c_j)(z_j - z_{j-1})$$

$$= \sum_{j=1}^{n} \left(u(c_j) + iv(c_j) \right) \left(x_j + iy_j - (x_{j-1} + iy_{j-1}) \right)$$

$$= \sum_{j=1}^{n} u(c_j)\Delta x_j - v(c_j)\Delta y_j + i \sum_{j=1}^{n} \left(u(c_j)\Delta y_j + v(c_j)\Delta x_j \right).$$

The convergence of the Riemann sums $R(f, p, c)$ depends on the convergence of the four sums

$$\sum u(c_j)\Delta x_j, \sum v(c_j)\Delta y_j, \sum u(c_j)\Delta y_j, \sum v(c_j)\Delta x_j. \tag{3}$$

Notice that here the Δx_j and Δy_j are determined by the partition $\{z_0, \ldots, z_n\}$ of γ. In particular, Δx_j and Δy_j are not always positive. The four sums in (3) converge if u and v are continuous on γ (i.e., f is continuous on γ), and γ is a smooth curve. We prove this for the first sum in the following proposition; the other cases are similar.

PROPOSITION 1: *If $u(z)$ is continuous on the smooth curve γ, and $\zeta(t) = x(t) + iy(t)$ is a smooth map of $[a, b]$ onto γ, then*

$$\lim_{P} \sum_{j=1}^{n} u(c_j)\Delta x_j = \int_{a}^{b} u\big(\zeta(t)\big)x'(t)\, dt. \tag{4}$$

Proof: Each partition $p = \{z_0, \ldots, z_n\}$ corresponds to a specific partition $P = \{t_0, \ldots, t_n\}$ of $[a, b]$, with $z_j = \zeta(t_j)$. The integrand $u(\zeta(t))x'(t)$ is continuous on $[a, b]$, and so the integral exists, and is the limit of sums $\sum u(\zeta(s_j))x'(s_j)\Delta t_j$. We rewrite the sum $\sum u(c_j)\Delta x_j$ in terms of the parameterizing function ζ: Let $c_j = \zeta(s_j)$ for each j, with $t_{j-1} \leq s_j \leq t_j$, so that

$$\sum_{j=1}^{n} u(c_j)\Delta x_j = \sum_{j=1}^{n} u\big(\zeta(s_j)\big) \big(x(t_j) - x(t_{j-1})\big)$$

$$= \sum_{j=1}^{n} u\big(\zeta(s_j)\big) x'(r_j)\Delta t_j,$$

where r_j is chosen in (t_{j-1}, t_j) in accordance with the Mean Value Theorem. If we let

$$F(s, r) = u\big(\zeta(s)\big)x'(r),$$

then F is continuous on $[a, b] \times [a, b]$ and so uniformly continuous. Hence, given $\varepsilon > 0$, there is $\delta > 0$ so that $|F(s, r) - F(s', r')| < \varepsilon/(b - a)$ whenever $|s - s'| < \delta$ and $|r - r'| < \delta$. Now assume that $P = \{t_0, \ldots, t_n\}$ is any partition of $[a, b]$ with $|t_j - t_{j-1}| < \delta$ for all j. Assume also that P is sufficiently refined that, in addition,

$$\left| \sum_{j=1}^{n} u\big(\zeta(s_j)\big) x'(s_j) \Delta t_j - \int_a^b u\big(\zeta(t)\big) x'(t)\, dt \right| < \varepsilon$$

for any choice of $s_j \in (t_{j-1}, t_j)$. Because all $\Delta t_j < \delta$,

$$\left| \sum_{j=1}^{n} u\big(\zeta(s_j)\big) x'(s_j) \Delta t_j - \sum_{j=1}^{n} u\big(\zeta(s_j)\big) x'(r_j) \Delta t_j \right| < \varepsilon.$$

Since one of these sums is an ε-approximation to the integral

$$\int_a^b u\big(\zeta(t)\big) x'(t)\, dt,$$

and the other is equal to the left-hand sum in (4), we have the equality (4).   ∎

We will use the notation

$$\int_\gamma u(z)\, dx = \lim_P \sum u(c_i) \Delta x_i,$$

so that Proposition 1 is the statement that

$$\int_\gamma u(z)\, dx = \int_a^b u\big(\zeta(t)\big) x'(t)\, dt$$

for *any* smooth function $\zeta(t) = x(t) + iy(t)$ that maps $[a, b]$ onto γ. A proof similar to that of Proposition 1 shows that

$$\lim_P \sum_{j=1}^{n} u(c_j) \Delta y_j = \int_\gamma u(z)\, dy = \int_a^b u\big(\zeta(t)\big) y'(t)\, dt$$

for any smooth mapping ζ of $[a, b]$ onto γ. Finally, for γ a smooth curve, $\zeta(t) = x(t) + iy(t)$ any smooth mapping of $[a, b]$ onto γ, and $f(z) = u(z) + iv(z)$ any continuous function on γ, we have

$$\int_\gamma f(z)\,dz = \int_{\gamma_v} (u + iv)(dx + i\,dy)$$

$$= \int_\gamma (u\,dx - v\,dy) + i\int_\gamma u\,dy + v\,dx$$

$$= \int_a^b \left[u(\zeta(t))x'(t) - v(\zeta(t))y'(t)\right]\,dt$$

$$+i\int_a^b \left[u(\zeta(t))y'(t) + v(\zeta(t))x'(t)\right]\,dt. \tag{5}$$

Notice that we can also write the line integral as follows:

$$\int_\gamma f(z)\,dz = \int_a^b f(\zeta(t))\zeta'(t)\,dt, \tag{6}$$

which simply collects the terms on the right side of (5) into a single complex formula.

PROBLEM 4: Write out the proof that

$$\int_\gamma u(z)\,dy = \lim_P \sum_{j=1}^n u(c_j)\,\Delta y_j = \int_a^b u(\zeta(t))y'(t)\,dt$$

for any smooth map $\zeta = x + iy$ on $[a, b]$ onto γ.   •

PROPOSITION 2: *If f is a continuous function on a smooth curve γ, then*

$$\left|\int_\gamma f(z)\,dz\right| \le \left(\max_{z\in\gamma} |f(z)|\right)\ell(\gamma).$$

Proof: For any partition p of γ, we have

$$\left|\sum f(c_i)(z_i - z_{i-1})\right| \le \sum |f(c_i)|\,|z_i - z_{i-1}|$$

$$\le \left(\max_{z\in\gamma} |f(z)|\right)\sum |z_i - z_{i-1}|$$

$$\le \left(\max_{z\in\gamma} |f(z)|\right)\ell(\gamma).$$

Since $\int_\gamma f(z)\,dz$ is the limit of the sums on the left of this inequality, we have the desired inequality.   ∎

PROBLEM 5: If $\{f_n(z)\}$ is a sequence of continuous complex-valued functions on a smooth curve γ, and $f_n(z) \longrightarrow f(z)$ uniformly on γ, then f is integrable over γ, and $\int_\gamma f_n(z)\,dz \longrightarrow \int_\gamma f(z)\,dz$.   •

PROPOSITION 3: *If $f(z)$ and $g(z)$ are continuous on a smooth curve γ, and k is a complex constant, then*

$$\int_\gamma kf(z)\, dz = k \int_\gamma f(z)\, dz,$$

$$\int_\gamma \big(f(z) + g(z)\big)\, dz = \int_\gamma f(z)\, dz + \int_\gamma g(z)\, dz.$$

Proof: For any partition $p = \{z_0, \ldots, z_n\}$ of γ, and any choice c_j with each c_j a point of γ between z_{j-1}, and z_j, we have

$$R(kf, p, c) = \sum kf(c_j)(z_j - z_{j-1})$$
$$= k \sum f(c_j)(z_j - z_{j-1})$$
$$= kR(f, p, c).$$

Since the net $\{R(f, p, c)\}$ converges to $\int_\gamma f(z)\, dz$, the net $\{kR(f, p, c)\}$ converges to $k \int_\gamma f(z)\, dz$, which gives the first equality. Similarly,

$$R(f + g, p, c) = R(f, p, c) + R(g, p, c).$$

Since both nets on the right converge, the net on the left converges, which gives the sum formula. ∎

If γ is a smooth curve with initial point α and terminal point β, we will let $-\gamma$ denote the same curve oriented in the opposite direction, from β to α. If ζ maps $[a, b]$ onto γ, and $\zeta_0(t) = \zeta(ta + (1 - t)b)$, then ζ_0 maps $[0, 1]$ smoothly onto $-\gamma$.

PROBLEM 6: If $f(z)$ is continuous on the smooth curve γ, then $\int_{-\gamma} f(z)\, dz = -\int_\gamma f(z)\, dz$. Hint: It is easy to prove this using Riemann sums and the definition. •

We will make extensive use of the following proposition in the study of analytic functions of a complex variable.

PROPOSITION 4: *If γ is any smooth curve from α to β, then for any nonnegative integer n,*

$$\int_\gamma z^n\, dz = \frac{1}{n + 1} \left[\beta^{n+1} - \alpha^{n+1} \right].$$

If n is a negative integer other than -1, the same result holds for every smooth curve γ not through 0.

Proof: Let $\zeta(t) = x(t) + iy(t)$ map $[a, b]$ onto γ, with $\zeta(a) = \alpha$, $\zeta(b) = \beta$. Assume $0 \notin \gamma$ if $n < 0$, and assume $n \neq -1$. Then

$$\int_\gamma z^n \, dz = \int_a^b \left(\zeta(t)\right)^n \zeta'(t) \, dt.$$

Now we use Problem 3(ii), and we get

$$\int_\gamma z^n \, dz = \frac{1}{n + 1} \left[\zeta(b)^{n+1} - \zeta(a)^{n+1}\right]$$

$$= \frac{1}{n + 1} \left[\beta^{n+1} - \alpha^{n+1}\right]. \qquad \blacksquare$$

PROBLEM 7: Let γ_1 and γ_2 be two different smooth curves from $\alpha = 0$ to $\beta = 1 + i$. For instance, let γ_1 be the line segment $\zeta_1(t) = t + it, 0 \leq t \leq 1$, and γ_2 the parabola $\zeta_2(t) = t + it^2, 0 \leq t \leq 1$. Calculate $\int_{\gamma_1} z^2 \, dz$ and $\int_{\gamma_2} z^2 \, dz$ using (6) with the two parameterizations and $u(z) = x^2 - y^2$, $v(z) = 2xy$.   •

PROBLEM 8: (i) If $f(z)$ is a polynomial, then there is a polynomial $F(z)$ such that for any smooth curve γ from α to β, $\int_\gamma f(z) \, dz = F(\beta) - F(\alpha)$.

(ii) If $f(z)$ is a function on a domain G such that for every smooth curve γ in G there is a sequence $\{f_n(z)\}$ of polynomials such that $f_n(z) \longrightarrow f(z)$ uniformly on γ, then there is a function $F(z)$ on G such that for every smooth curve γ from α to β, $\int_\gamma f(z) \, dz = F(\beta) - F(\alpha)$.   •

PROPOSITION 5: *If Γ is the positively oriented unit circle, then $\int_\Gamma \dfrac{dz}{z} = 2\pi i$.*

Proof: Let $f(z) = z^{-1}$, so that $f(z) = \bar{z}$ for $z \in \Gamma$. We parameterize Γ as follows:

$$\Gamma = \{\cos \theta + i \sin \theta : 0 \leq \theta \leq 2\pi\}.$$

Thus $\zeta(\theta) = \cos \theta + i \sin \theta$ is a smooth function mapping $[0, 2\pi]$ onto Γ. With this parameterization,

$$f\left(\zeta(\theta)\right) = \overline{\zeta(\theta)} = \cos \theta - i \sin \theta,$$

and

$$\int_\Gamma z^{-1} \, dz = \int_0^{2\pi} \overline{\zeta(\theta)} \zeta'(\theta) \, d\theta$$

$$= \int_0^{2\pi} (\cos \theta - i \sin \theta)(- \sin \theta + i \cos \theta) \, d\theta$$

$$= \int_0^{2\pi} i(\cos^2 \theta + \sin^2 \theta) \, d\theta$$

$$= 2\pi i. \qquad \blacksquare$$

The function $f(z) = z^{-1}$ is of course not a uniform limit of polynomials on every smooth curve γ, since z^{-1} is not defined at 0. However, z^{-1} is the limit of polynomials on curves in some domains.

PROPOSITION 6: *If $z_0 \neq 0$, then z^{-1} is the uniform limit of polynomials on any smooth curve in $D\left(z_0, |z_0|\right)$.*

Proof: Each smooth curve γ in $D(z_0, |z_0|)$ is a compact set, so there is $r < 1$ such that $|z - z_0| \leq r|z_0|$ for all $z \in \gamma$. Now write z^{-1} as a geometric series in powers of $z - z_0$.

$$
\begin{aligned}
\frac{1}{z} &= \frac{1}{z_0 + (z - z_0)} \\
&= \frac{1}{z_0} \frac{1}{1 + \left(\dfrac{z - z_0}{z_0}\right)}.
\end{aligned}
$$

If $z \in \gamma$, then $|z - z_0|/|z_0| \leq r < 1$. Therefore,

$$
\frac{1}{z} = \frac{1}{z_0} \sum_{n=0}^{\infty} (-1)^n \left(\frac{z - z_0}{z_0}\right)^n, \tag{7}
$$

and the series converges uniformly on γ. The partial sums of (7) are polynomials in z that uniformly approximate z^{-1} on γ. ∎

PROBLEM 9: (i) Let $F(z) = \int_\gamma \dfrac{dw}{w}$ where γ is any smooth curve in $D(1, 1)$ from 1 to z. Show that if z is real and $0 < z < 2$, then $F(z) = \log z$.
(ii) Let $z = r \cos \theta + ir \sin \theta$ be any point in $D(1, 1)$. Let γ be the curve from 1 to z consisting of the real interval from 1 to r, followed by the arc of the circle of radius r from r to z. Show that $\int_\gamma z^{-1} dz = \log r + i\theta$. Hint: The second piece of γ can be parameterized by $\zeta(t) = r \cos t + ir \sin t$, $0 \leq t \leq \theta$, if $\theta > 0$. •

PROBLEM 10: Let γ_1 be the segment from $-i$ to i, and let γ_2 be the right half of the unit circle (from $-i$ to i). Compute $\int |z|\, dz$, $\int \bar{z}\, dz$, and $\int z\, dz$ over both curves. •

PROBLEM 11: Let γ be the segment $[1, 1+i]$ followed by the segment $[1+i, i]$. Set up the real integrals for $\int_\gamma z^2\, dz$, and verify that the answer (from Proposition 4) is $\dfrac{1}{3}i^3 - \dfrac{1}{3}$. •

PROBLEM 12: Let $f(z)$ be continuous in $|z| \geq R$, and let $M(r) = \max \{|f(z)| : |z| = r\}$ for $r \geq R$. Show that if $rM(r) \longrightarrow 0$ as $r \longrightarrow \infty$, then $\int_{|z|=r} f(z)\, dz \longrightarrow 0$ as $r \longrightarrow \infty$. State a similar result for a function f such that $rM(r) \longrightarrow 0$ as $r \longrightarrow 0$. •

PROBLEM 13: Let γ_r be the closed curve consisting of the real interval $[-r, r]$ followed by the top half of the circle $|z| = r$. Show that $\lim_{r \to \infty} \int_{\gamma_r} \dfrac{dz}{1 + z^2} = \pi$. Hint: First show that $\frac{d}{dx} \tan^{-1} x = \frac{1}{1+x^2}$, using $\frac{d}{dx} \tan(\tan^{-1} x) = 1$ and $\sec^2 x = \tan^2 x + 1$. Then compare with Problem 12.   •

PROBLEM 14: Let $z^{1/2} = |z|^{1/2} e^{i\theta/2}$ where $\theta = \arg z$, $-\pi < \theta \le \pi$. Compute $\int z^{1/2}\, dz$ over the top half of the unit circle from 1 to -1, and then over the bottom half—i.e., the curve $z = e^{i\theta}$ as θ runs from 0 to $-\pi$.   •

XXVI

POWER SERIES

We have already treated series of real numbers, $\sum x_n$, and series of real functions, $\sum f_n(x)$. Series of complex numbers and complex functions offer little new in the way of excitement until we get to power series and analytic functions. If $\{z_n\}$ is a sequence of complex numbers, with $z_n = x_n + iy_n$, then $\sum z_n = Z$ provided $s_n \longrightarrow Z$, where

$$s_n = z_1 + \cdots + z_n$$
$$= (x_1 + \cdots + x_n) + i(y_1 + \cdots + y_n).$$

Clearly $s_n \longrightarrow Z = X + iY$ if and only if

$$x_1 + \cdots + x_n \longrightarrow X \qquad \text{and} \qquad y_1 + \cdots + y_n \longrightarrow Y.$$

Hence $\sum z_n$ converges if and only if the two real series $\sum \operatorname{Re} z_n$, $\sum \operatorname{Im} z_n$ converge. It is clear that convergent series of complex numbers can be added term by term, or multiplied by a constant term by term. That is, if $\sum z_n = \alpha$ and $\sum w_n = \beta$, then $\sum(z_n + w_n) = \alpha + \beta$ and $\sum k z_n = k\alpha$ for every complex number k.

A complex series $\sum z_n$ **converges absolutely** provided the real positive series $\sum |z_n|$ converges. Absolute convergence of complex series implies convergence, just as for real series.

PROBLEM 1: (i) Show that $\sum z_n$ converges if $\sum |z_n|$ converges.
(ii) If $\sum(x_n + iy_n)$ converges absolutely, then $\sum x_n$ and $\sum y_n$ converge absolutely, and conversely. •

Recall that an absolutely convergent real series can be rearranged and grouped at will. It follows (see Problem 2) that absolutely convergent complex series can also be rearranged and grouped as needed. We will need that fact, for instance, to multiply two series.

PROBLEM 2: Show that if $\sum_{n=1}^{\infty} z_n = Z$, the series converges absolutely, and $\{z_{n_k}\}$ is a rearrangement of $\{z_n\}$ (i.e., $\{n_k\}$ is one-to-one on $\mathbb{N}$ onto $\mathbb{N}$), then $\sum_{k=1}^{\infty} z_{n_k} = Z$. Hint: Use Problem 1.   •

Series of complex-valued functions defined on a set in $\mathbb{R}$ or $\mathbb{C}$ are treated in the obvious way. The series $\sum f_n(z)$ **converges to** $F(z)$ **pointwise** on the set $S \subset \mathbb{C}$ provided that, for each $z \in S$ and each $\varepsilon > 0$, there is N, which *a priori* depends on both ε and z, such that $\left| \sum_{k=1}^{n} f_k(z) - F(z) \right| < \varepsilon$ for all $n \geq N$. If, for each ε, there is N such that $\left| \sum_{k=1}^{n} f_k(z) - F(z) \right| < \varepsilon$ for all $z \in S$, then $\sum f_k(z)$ **converges to** $F(z)$ **uniformly** on S. Finite sums of complex functions are continuous, and uniform limits of continuous complex functions are continuous. The proofs are exactly as for real functions. Therefore, if $\{f_n(z)\}$ is a sequence of complex-valued functions, all of which are continuous on some set $S \subset \mathbb{C}$, and if $\sum f_n(z)$ converges uniformly on S to $F(z)$, then F is continuous on S.

PROBLEM 3: Verify that if each complex-valued function $s_n(z)$ is continuous on a set $S \subset \mathbb{C}$, and $s_n(z) \longrightarrow F(z)$ uniformly on S, then F is continuous on S.   •

We will be concerned in this chapter with **power series**, that is, series of the form

$$\sum_{n=0}^{\infty} a_n z^n \quad \text{or} \quad \sum_{n=0}^{\infty} a_n(z - z_0)^n, \tag{1}$$

where $\{a_n\}$ is a sequence of complex (or real) numbers. We have already seen some real power series in the form of Taylor series. Recall that if f has derivatives of all orders on some interval $(-r, r)$, then f has the following Taylor series:

$$\sum_{n=0}^{\infty} \frac{1}{n!} f^{(n)}(0) x^n = f(0) + f'(0)x + \frac{1}{2!} f''(0)x^2 + \cdots.$$

The Taylor series for f does not always converge to f, but it does if the functions $f^{(n)}(x)$ are suitably bounded. For instance, we showed with Taylor's Theorem that for all real x,

$$e^x = 1 + x + \frac{x^2}{2!} + \frac{x^3}{3!} + \cdots + \frac{x^n}{n!} + \cdots,$$

$$\cos x = 1 - \frac{x^2}{2!} + \frac{x^4}{4!} - \cdots + \frac{(-1)^n x^{2n}}{(2n)!} + \cdots,$$

$$\sin x = x - \frac{x^3}{3!} + \frac{x^5}{5!} - \cdots + \frac{(-1)^{n-1} x^{2n-1}}{(2n-1)!} + \cdots.$$

It is the nature of the beasts that power series should be considered in the complex domain. We may write x in a power series $\sum a_n x^n$, and think of x as real, but we cannot stop x from thinking of itself as complex. Complex algebra is the same as real algebra, and so a complex x can do anything a real x can. For example, if $f(x) = 1/(1 + x^2)$, then f has derivatives of all orders on all of $\mathbb{R}$. The Taylor series for $f(x)$ is the geometric series

$$f(x) = \frac{1}{1 + x^2} = 1 - x^2 + x^4 - x^6 + \cdots, \tag{2}$$

which converges only on $(-1, 1)$. The reason we should not be surprised that the series (2) does not converge to $f(x)$ on the whole line is that x could be complex; that is, the same series (2), with x replaced by z, is the series for $f(z) = 1/(1 + z^2)$. If (2) converged for all real x, then (2) would also converge for all complex x by Proposition 1 below. Of course, $f(z) = 1/(1 + z^2)$ is not defined for $z = i$, and $|f(z)| \longrightarrow \infty$ as $z \longrightarrow i$.

We will treat series in powers of z for simplicity. Any statement about $\sum a_n z^n$ has an obvious consequence for $\sum a_n (z - z_0)^n$, and vice versa.

The simplest and most versatile power series is the geometric series $\sum z^n$. The same algebra that shows that for $x \neq 1$,

$$1 + x + x^2 + \cdots + x^n = \frac{1}{1 - x} - \frac{x^{n+1}}{1 - x},$$

works equally well in any field, and particular in $\mathbb{C}$. Therefore, for any complex number $z \neq 1$,

$$1 + z + z^2 + \cdots + z^n = \frac{1}{1 - z} - \frac{z^{n+1}}{1 - z}. \tag{3}$$

The sequence $\{z^{n+1}\}$ converges if and only if $|z| < 1$, or $z = 1$, which is ruled out in (3). Therefore, if $|z| < 1$,

$$\sum_{n=0}^{\infty} z^n = \frac{1}{1 - z}, \tag{4}$$

and the series diverges ($z^n \nrightarrow 0$) if $|z| \geq 1$. Notice that (2) is a special case of (4), since $|x| < 1$ if and only if $|-x^2| < 1$.

Some power series converge only at $z = 0$; some power series converge for all z. Barring these two possibilities, a power series converges inside some disc and diverges outside this disc. That is the content of the next proposition.

PROPOSITION 1: *If $\sum a_n z^n$ converges at $z_0 \neq 0$, then $\sum a_n z^n$ converges absolutely for all z such that $|z| < |z_0|$. Consequently, if $\sum a_n z^n$ diverges at some number z_1, then $\sum a_n z^n$ diverges for all z such that $|z| > |z_1|$.*

Proof: Suppose $z_0 \neq 0$ and $\sum a_n z_0^n$ converges. This implies that $a_n z_0^n \longrightarrow 0$, and so in particular $\{a_n z_0^n\}$ is a bounded sequence. Let $|a_n z_0^n| \leq M$ for all n. If

$|z| < |z_0|$, then $|z/z_0| = r < 1$, and

$$|a_n z^n| = |a_n z_0^n| \left| \frac{z}{z_0} \right|^n \leq M r^n.$$

Since $\sum M r^n$ converges if $0 \leq r < 1$, it follows that $\sum |a_n z^n|$ converges if $|z| < |z_0|$; that is, $\sum a_n z^n$ converges absolutely for all $|z| < |z_0|$.   ■

It follows from Proposition 1 that $\sum a_n z^n$ converges only for $z = 0$, or converges for all z, or there is some $r > 0$ such that $\sum a_n z^n$ converges for $z \in D(0, r)$ and diverges for $z \notin \overline{D}(0, r)$. The first two cases are real possibilities (see Problem 4). In the last case (see Problem 5), the open disc $D(0, r)$ is called the **circle of convergence** of the series. This unfortunate nomenclature arises from the confusion in English of circle-as-curve and circle-as-inside-of-curve. We will write "z is inside the circle of convergence" to avoid the confusion. The number r is called the **radius of convergence** of the series.

PROBLEM 4: (i) Show that $\sum n! z^n$ converges only for $z = 0$.

(ii) Show that $\sum z^n / n!$ converges for all z. Hint: Show that if $\sum a_n x^n$ converges for all real x, then $\sum a_n z^n$ converges for all complex z. We know the series for e^x converges for all real x by Taylor's Theorem.   ●

PROBLEM 5: Assume that $\sum a_n z^n$ converges for some nonzero number z_0, but does not converge for all z. Show there is $r > 0$ such that the series converges in $D(0, r)$ and diverges outside $\overline{D}(0, r)$.   ●

PROBLEM 6: Assume that the series $\sum_{n=1}^{\infty} a_n z^{-n} = a_1 z^{-1} + a_2 z^{-2} + a_3 z^{-3} + \cdots$ converges for at least one z and diverges for at least one z (other than 0 of course).

(i) Show that if the series converges at z_1, then the series converges for all z with $|z| > |z_1|$.

(ii) If the series diverges at z_2, then the series diverges for all z with $|z| < |z_2|$.

(iii) There is $r > 0$ such that the series converges absolutely for $|z| > r$, and the series diverges on $\{z : 0 < |z| < r\}$. Hint: Do this problem two ways. First use the method of Proposition 1 and compare the series with an appropriate geometric series. Then let $z = 1/w$ and use what you know about the series $\sum_{n=1}^{\infty} a_n w^n$ to describe the convergence of $\sum_{n=1}^{\infty} a_n z^{-n}$.   ●

The root and ratio tests for real series automatically become tests for the absolute convergence of complex series and, in particular, for power series.

PROPOSITION 2: *If $a_n \neq 0$ for all n and $\lim |a_{n+1}/a_n| = s$, then $\sum a_n z^n$ converges absolutely if $|z| < 1/s$ and diverges if $|z| > 1/s$; in other words, $1/s$ is the radius of convergence.*

Proof: Apply the ratio test to the real series $\sum |a_n z^n|$: since $\lim |a_{n+1}/a_n| = s$,

$$\lim_{n \to \infty} \frac{|a_{n+1} z^{n+1}|}{|a_n z^n|} = s|z|.$$

If $|z| < 1/s$, so that $s|z| < 1$, the series converges, and if $|z| > 1/s$, the series diverges. Of course if $s = 0$, the series converges for all z. ∎

As we noted earlier, the ratio test has the weakness that $\lim |a_{n+1}/a_n|$ may not converge, whereas $\limsup \sqrt[n]{|a_n|}$ always either converges or diverges to $+\infty$.

PROPOSITION 3: *If* $\limsup \sqrt[n]{|a_n|} = s$, *then* $1/s$ *is the radius of convergence of the series* $\sum a_n z^n$. *If* $s = 0$, *the series converges for all* z. *If* $s = \infty$, *the series converges only for* $z = 0$.

Proof: Assume $0 < \limsup \sqrt[n]{|a_n|} = s < \infty$. Then applying the root test to $\sum a_n z^n$, we get

$$\limsup \sqrt[n]{|a_n z^n|} = |z| \limsup \sqrt[n]{|a_n|} = s|z|.$$

Therefore, $\sum a_n z^n$ converges absolutely if $s|z| < 1$, and the series diverges if $s|z| > 1$; that is, $1/s$ is the radius of convergence if $s \neq 0$. ∎

PROBLEM 7: Show that if $\limsup \sqrt[n]{|a_n|} = \infty$, then $\sum a_n z^n$ converges only for $z = 0$, and if $\limsup \sqrt[n]{|a_n|} = 0$, then $\sum a_n z^n$ converges for all z. •

PROBLEM 8: Show that $\sum_{n=0}^{\infty} a_n z^n$, $\sum_{n=1}^{\infty} n a_n z^{n-1}$, and $\sum_{n=0}^{\infty} \frac{1}{n+1} a_n z^{n+1}$ all have the same radius of convergence (including the cases $r = 0$ or $r = \infty$). Hint: Explain why it suffices to show that the first two have the same radius. •

The Weierstrass M-test is a standard tool for showing uniform convergence for both real and complex series.

PROBLEM 9 (Weierstrass M-test): If $\{u_n(z)\}$ is a sequence of complex-valued functions on a set $S \subset \mathbb{C}$, $|u_n(z)| \leq M_n$ for all n and all $z \in S$, and $\sum M_n$ converges, then $\sum u_n(z)$ converges uniformly on S. •

PROPOSITION 4: *If* r *is the radius of convergence of the power series* $\sum a_n z^n$, *then the series converges uniformly on every closed disc* $\overline{D}(0, s)$ *with* $s < r$. *If* $f(z) = \sum a_n z^n$ *for* $|z| < r$, *then* f *is continuous on* $D(0, r)$.

Proof: Assume that r is the radius of convergence of $\sum a_n z^n = f(z)$, and let $0 < s < r$. Then $\sum |a_n s^n| = \sum |a_n| s^n$ converges by Proposition 1. If $|z| \leq s$, then $|a_n z^n| \leq |a_n| s^n$, so that the Weierstrass M-test with $M_n = |a_n| s^n$ shows that the series converges uniformly on $\overline{D}(0, s)$. Since $f(z)$ is continuous on every $\overline{D}(0, s)$ with $s < r$, $f(z)$ is continuous on $D(0, r)$. ∎

PROBLEM 10: (i) If $\sum a_n z^n$ has radius of convergence r, then $\sum a_n z^n$ converges uniformly on every compact set $K \subset D(0, r)$ and, in particular, on every curve $\gamma \subset D(0, r)$.

(ii) If $\sum_{n=1}^{\infty} a_n z^{-n}$ converges for $|z| > r$, then the series converges uniformly on every compact subset of $\{z : |z| > r\}$. •

Now consider the question of whether a function can have two different power series representations; that is, is there $f(z)$ such that

$$f(z) = \sum_{n=0}^{\infty} a_n z^n = \sum_{n=0}^{\infty} b_n z^n \tag{5}$$

for all z in some disc, where $\{a_n\}$, $\{b_n\}$ are different sequences? It would be unfortunate if this were the case, but it is not obvious that it isn't. We can rephrase the question as follows: If $\sum a_n z^n = 0$ for all z in some disc, does it follow that all $a_n = 0$? The answer is yes, but even more is true.

PROPOSITION 5 (Identity Theorem for Series): *If $f(z) = \sum a_n z^n$ for $|z| < r$, and there is a sequence $\{z_n\}$ in $D(0, r) \sim \{0\}$ such that $z_n \longrightarrow 0$ and $f(z_n) = 0$ for all n, then $a_n = 0$ for all n, and $f(z) \equiv 0$.*

Proof: If $f(z_n) = 0$ for all n, and $z_n \longrightarrow 0$, then $f(0) = 0$ since f is continuous on $D(0, r)$ by Proposition 4. Therefore, $a_0 = f(0) = 0$, and

$$f(z) = z(a_1 + a_2 z + a_3 z^2 + \cdots).$$

The series

$$f_1(z) = a_1 + a_2 z + a_3 z^2 + \cdots$$

also has radius of convergence r (see Problem 11), and $f_1(z_n) = 0$ for all n, so that $f_1(0) = a_1 = 0$. Therefore,

$$f(z) = z^2(a_2 + a_3 z + a_4 z^4 + \cdots),$$

and the same argument shows $a_2 = 0$, and so forth. ∎

PROBLEM 11: Show that if $\sum a_n z^n$ has radius of convergence r, and $a_0 = a_1 = \cdots = a_{n-1} = 0$, then

$$a_n + a_{n+1} z + a_{n+2} z^2 + \cdots$$

also has radius of convergence r.   •

We have seen (Problem 8) that if

$$f(z) = a_0 + a_1 z + a_2 z^2 + \cdots + a_n z^n + \cdots, \tag{6}$$

and the radius of convergence is r, then

$$g(z) = a_1 + 2a_2 z + 3a_3 z^2 + \cdots + na_n z^{n-1} + \cdots \tag{7}$$

also has radius of convergence r. Both series therefore converge absolutely for $|z| < r$, and converge uniformly on every disc $\bar{D}(0, s)$ with $s < r$. We will show in Chapter 29 that if f is given by (6), then f has a complex derivative (to be defined later), and that derivative is represented by the series (7); that is, $f'(z) = g(z)$.

It is a remarkable theorem that if $f'(z_0) \neq 0$ for a complex function, then f is one-to-one on some disc around z_0. (The analogous statement for real functions is not true — see Problem 12.) Without worrying yet about what $f'(z)$ means, we will prove next that the series (6) represents a function that is locally one-to-one near zero whenever $a_1 \neq 0$.

PROPOSITION 6: *Let*

$$f(z) = a_0 + a_1 z + a_2 z^2 + \cdots + a_n z^n + \cdots, \tag{8}$$

and suppose r is the radius of convergence of the series. If $a_1 \neq 0$, then f is one-to-one on some disc around 0.

Proof: We prove the contrapositive statement: if $f(z_n) = f(w_n)$ for pairs z_n, w_n with $z_n \neq w_n$ and z_n, w_n approaching zero, then $a_1 = 0$. Accordingly, let $z_n \longrightarrow 0$, $w_n \longrightarrow 0$, $z_n \neq w_n$ for all n, with $f(z_n) = f(w_n)$ for all n. Then

$$f(z_n) - f(w_n) = 0 = a_1(z_n - w_n) + a_2(z_n^2 - w_n^2) + \cdots,$$

and so for all n,

$$0 = (z_n - w_n)\left[a_1 + a_2 \left(\frac{z_n^2 - w_n^2}{z_n - w_n} \right) + a_2 \left(\frac{z_n^3 - w_n^3}{z_n - w_n} \right) + \cdots \right].$$

Hence, since $z_n \neq w_n$,

$$0 = a_1 + \sum_{k=2}^{\infty} a_k \left(\frac{z_n^k - w_n^k}{z_n - w_n} \right).$$

We define $P_k(z, w)$ for $z \neq w$ by

$$P_k(z, w) = \frac{z^k - w^k}{z - w} = z^{k-1} + z^{k-2}w + z^{k-3}w^2 + \cdots + w^{k-1}.$$

Hence if $|z| \leq s < r$ and $|w| \leq s < r$,

$$|P_k(z, w)| \leq k s^{k-1},$$

and by Problem 8, $\sum_{k=2}^{\infty} k|a_k|s^{k-1}$ converges. Now we have, for all n,

$$0 = a_1 + \sum_{k=2}^{\infty} a_k P_k(z_n, w_n), \tag{9}$$

and if $|z_n| \leq s$, $|w_n| \leq s$, then

$$\left| \sum_{k=2}^{\infty} a_k P_k(z_n, w_n) \right| \leq \sum_{k=2}^{\infty} k|a_k|s^k.$$

The last series is a function of s that converges to 0 as $s \longrightarrow 0$, and so $\sum_{k=2}^{\infty} a_k P_k(z_n, w_n)$ approaches 0 as z_n and w_n approach 0. Therefore, from (9), $a_1 = 0$. ∎

PROBLEM 12: Exhibit a real function $f(x)$ such that $f'(x)$ exists for all x, $f'(0) \neq 0$, and f is not one-to-one on any interval around 0. Hint: Any function f such that $x \le f(x) \le e^x - 1$ for all x satisfies $f'(0) = 1$. Show that $\frac{1}{2}(x + e^x - 1) + \frac{1}{2}(e^x - 1 - x) \sin \frac{1}{x}$ is such a function.   •

PROBLEM 13: Extend Proposition 6 as follows: If $f(z) = a_0 + a_1 z + a_2 z^2 + \cdots$ for $|z| < r$, and in every disc around 0 there are three distinct points s, z, w such that $f(s) = f(z) = f(w)$, then $a_1 = a_2 = 0$. Hint: The same techniques as in Proposition 6 will work. Show that

$$\sum_{k=2}^{\infty} a_k \left(P_k(s, z) - P_k(s, w) \right) = 0,$$

and for $k \ge 2$,

$$P_k(s, z) - P_k(s, w) = (z - w) \left[s^{k-2} + s^{k-3} P_2(z, w) + \cdots + P_{k-1}(z, w) \right]. \quad •$$

XXVII

THE TRANSCENDENTAL FUNCTIONS

The Taylor series for e^x, $\cos x$, and $\sin x$ converge for all real x, and so these series converge for all $z \in \mathbb{C}$. We use these series to extend the domains of e^x, $\cos x$, and $\sin x$ to all of $\mathbb{C}$. We accordingly define, for all complex numbers z,

$$e^z = 1 + z + \frac{z^2}{2!} + \frac{z^3}{3!} + \cdots,$$

$$\cos z = 1 - \frac{z^2}{2!} + \frac{z^4}{4!} - \frac{z^6}{6!} + \cdots, \tag{1}$$

$$\sin z = z - \frac{z^3}{3!} + \frac{z^5}{5!} - \frac{z^7}{7!} + \cdots.$$

From the formulas (1) we get immediately the lovely identity known as **Euler's formula**:

$$e^{iz} = \cos z + i \sin z. \tag{2}$$

187

If z is real, say $z = \theta$, then $e^{i\theta} = \cos \theta + i \sin \theta$, and $|e^{i\theta}| = 1$. The function $\zeta(\theta) = e^{i\theta}$ therefore maps the interval $[-\pi, \pi]$ smoothly onto the unit circle Γ. For any $z \neq 0$, $z/|z| = e^{i\theta}$ for one $\theta \in (-\pi, \pi]$; this θ is denoted $\arg z$. Therefore, we have the following **polar representation** for any $z \neq 0$:

$$z = |z|e^{i \arg z}. \tag{3}$$

PROBLEM 1: Show that $e^z = 1$ if and only if $z = 2n\pi i$ for some integer n. Hint: If $e^z = 1$, then $e^z = \cos \theta + i \sin \theta = e^{i\theta}$ for some real θ with $\cos \theta = 1$, $\sin \theta = 0$. Show that for $\theta \in [-\pi, \pi]$ this only happens if $\theta = 0$, and then use the periodicity of $\sin \theta$ and $\cos \theta$ for real θ.   ●

Now we have to show that e^z still behaves like an exponential for complex numbers z (i.e., that $e^{z_1} e^{z_2} = e^{z_1 + z_2}$), and that the familiar trigonometric identities hold for complex values of the variable. We will show that $e^{z_1} e^{z_2} = e^{z_1 + z_2}$ by multiplying the series for e^{z_1} and e^{z_2}, and so we start with a proposition that legitimizes the multiplication of series.

PROPOSITION 1: *Let A and B be two index sets and x_α, y_β be real or complex functions on A and B, respectively, with $\sum_{\alpha \in A} x_\alpha = x_0$, $\sum_{\beta \in B} y_\beta = y_0$. Then $\sum_{(\alpha, \beta) \in A \times B} x_\alpha y_\beta = x_0 y_0$.*

Proof: Since these are unordered sums, they converge absolutely. Let $\sum |x_\alpha| \leq M$, $\sum |y_\beta| \leq M$, so that any finite sum of elements $|x_\alpha|$ or $|y_\beta|$ is bounded by M. Let F_0, G_0 be finite subsets of A and B, respectively, such that if $F \supset F_0, G \supset G_0$; then

$$\sum_{F - F_0} |x_\alpha| < \varepsilon \quad \text{and} \quad \left| \sum_F x_\alpha - x_0 \right| \leq \varepsilon,$$

$$\sum_{G - G_0} |y_\beta| < \varepsilon \quad \text{and} \quad \left| \sum_G y_\beta - y_0 \right| \leq \varepsilon.$$

Let H be a finite subset of $A \times B$ with $H \supset F_0 \times G_0$. Let

$$F = \{\alpha : (\alpha, \beta) \in H \text{ for some } \beta\},$$

$$G = \{\beta : (\alpha, \beta) \in H \text{ for some } \alpha\},$$

so that $H \subset F \times G$, and

$$H - F_0 \times G_0 \subset [(F - F_0) \times G_0] \cup [F_0 \times (G - G_0)] \cup [(F - F_0) \times (G - G_0)].$$

The three sets on the right are finite and disjoint. The sum of $|x_\alpha y_\beta|$ over $(F - F_0) \times G_0$ is less than $\varepsilon \cdot M$, and similarly for the other sets, so that

$$\left| \sum_{H - F_0 \times G_0} x_\alpha y_\beta \right| \leq \sum_{H - F_0 \times G_0} |x_\alpha y_\beta| \leq \varepsilon M + \varepsilon M + \varepsilon^2.$$

Moreover,

$$\left| \sum_{F_0 \times G_0} x_\alpha y_\beta - x_0 y_0 \right| = \left| \sum_{F_0} x_\alpha \sum_{G_0} y_\beta - x_0 y_0 \right|$$

$$\leq \left| \sum_{F_0} x_\alpha \left(\sum_{G_0} y_\beta - y_0 \right) \right| + \left| y_0 \left(\sum_{F_0} x_\alpha - x_0 \right) \right|$$

$$\leq M\varepsilon + |y_0|\varepsilon.$$

Therefore, if H is a finite subset of $A \times B$ and $H \supset F_0 \times G_0$, then

$$\left| \sum_{(\alpha,\beta) \in H} x_\alpha y_\beta - x_0 y_0 \right| \leq \left| \sum_H x_\alpha y_\beta - \sum_{F_0 \times G_0} x_\alpha y_\beta \right|$$

$$+ \left| \sum_{F_0 \times G_0} x_\alpha y_\beta - x_0 y_0 \right|$$

$$< 3M\varepsilon + \varepsilon^2 + |y_0|\varepsilon. \qquad \blacksquare$$

Now we check that the product of the series for e^{z_1} and e^{z_2} gives the series for $e^{z_1 + z_2}$.

PROPOSITION 2: *For all z_1, z_2, $e^{z_1} \cdot e^{z_2} = e^{z_1 + z_2}$.*

Proof:

$$e^{z_1 + z_2} = \sum_{n=0}^{\infty} \frac{1}{n!}(z_1 + z_2)^n$$

$$= \sum_{n=0}^{\infty} \frac{1}{n!} \sum_{k=0}^{n} \frac{n!}{k!(n-k)!} z_1^k z_2^{n-k}$$

$$= \sum_{n=0}^{\infty} \sum_{k=0}^{n} \frac{1}{k!(n-k)!} z_1^k z_2^{n-k}.$$

So far we have used nothing but the convergence of the series for e^z at $z_1 + z_2$, and the binomial expansion of $(z_1 + z_2)^n$. Now multiply the series for e^{z_1} and e^{z_2}. These series converge absolutely, and so can be considered unordered sums as in Proposition 1:

$$e^{z_1} e^{z_2} = \sum_{k=0}^{\infty} \frac{1}{k!} z_1^k \sum_{j=0}^{\infty} \frac{1}{j!} z_2^j$$

$$= \sum_{(k,j)} \frac{1}{k!j!} z_1^k z_2^j.$$

Now group all terms $z_1^k z_2^j$ with $j + k = n$ (see Problem 6, Chapter 17):

$$e^{z_1} e^{z_2} = \sum_{n=0}^{\infty} \sum_{k=0}^{n} \frac{1}{k!(n-k)!} z_1^k z_2^{n-k}$$

$$= e^{z_1 + z_2}.$$ ∎

The trigonometric identities now follow easily from Proposition 2 and the following easy consequences of Euler's formula (2):

$$\cos z = \frac{e^{iz} + e^{-iz}}{2},$$

$$\sin z = \frac{e^{iz} - e^{-iz}}{2i}.$$
(4)

PROBLEM 2: Prove the following identities:

(i) $\cos^2 z + \sin^2 z = 1$.
(ii) $\cos(z_1 + z_2) = \cos z_1 \cos z_2 - \sin z_1 \sin z_2$.
(iii) $\sin(z_1 + z_2) = \sin z_1 \cos z_2 + \cos z_2 \sin z_1$. •

PROBLEM 3: Define $\cosh z = (e^z + e^{-z})/2$, $\sinh z = (e^z - e^{-z})/2$. Show that for all z:

(i) $\cosh iz = \cos z$, $\cos iz = \cosh z$.
(ii) $\sinh iz = -i \sin z$, $\sin iz = -i \sinh z$.
(iii) $\cosh^2 z - \sinh^2 z = 1$.
(iv) Show that $\cos z$ and $\sin z$ are not bounded functions on $\mathbb{C}$. •

PROBLEM 4: Show that multiplication by $z_0 = |z_0| e^{i \arg z_0}$ rotates the plane through an angle $\arg z_0$; that is, if $f(z) = z_0 z$, then $\arg f(z) = \arg z + \arg z_0$. (If the sum is not in $(-\pi, \pi]$, then $\arg z + \arg z_0$ must be interpreted modulo 2π.) •

Now consider the geometry of the mapping $f(z) = e^z$. If $z = x + iy$, then by Proposition 2,

$$e^z = e^x e^{iy} = e^x(\cos y + i \sin y).$$

As z goes from left to right along any horizontal line $y =$ constant, $f(z)$ goes outward from the origin along the ray through e^{iy}. Clearly $e^z \neq 0$ for all z, and so $f(z) = e^z$ maps any strip $(y_0, y_0 + 2\pi]$ one-to-one onto $\mathbb{C} - \{0\}$. In particular, e^z maps the strip $-\pi < \operatorname{Im} z \leq \pi$ one-to-one onto $\mathbb{C} - \{0\}$, and we choose this part of the mapping to define the inverse function $\log z$. Thus $\log z$ will be *the* complex number w such that $-\pi < \operatorname{Im} w \leq \pi$ and $e^w = z$. If $e^{u+iv} = z$, so that $e^u e^{iv} = z$, then $e^u = |z|$ and $v = \arg z$; that is, for $z \neq 0$,

$$\log z = \log |z| + i \arg z.$$
(5)

With the definition (5), $\log z$ is defined on all of $\mathbb{C} - \{0\}$, but is discontinuous at all points of the negative real axis. If we restrict z to the plane minus the negative real axis, $\log z$ is continuous. Notice that if z is a real positive number, $\log z$ agrees with our previous restricted definition. Of course e^z, $\cos z$, and $\sin z$ also agree with the earlier definitions if z is real.

PROBLEM 5: Show that if $\log z$ has a power series expansion $\sum_{n=0}^{\infty} a_n(z-1)^n$ that converges in $D(1, 1)$ (it does, but we do not know that yet), then $\log z = \sum_{n=0}^{\infty} \frac{(-1)^n}{n+1}(z-1)^n$. Hint: $\dfrac{1}{z} = \dfrac{1}{1-(1-z)} = \sum_{n=0}^{\infty}(1-z)^n$ if $|1-z| < 1$. •

PROBLEM 6: Show that $(e^{z_1})^{z_2} \neq e^{(z_1 z_2)}$ in general; for example, if $z_1 = 2\pi i$, $z_2 = 1/\pi i$. Hint: Define $a^z = e^{z \log a}$ for $a \neq 0$. In particular, $1^z = e^{z \log 1} = e^0 = 1$. •

PROBLEM 7: Show that any nonzero complex number $z = re^{i \arg z}$ has k distinct kth roots $r^{1/k}e^{i\theta_j}$, where $\theta_1 = \arg z/k$, $\theta_2 = \arg z/k + 2\pi/k$, and so on. •

PROBLEM 8: Show that the function $f(z) = z^k$, $k \in \mathbb{N}$, maps $D(0, R) - \{0\}$ exactly k-to-one onto $D(0, R^k) - \{0\}$. •

PROBLEM 9: Show that if f is a function on the disc $D(0, R)$ such that $f(0) = 0$ and $f[D(0, R)]$ contains some disc around 0, then $g[D(0, R)]$ also contains some disc around 0, where $g(z) = z^k f(z)$. Hint: Do not confuse $g(z)$ with $f(z)^k$, which also has this property. •

XXVIII

ANALYTIC FUNCTIONS

The derivative of a complex-valued function of a complex variable is defined just like the derivative with respect to a real variable: if $f(z)$ is defined in some disc $D(z_0, r)$, then

$$f'(z_0) = \lim_{z \to z_0} \frac{f(z) - f(z_0)}{z - z_0}$$

$$= \lim_{h \to 0} \frac{f(z_0 + h) - f(z_0)}{h}. \tag{1}$$

In the definition for the derivative with respect to a real variable, x can approach x_0 from only two directions. In the derivative with respect to a complex variable, z can approach z_0 from an infinity of directions, and along a multitude of nonlinear paths. Recall (Problem 4 of Chapter 21) that a real-valued function $f(x, y)$ can have a directional derivative along every line through (x_0, y_0) and still be discontinuous at (x_0, y_0). If all directional derivatives exist at (x_0, y_0), then $f(x, y) \longrightarrow f(x_0, y_0)$ along every line through (x_0, y_0), but the rate at which $f(x, y)$ approaches $f(x_0, y_0)$ need not be uniform along all lines. In contrast, if $f'(z_0)$ exists, then $(f(z) - f(z_0))/(z - z_0)$ must be close to $f'(z_0)$ for all z close to z_0, which implies that $f(z)$ must be uniformly close to $f(z_0)$ along all lines through z_0. The existence of $f'(z_0)$ does imply (see Problem 1) that f is continuous at z_0. The existence of the complex derivative is a very much stronger assumption than the existence of all directional derivatives.

PROBLEM 1: Show that if $f'(z_0)$ exists, then f is continuous at z_0.   •

The usual rules for the derivative of a sum, difference, product, or quotient hold for complex functions of a complex variable, and the proofs are the same as for real functions of a real variable (see Problem 2). Thus if $f'(z_0)$, $g'(z_0)$ exist, and $s(z) = f(z) + g(z)$, $p(z) = f(z)g(z)$, $q(z) = f(z)/g(z)$, then

$$s'(z_0) = f'(z_0) + g'(z_0),$$

$$p'(z_0) = f(z_0)g'(z_0) + f'(z_0)g(z_0),$$

$$q'(z_0) = \frac{g(z_0)f'(z_0) - f(z_0)g'(z_0)}{g(z_0)^2} \quad \text{if } g(z_0) \neq 0.$$

(2)

PROBLEM 2: Prove the identities (2).   •

PROPOSITION 1 (Chain Rule): *If $f'(w_0)$ exists, $g'(z_0)$ exists, and $g(z_0) = w_0$, then $h(z) = f(g(z))$ is defined in some disc $D(z_0, r)$, and $h'(z_0) = f'(g(z_0))g'(z_0)$.*

Proof: Since $f'(w_0)$ and $g'(z_0)$ exist, f is continuous at w_0 and g is continuous at z_0. We know that $f(w)$ is defined in some disc $D(w_0, r)$. Since $g(z_0) = w_0$ and g is continuous at z_0, there is $s > 0$ (this is the δ that corresponds to $\varepsilon = r$) such that $|g(z) - g(z_0)| < r$ if $|z - z_0| < s$. That is, $g(z) \in D(w_0, r)$ if $z \in D(z_0, s)$, and hence $h(z) = f(g(z))$ is defined for all $z \in D(z_0, s)$. For any $w \in D(w_0, r)$, let $\varepsilon(w)$ be defined by

$$f(w) - f(w_0) = f'(w_0)(w - w_0) + \varepsilon(w)(w - w_0).$$ (3)

Dividing both sides of (3) by $w - w_0$ shows that $\varepsilon(w) \longrightarrow 0$ as $w \longrightarrow w_0$. Let $z \in D(z_0, s)$, and substitute $g(z)$ and $g(z_0)$ for w and w_0 in (3) to get

$$h(z) - h(z_0) = f(g(z)) - f(g(z_0))$$

$$= f'(g(z_0))(g(z) - g(z_0)) + \varepsilon(g(z))(g(z) - g(z_0)).$$ (4)

Divide both sides of (4) by $z - z_0$ and let $z \longrightarrow z_0$ to obtain

$$\frac{h(z) - h(z_0)}{z - z_0} = f'(g(z_0))\frac{(g(z) - g(z_0))}{z - z_0}$$

$$+ \varepsilon(g(z))\frac{(g(z) - g(z_0))}{z - z_0}$$

$$\longrightarrow f'(g(z_0))g'(z_0) + 0 \cdot g'(z_0).$$

In the last line we used $g(z) \longrightarrow g(z_0) = w_0$ as $z \longrightarrow z_0$, so that $\varepsilon(g(z)) \longrightarrow 0$ as $z \longrightarrow z_0$.   ∎

We will also use the usual Leibniz notation for the complex derivative: $\frac{d}{dz}f(z) = f'(z)$. Thus (2) can be rewritten, with some lack of precision, as

$$\frac{d}{dz}\big(f(z) + g(z)\big) = \frac{d}{dz}f(z) + \frac{d}{dz}g(z),$$

$$\frac{d}{dz}\big(f(z)g(z)\big) = f(z)\frac{d}{dz}g(z) + g(z)\frac{d}{dz}f(z),$$

$$\frac{d}{dz}\left(\frac{f(z)}{g(z)}\right) = \frac{g(z)\dfrac{d}{dz}f(z) - f(z)\dfrac{d}{dz}g(z)}{g(z)^2}.$$

PROPOSITION 2 (The Cauchy–Riemann Equations): *If $f(z) = u(x, y) + iv(x, y)$, with $z = x + iy$, and $f'(z_0)$ exists, then u_x, u_y, v_x, v_y exist at (x_0, y_0) and*

$$f'(z_0) = u_x(x_0, y_0) + iv_x(x_0, y_0)$$

$$= v_y(x_0, y_0) - iu_y(x_0, y_0). \tag{5}$$

Consequently

$$u_x(x_0, y_0) = v_y(x_0, y_0); \qquad u_y(x_0, y_0) = -v_x(x_0, y_0). \tag{6}$$

*(Equations (6) are the **Cauchy–Riemann equations**.)*

Proof: We first consider the derivative along the horizontal line through z_0, so that $z - z_0$ is a real number h:

$$f'(z_0) = \lim_{h \to 0} \frac{f(x_0 + h, y_0) - f(x_0, y_0)}{h}$$

$$= \lim_{h \to 0} \left[\frac{u(x_0 + h, y_0) - u(x_0, y_0)}{h} + i\,\frac{v(x_0 + h) - v(x_0, y_0)}{h} \right]$$

$$= u_x(x_0, y_0) + iv_x(x_0, y_0). \tag{7}$$

Now let $z - z_0 = ih$ for real h; in other words, let $z \longrightarrow z_0$ in a vertical direction. Then

$$f'(z_0) = \lim_{h \to 0} \frac{f(x_0, y_0 + h) - f(x_0, y_0)}{ih}$$

$$= \left[\lim_{h \to 0} \frac{u(x_0, y_0 + h) - u(x_0, y_0)}{ih} + i\,\frac{v(x_0, y_0 + h) - v(x_0, y_0)}{ih} \right]$$

$$= \frac{1}{i}u_y(x_0, y_0) + v_y(x_0, y_0)$$

$$= v_y(x_0, y_0) - iu_y(x_0, y_0). \tag{8}$$

Comparison of (7) and (8) gives the result. ∎

For complex functions the existence of $f'(z_0)$ for a single point z_0 is not very interesting. The usual, and very powerful, assumption is that $f'(z)$ exists for all z in some domain. If $f'(z)$ exists for $z \in G$ we say f is **analytic in** G (or **regular**, or

holomorphic in G). The derivative of f determines f up to an additive constant, just as for real functions. The proof depends on the following problem.

PROBLEM 3: Show that any two points in a domain G can be connected by a path consisting of a finite number of horizontal and vertical segments. Hint: You can show that the set of points z that can be connected to a given point α by such a path is open, and the set of z that cannot be connected to α is also open. Or, use the fact that a domain is arcwise connected. If γ is a curve in G, then the compact set γ can be covered by a finite number of overlapping discs all of which are in G. Any point of a disc can be connected to the center by a horizontal and a vertical segment.   •

PROPOSITION 3: *If f is analytic in a domain G, and $f'(z) = 0$ for $z \in G$, then f is constant in G.*

Proof: If $f(z) = u(x, y) + iv(x, y)$, then u and v are continuous in G, and u_x, u_y, v_x, v_y all exist in G, and from (7) and (8) we see that

$$u_x(x, y) = u_y(x, y) = v_x(x, y) = v_y(x, y) = 0$$

for all $x + iy \in G$. It follows from the Mean Value Theorem that u and v are constant along each horizontal or vertical segment in G. Since any $z \in G$ can be connected to a fixed α by a path consisting of such horizontal and vertical segments, $f(z) = f(\alpha)$ for all $z \in G$.   ∎

Now let us see whether there are enough analytic functions to make the subject worthwhile. It turns out that very few complex functions are analytic — in fact, only polynomials and functions that are locally the uniform limit of polynomials are analytic. However, there are nevertheless enough analytic functions to map the unit disc one-to-one onto any bounded domain the complement of which has only one component. For example, there is a function analytic on $D(0, 1)$ that maps $D(0, 1)$ one-to-one onto the domain G consisting of $(0, 1) \times (0, 1)$ with all the following segments removed: $\{(x, y) : x = 1/2^n, 0 \le y \le 1 - 1/2^n\}$. See Figure I. If that seems unbelievable — and it should — then consider also that the inverse mapping is also analytic!

If $f(z) = z$, then $f'(z) = 1$, and so z is analytic in the whole plane. It follows from the product rule that z^n is analytic for each $n \in \mathbb{N}$, and

$$\frac{d}{dz} z^n = nz^{n-1}. \tag{9}$$

The quotient formula gives us

$$\frac{d}{dz} \frac{1}{g(z)} = \frac{g(z) \cdot 0 - 1g'(z)}{z^2}.$$

Hence

$$\frac{d}{dz} \frac{1}{z} = \frac{-1}{z^2}.$$

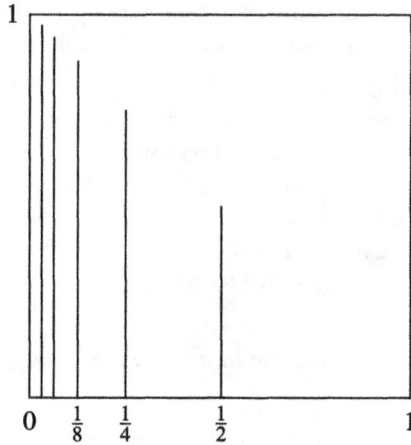

FIGURE I
An Analytic Image of the Unit Disc

The chain rule then shows that for $n \in \mathbb{N}$,

$$\frac{d}{dz} z^{-n} = \frac{d}{dz} \left(z^{-1} \right)^n$$

$$= n \left(z^{-1} \right)^{n-1} \left(\frac{-1}{z^2} \right)$$

$$= -nz^{-n-1}.$$

Thus integer powers of z are analytic everywhere, except for $z = 0$ when the exponent is negative. Recall from Chapter 25 that if γ is any smooth curve from α to β, and $n \in \mathbb{N}$, then

$$\int_\gamma z^n \, dz = \frac{1}{n+1} \left[\beta^{n+1} - \alpha^{n+1} \right], \tag{10}$$

and the same formula holds for negative powers of z other than -1, if $0 \notin \gamma$. This is a special case of the following proposition.

PROPOSITION 4: *If $F(z)$ is analytic in a domain G, and $F'(z)$ is continuous in G, then for any smooth curve γ from α to β in G, $\int_\gamma F'(z) \, dz = F(\beta) - F(\alpha)$. Consequently, if γ is a smooth closed curve in G, $\int_\gamma F'(z) \, dz = 0$.*

Proof: Let $\zeta(t)$ be a smooth mapping of $[a, b]$ onto γ, with $\zeta(a) = \alpha$, $\zeta(b) = \beta$. The chain rule for differentiating $F(\zeta(t))$ works for the special case that $\zeta(t)$ is a function of the real variable t, and so

$$\frac{d}{dt} F\left(\zeta(t) \right) = F'\left(\zeta(t) \right) \zeta'(t)$$

$$= f\left(\zeta(t) \right) \zeta'(t).$$

Therefore,

$$\int_\gamma f(z)\,dz = \int_a^b f\big(\zeta(t)\big)\zeta'(t)\,dt$$

$$= F\big(\zeta(b)\big) - F\big(\zeta(a)\big)$$

$$= F(\beta) - F(\alpha).$$

If $\int_\alpha^\beta f(z)\,dz$ does not depend on the curve in G from α to β, then

$$\int_\beta^\alpha f(z)\,dz = -\int_\alpha^\beta f(z)\,dz$$

for any curve from β to α. It follows that if γ is a closed curve, then $\int_\gamma f(z)\,dz = 0$, since we can let α and β be any two points on γ, and think of γ as a curve from α to β followed by a curve from β to α. ∎

From Proposition 4 we know that if $f(z) = F'(z)$ in a domain G, then $\int_\gamma f(z)\,dz$ does not depend on which curve γ joins any two points α and β of G; that is, the integral depends only on the endpoints of a given curve. If $f(z)$ is any continuous function on G such that $\int_\gamma f(z)\,dz$ depends only on the endpoints of γ, we will say that $\int_\gamma f(z)\,dz$ is **independent of path in** G, and write $\int_\alpha^\beta f(z)\,dz$ to indicate the integral over *any* smooth curve from α to β in G. We know that all polynomials are derivatives, and so the integral of a polynomial is independent of path in any domain, including $\mathbb{C}$. In addition, functions of the form $c(z - z_0)^{-n}$, for $n \in \mathbb{N}$, $n \neq -1$, are derivatives in any domain not containing z_0. Therefore, any finite sum of such terms is a function the integral of which is independent of path in $\mathbb{C} - \{z_0\}$.

We have noticed that an integral $\int f(z)\,dz$ is independent of path in G if and only if $\int_\gamma f(z)\,dz = 0$ for every smooth *closed* curve in G. This property is critically important in the theory of analytic functions. The next proposition shows that $\int f(z)\,dz$ is independent of path in G only if f is a derivative in G; that is, for the existence of an antiderivative of f in G, it is necessary and sufficient that $\int f(z)\,dz$ be independent of path, or equivalently, that $\int_\gamma f(z)\,dz = 0$ for all closed curves γ in G.

Recall that every continuous function f on the line is a derivative, since if

$$F(x) = \int_a^x f(t)\,dt,$$

then $F'(x) = f(x)$. The same argument applies to complex functions, provided the integral from z_0 to z does not depend on the path. For functions of a real variable there is only one integration path from a to x, and so there is no such problem. For functions of a complex variable, independence of path is the critical assumption.

PROPOSITION 5: *If $f(z)$ is a continuous function in G and $\int f(z)\,dz$ is independent of path in G, and $F(z)$ is defined in G by $F(z) = \int_\alpha^z f(w)\,dw$ for some fixed $\alpha \in G$, then $F'(z) = f(z)$ for all $z \in G$.*

Proof: Fix $\alpha \in G$ and let $F(z) = \int_\alpha^z f(w)\, dw$. Let $z_1 \in G$, and compute the difference quotient for F at z_1:

$$\frac{F(z) - F(z_1)}{z - z_1} = \frac{1}{z - z_1}\left[\int_\alpha^z f(w)\, dw - \int_\alpha^{z_1} f(w)\, dw\right]. \tag{11}$$

The integrals in (11) can be taken over any smooth curves from α to z and α to z_1. We consider any curve γ from α to z_1, and take for the curve in the first integral the curve γ followed by the segment from z_1 to z for z close to z_1. Then the two integrals from α to z_1 cancel, and (11) becomes

$$\frac{F(z) - F(z_1)}{z - z_1} = \frac{1}{z - z_1}\left[\int_{z_1}^z f(w)\, dw\right]. \tag{12}$$

Let $f(w) = f(z_1) + \varepsilon(w)$ for $w \in [z_1, z]$. Then $\varepsilon(w) \longrightarrow 0$ as $w \longrightarrow z_1$, since f is continuous. Now we have

$$\frac{F(z) - F(z_1)}{z - z_1} = \frac{1}{z - z_1}\int_{z_1}^z \left[f(z_1) + \varepsilon(w)\right]\, dw$$

$$= f(z_1) + \frac{1}{z - z_1}\int_{z_1}^z \varepsilon(w)\, dw.$$

The last term tends to zero as $z \longrightarrow z_1$, since

$$\left|\int_{z_1}^z \varepsilon(w)\, dw\right| \le \max_{w \in [z_1, z]} |\varepsilon(w)| \cdot |z - z_1|,$$

and $\max\{|\varepsilon(w)| : w \in [z_1, z]\}$ tends to zero as $z \longrightarrow z_1$. Notice that z approaches z_1 any which way in the limit for $f'(z_1)$, but for any fixed z we calculate the difference quotient by integrating over the segment $[z_1, z]$, and the estimate on the sup of $|\varepsilon(w)|$ is uniform over all segments $[z_1, z]$ with $|z - z_1| < \delta$.   ∎

To sum up, functions with integrals that are independent of path have antiderivatives, and any function with an antiderivative in G has integrals that are independent of path in G. Equivalently, functions that are derivatives in G are exactly those functions whose integrals around all closed paths in G are zero.

Now we can prove the important theorem that all power series represent analytic functions.

PROPOSITION 6: *If $f(z) = \sum_{n=0}^\infty a_n(z - z_0)^n$ for $z \in D(z_0, r)$, then $f(z)$ is analytic in $D(z_0, r)$, and $f'(z) = \sum_{n=1}^\infty na_n(z - z_0)^{n-1}$ in $D(z_0, r)$. By applying the result to $f'(z)$, which is just another power series converging in $D(z_0, r)$, we see that $f'(z)$ is analytic, and $f''(z) = \sum_{n=2}^\infty n(n - 1)a_n(z - z_0)^{n-2}$. Hence if $f(z)$ is the sum of a power series, then $f(z)$ has derivatives of all orders, and $f^{(n)}(z)$ is gotten by differentiating the series for $f(z)$ term by term n times. All the series for the derivatives $f^{(n)}(z)$ have the same radius of convergence.*

Proof: Let $g(z) = \sum_{n=1}^{\infty} n a_n (z - z_0)^{n-1}$. It is easy to check that the series for $g(z)$ has the same radius of convergence as that for $f(z)$, and so $g(z)$ is defined and continuous in $D(z_0, r)$. Let $\alpha, \beta \in D(z_0, r)$. We show that $\int g(z) \, dz$ is independent of path in $D(z_0, r)$. If γ is a smooth curve in $D(z_0, r)$ from α to β, then γ is a compact set, and so $\gamma \subset \overline{D}(z_0, s)$ for some $s < r$, and the series for $g(z)$ converges uniformly on γ. Therefore, the series can be integrated term by term:

$$\int_{\gamma} g(z) \, dz = \int_{\gamma} \sum_{n=1}^{\infty} n a_n (z - z_0)^{n-1} \, dz$$

$$= \sum_{n=1}^{\infty} \int_{\gamma} n a_n (z - z_0)^{n-1} \, dz$$

$$= \sum_{n=1}^{\infty} \left[a_n (\beta - z_0)^n - a_n (\alpha - z_0)^n \right]$$

$$= f(\beta) - f(\alpha). \tag{13}$$

The foregoing integral obviously depends only on α and β, and not on the particular curve γ from α to β. Moreover, we see from (13) that for any $z \in D(z_0, r)$,

$$f(z) = \int_{\alpha}^{z} g(w) \, dw + f(\alpha),$$

and so $f(z)$ is analytic with $f'(z) = g(z)$ by Proposition 5 (the f here is the F of Proposition 5). ∎

Analyticity — the existence of a derivative — is a local property. Therefore, if $f(z)$ has a power series representation on *some* disc $D(z_0, r)$ at each $z_0 \in G$, then $f(z)$ is analytic in each $D(z_0, r)$ and therefore analytic in G. The series for e^z, $\cos z$, and $\sin z$ converge for all z, and so these functions are analytic in the whole plane.

PROBLEM 4: Let $f(z) = \sum_{n=1}^{\infty} a_n z^{-n}$, and assume the series converges for $|z| > r$. (cf. Problems 6 and 10 of Chapter 26). Let $g(z) = \sum_{n=1}^{\infty} -n a_n z^{-n-1}$.

(i) Show that the series for $g(z)$ converges for $|z| > r$, and converges uniformly on any curve γ in $|z| > r$.

(ii) Show that $\int_{\gamma} g(z) \, dz$ is independent of path in $|z| > r$, and that $\int_{\alpha}^{\beta} g(z) \, dz = f(\beta) - f(\alpha)$ for all α, β with $|\alpha| > r$, $|\beta| > r$.

(iii) Fix α with $|\alpha| > r$, so that

$$f(z) = f(\alpha) + \int_{\alpha}^{z} g(w) \, dw,$$

for all $|z| > r$. Show f is analytic in $|z| > r$. •

PROBLEM 5: Show that $\frac{d}{dz}e^z = e^z$, $\frac{d}{dz}\cos z = -\sin z$, and $\frac{d}{dz}\sin z = \cos z$ for all z.   •

The fact that $\log z$ is analytic on $\mathbb{C}$ minus the negative real axis will follow from the following problem.

PROBLEM 6: Assume $f(z)$ is an analytic function that maps a domain G one-to-one onto a domain H. Let $g(w)$ be the inverse function, so that $g(f(z)) = z$ and $f(g(w)) = w$ for all $z \in G$, and all $w \in H$, and assume that $g(w)$ is continuous on H. If $f'(z) \neq 0$ for all $z \in G$, then g is analytic on H. Hint: Let $f(z_1) = w_1$ and let Δz and Δw be the corresponding changes in z from z_1 and w from w_1. Since $\Delta w/\Delta z \longrightarrow f'(z_1)$, $\Delta z/\Delta w \longrightarrow g'(w_1) = 1/f'(z_1)$. Of course $g'(w_1)$ is the limit as $\Delta w \longrightarrow 0$ and $f'(z)$ is the limit as $\Delta z \longrightarrow 0$, so that there are some details to fill in.   •

Since $\frac{d}{dz}e^z = e^z \neq 0$, and e^z maps the domain $-\pi < \operatorname{Im} z < \pi$ one-to-one onto $\mathbb{C}$ minus the negative real axis, and the inverse function $\log z$ is continuous on $\mathbb{C}$ minus the negative real axis, $\log z$ is analytic on the plane minus the negative real axis.

PROBLEM 7: Show that if z is in the domain of $\log z$ (i.e., not on the negative real axis) then $\frac{d}{dz}\log z = \frac{1}{z}$. Hint: Use the chain rule and the identity $e^{\log z} = z$.   •

It is a simple translation of the result of Problem 7 that $\log(z - z_0)$ is analytic in the plane minus the ray $\{z_0 - t : t \geq 0\}$, and that $\frac{d}{dz}\log(z - z_0) = \frac{1}{z-z_0}$ in this domain. We will need the following result in the derivation of Cauchy's formula in the next chapter.

PROPOSITION 7: *If γ is any positively oriented circle containing z_0 in its interior, then*

$$\int_\gamma \frac{dz}{z - z_0} = 2\pi i.$$

Proof: Let $\rho = \{z_0 - t : t \geq 0\}$ be the horizontal ray to the left of z_0, so that $\log(z - z_0)$ is analytic in $\mathbb{C} - \rho$, and $\frac{d}{dz}\log(z - z_0) = 1/(z - z_0)$ in $\mathbb{C} - \rho$. Therefore, the integral of $1/(z - z_0)$ is independent of path in $\mathbb{C} - \rho$. Let γ intersect ρ at $\alpha \neq z_0$, and let z_1 be a point of γ just below α but close, with $\arg(z_1 - \alpha) = -\pi + \varepsilon$, and let z_2 be a point of γ above α but close, with $\arg(z_2 - z_0) = \pi - \varepsilon$.

$$\int_{z_1}^{z_2} \frac{1}{z - z_0}\,dz = \log(z_2 - z_0) - \log(z_1 - z_0)$$

$$= i\big(\arg(z_2 - z_0) - \arg(z_1 - z_0)\big)$$

$$= i\big(\pi - \varepsilon - (-\pi + \varepsilon)\big) = (2\pi - 2\varepsilon)i.$$

The integral of $1/(z - z_0)$ around γ is the limit of the integrals from z_1 to z_2 as both z_1 and z_2 approach α, that is, as $\varepsilon \longrightarrow 0$. Notice that $1/(z - z_0)$ is continuous on γ even though $\log(z - z_0)$ is discontinuous at points of ρ.   ∎

PROBLEM 8: Modify the proof of Proposition 7 to show that if γ is any smooth curve around z_0 such that γ intersects $\rho = \{z_0 - t : t \geq 0\}$ exactly once, at $\alpha \neq z_0$, then $\int_\gamma \frac{dz}{z-z_0} = 2\pi i$. Here we define "curve around z_0" to mean that $\gamma = \{\zeta(t) : a \leq t \leq b\}$ with $\zeta(a) = \zeta(b) = \alpha$, and $\operatorname{Im}\zeta(t) < \operatorname{Im}z_0$ on some interval $(a, a + \delta)$ and $\operatorname{Im}\zeta(t) > \operatorname{Im}z_0$ on some interval $(b - \delta, b)$.   •

PROBLEM 9: (i) Let $f(z) = z^2$. Show that f maps the right half-plane $H = \{z : \operatorname{Re}z > 0\}$ one-to-one onto the plane minus the negative real axis: $G = \mathbb{C} - \{x : x \leq 0\}$. Hint: Write $z = |z|e^{i\arg z}$, with $-\frac{\pi}{2} < \arg z < \frac{\pi}{2}$ for $z \in H$, and use the fact (Chapter 27) that $e^{z_1} \cdot e^{z_2} = e^{z_1+z_2}$ to deduce that $z^2 = |z|^2 e^{i2\arg z}$.

(ii) Let $\sqrt{z}$ denote the inverse function $f^{-1}(z)$, which maps G onto H. Show that $\sqrt{z}$ is continuous on G, and conclude from Problem 6 that $\sqrt{z}$ is analytic on the slit plane G.   •

PROBLEM 10: Fill in the details of the hint and give an alternative proof that if $f'(z) = 0$ on a disc $D(0, r)$, or on a domain, then f is constant. Hint: Assume $f(0) = 0$ and fix a direction θ. Let $g(t) = |f(te^{i\theta})|$, and show that $g'(t) = 0$ for all t and all θ.   •

XXIX

CAUCHY'S INTEGRAL THEOREMS

The basic theorems of complex analysis are Cauchy's Theorem and Cauchy's integral formula. Cauchy's Theorem says that the integral of an analytic function is independent of path in any domain with no holes in it. Cauchy's integral formula is

$$f(z) = \frac{1}{2\pi i} \int_\gamma \frac{f(\zeta)}{\zeta - z} \, d\zeta,$$

where γ is any closed curve that goes around z once, and f is analytic on a domain containing γ and its inside. Most of the difficulty in proving these theorems involves the problem of defining "domain with no holes" and "a curve that goes around z once." We get around these topological problems by restricting our attention to convex domains, for these are easy to define and they have no holes. We will construct an antiderivative for any function that is analytic on a convex domain. The integral of a derivative is independent of path, as we have seen, and that proves Cauchy's Theorem.

A domain G is **convex** if whenever z_1 and z_2 are points of G, the whole segment $[z_1, z_2]$ lies in G. Recall that

$$[z_1, z_2] = \{(1 - \lambda) z_1 + \lambda z_2 : 0 \le \lambda \le 1\}.$$

The **triangle** determined by three points z_0, z_1, z_2 is the set

$$T = \{\lambda_0 z_0 + \lambda_1 z_1 + \lambda_2 z_2 : \lambda_i \ge 0, \lambda_1 + \lambda_2 + \lambda_3 = 1\}.$$

PROBLEM 1: Show that the triangle T determined by z_0, z_1, z_2 is compact and convex. Hint: Convexity is easy — just write out the algebra. To prove compactness, show that every sequence $\{w_n\}$ in T has a subsequence that converges to a point of T. Let

$$w_n = \mu_n z_0 + \nu_n z_1 + (1 - \mu_n - \nu_n)z_2$$

with $\mu_n \in [0, 1]$, $\nu_n \in [0, 1]$, $\mu_n + \nu_n \in [0, 1]$. Clearly,

$$w_{n_k} \longrightarrow w_0 = \mu_0 z_0 + \nu_0 z_1 + (1 - \mu_0 - p_0)z_2$$

if $\{\mu_{n_k}\}$ and $\{\nu_{n_k}\}$ converge. What you use here is the fact that $[0, 1] \times [0, 1]$ is compact, which is a special case of our earlier result that every closed bounded set in $\mathbb{R}^2$ is compact.   ●

PROBLEM 2: (i) Show that the intersection of any family of convex sets is convex, and hence that there is a smallest convex set containing any given set.

(ii) Show that the triangle determined by z_0, z_1, z_2 is the smallest convex set containing the set $\{z_0, z_1, z_2\}$. Hint: If $\lambda_0, \lambda_1, \lambda_2 \in [0, 1]$ and $\lambda_0 + \lambda_1 + \lambda_2 = 1$, then there are numbers $\mu, \nu \in [0, 1]$ such that

$$\lambda_0 z_0 + \lambda_1 z_1 + \lambda_2 z_2 = (1 - \nu)\left[(1 - \mu)z_0 + \mu z_1\right] + \nu z_2.$$

Thus any convex set that contains z_0, z_1, z_2 must also contain (why?) $\lambda_0 z_0 + \lambda_1 z_1 + \lambda_2 z_2$.

(iii) If z_0, z_1, z_2 belong to a convex domain G, then the triangle T determined by z_0, z_1, z_2 is a subset of G.   ●

If T is the triangle determined by z_0, z_1, z_2, then the **boundary of** T, denoted ∂T, is the piecewise smooth curve consisting of the three segments $[z_0, z_1]$, $[z_1, z_2]$, and $[z_2, z_0]$. Our aim is to show that the integral of an analytic function around ∂T is zero.

PROPOSITION 1: *If $f(z)$ is analytic in a convex domain G, and T is the triangle determined by $z_0, z_1, z_2 \in G$, then $\int_{\partial T} f(z)\, dz = 0$.*

Proof: Form four new triangular closed curves from the points z_0, z_1, z_2 and the mid-points of the three segments, as indicated in Figure I. The sum of the integrals around the four new curves, with the indicated orientations, is the same as the integral around ∂T because of the pairwise cancellations over the new lines. It follows that for at least one of the four small triangles, call it T_1,

$$\left| \int_{\partial T_1} f(z)\, dz \right| \geq \frac{1}{4} \left| \int_{\partial T} f(z)\, dz \right|.$$

Now divide T_1 into four new triangles in the same way, and pick one, T_2, such that

$$\left| \int_{\partial T_2} f(z)\, dz \right| \geq \frac{1}{4} \left| \int_{\partial T_1} f(z)\, dz \right| \geq \frac{1}{4^2} \left| \int_{\partial T} f(z)\, dz \right|.$$

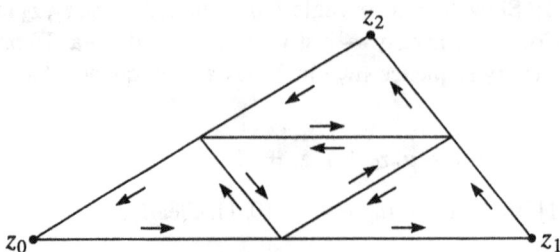

FIGURE I

Notice that the length of ∂T_1 is $1/2$ the length of ∂T, the length of T_2 is $1/4$ the length of ∂T, and so forth. Continue the division process and define a sequence of triangles T_n such that $\ell(\partial T_n) = (1/2^n)\ell(\partial T)$ and

$$\left| \int_{\partial T_n} f(z)\, dz \right| \geq \frac{1}{4^n} \left| \int_{\partial T} f(z)\, dz \right|. \tag{1}$$

The compact triangles T_n collapse to a single point a, which is in all T_n. We define $\varepsilon(z)$ by

$$\frac{f(z) - f(a)}{z - a} - f'(a) = \varepsilon(z),$$

so that

$$f(z) = f(a) + f'(a)(z - a) + \varepsilon(z)(z - a) \tag{2}$$

for all $z \neq a$. Clearly $\varepsilon(z) \longrightarrow 0$ as $z \longrightarrow a$. Since ∂T_n is a closed curve, polynomials (like $f(a) + f'(a)(z - a)$) have integral zero around ∂T_n. Therefore,

$$\left| \int_{\partial T_n} f(z)\, dz \right| = \left| \int_{\partial T_n} \varepsilon(z)(z - a)\, dz \right|$$

$$\leq \max_{z \in \partial T_n} \left[|\varepsilon(z)||z - a| \right] \ell(\partial T_n)$$

$$\leq \max_{z \in \partial T_n} |\varepsilon(z)| \ell(\partial T_n)^2$$

$$= \max_{z \in \partial T_n} |\varepsilon(z)| \frac{\ell(\partial T)^2}{4^n}. \tag{3}$$

Now from (1) and (3) we have, for all n,

$$\left| \int_{\partial T} f(z)\, dz \right| \leq 4^n \left| \int_{\partial T_n} f(z)\, dz \right|$$

$$\leq \ell(\partial T)^2 \max_{z \in \partial T_n} |\varepsilon(z)|.$$

Since $\varepsilon(z) \longrightarrow 0$ as $z \longrightarrow a$, and ∂T_n is uniformly close to a for large n, $\int_{\partial T} f(z)\,dz = 0$. ∎

PROBLEM 3: What goes wrong in the preceding proof if G is not convex? For example, why does the proof fail for the punctured disc $G = \{z : |z| < 1, z \neq 0\}$? •

If the integral of f around triangles is zero, then the integral of f over any side of a triangle can be replaced by the integral over the other two sides. That allows us to show that f has an antiderivative.

PROPOSITION 2 (Cauchy's Theorem for Convex Domains): *If f is analytic in a convex domain G, $\alpha \in G$, and $F(z)$ is defined for $z \in G$ by*

$$F(z) = \int_{[\alpha, z]} f(\zeta)\,d\zeta,$$

then $F'(z) = f(z)$ for all $z \in G$. Hence $\int f(z)\,dz$ is independent of path in G, and $\int_\gamma f(z)\,dz = 0$ for every smooth closed curve in G.

Proof: We let z_1 be any point in G and show that $F'(z_1) = f(z_1)$. We can represent $F(z)$ as the integral over $[\alpha, z_1]$ followed by the integral over $[z_1, z]$, because the integral over the triangular path $[\alpha, z_1] + [z_1, z] + [z, \alpha]$ is zero. Therefore,

$$F(z) = \int_{[\alpha, z_1]} f(\zeta)\,d\zeta + \int_{[z_1, z]} f(\zeta)\,d\zeta$$

$$= F(z_1) + \int_{[z_1, z]} f(\zeta)\,d\zeta.$$

Therefore

$$\frac{F(z) - F(z_1)}{z - z_1} = \frac{1}{z - z_1} \int_{[z_1, z]} f(\zeta)\,d\zeta.$$

Since f is continuous at z_1, the term on the right approaches $f(z_1)$ as $z \longrightarrow z_1$, and $F'(z_1) = f(z_1)$. Since f is a derivative in G, the integral of f is independent of path in G. ∎

Cauchy's Theorem is valid for any domain G such that the inside of any curve in G consists only of points of G — that is, G has no holes. To make these ideas precise — for instance to show that a simple closed curve has an inside — is very difficult, and has more to do with plane topology than with complex analysis. We will simplify the situation by restricting our attention to convex domains, or (as in Problem 4) to domains that are the careful union of convex domains.

The proof of Cauchy's integral formula goes like this. We write

$$\frac{f(z)}{z - z_0} = \frac{f(z_0)}{z - z_0} + \frac{f(z) - f(z_0)}{z - z_0}.$$

Let $f(z)$ be analytic on a domain containing $\overline{D}(a, r)$, where $z_0 \in D(a, r)$, and let γ be the circle $|z - a| = r$, positively oriented. Then

$$\int_\gamma \frac{f(z_0)}{z - z_0}\, dz = f(z_0) \int_\gamma \frac{dz}{z - z_0} = 2\pi i f(z_0),$$

and

$$\int_\gamma \frac{f(z) - f(z_0)}{z - z_0}\, dz = 0. \tag{4}$$

To show the last integral is zero, notice that if

$$g(z) = \begin{cases} \dfrac{f(z) - f(z_0)}{z - z_0} & \text{for } z \neq z_0, \\ f'(z_0) & \text{if } z = z_0, \end{cases}$$

then $g(z)$ is analytic wherever $f(z)$ is, except at z_0, but $g(z)$ is nevertheless continuous at z_0. We will therefore beef up Proposition 1 to include such functions — analytic except possibly at one point — to prove that the second integral (4) is zero.

PROPOSITION 1′: *If $g(z)$ is continuous in a convex domain G, and $g(z)$ is analytic in G except possibly at one point $a \in G$, then $\int_{\partial T} g(z)\, dz = 0$ for every triangle $T \subset G$.*

Proof: Let T be a triangle in G and assume $g(z)$ is analytic in G except at a, where g is nevertheless continuous. If $a \notin T$, the proof of Proposition 1 works without change. If a is a vertex of T, say $a = z_0$, with z_1 and z_2 the other vertices, then let b be a point of $[z_0, z_1]$ very close to a and let c be a point of $[z_0, z_2]$ very close to a. (See Figure II.) The integral of $g(z)$ around triangle a, b, c can be made arbitrarily small by picking b and c sufficiently close to a. The integral of g around ∂T can be written as the sum of the integrals around the three triangles $a, b, c; b, z_1, z_2;$ and b, z_2, c. The integrals around the last two triangles are zero, and so the integral around ∂T is zero. If a lies on one of the segments $[z_0, z_1], [z_1, z_2], [z_2, z_0],$

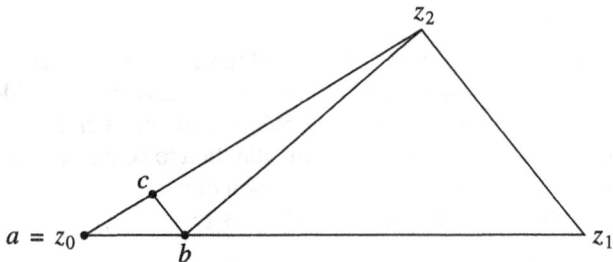

FIGURE II

or in the interior of T, then the integral around ∂T can similarly be written as the sum of integrals around triangles having a as a vertex.   ■

PROPOSITION 2': *If $g(z)$ is continuous in a convex domain G and analytic except possibly at one point a, then there is a function $F(z)$ such that $F'(z) = g(z)$ for all $z \in G$, and consequently $\int_\gamma g(z)\,dz = 0$ for every smooth closed curve in G.*

Proof: The proof of Proposition 2 works without change given the stronger form of Proposition 1'.   ■

Now the proof of Cauchy's integral formula is in hand. For any smooth closed curve γ in the convex domain G,

$$\int_\gamma \frac{f(z) - f(z_0)}{z - z_0}\,dz = 0.$$

If γ winds around z_0 once, then $\int_\gamma \dfrac{dz}{z - z_0} = 2\pi i$, and if γ winds around z_0 a total of n times, the integral is $2n\pi i$. We do not want to bog down in the definition of "winds around z_0 a total of n times," so we will stick to the following condition on γ, which effectively ensures that γ winds around z_0 once. Let ρ be the ray $\rho = \{z_0 - t : t \geq 0\}$ and let $\gamma = \{\zeta(t) : a \leq t \leq b\}$ be a smooth closed curve in G such that $\zeta(a) = \zeta(b) = \alpha$, where α is a point of ρ different from z_0, $\operatorname{Im} \zeta(t) < \operatorname{Im} z_0$ on some interval $(a, a + \delta)$ and $\operatorname{Im} \zeta(t) > \operatorname{Im} z_0$ on some interval $(b - \delta, b)$, and $\zeta(t) \notin \rho$ for t different from a and b. With these assumptions on γ, which certainly hold for any circle $|z - a| = r$ around z_0, we can use the antiderivative $\log(z - z_0)$ of $1/(z - z_0)$ to show $\int_\gamma \dfrac{dz}{z - z_0} = 2\pi i$.

Thus, if γ is a smooth curve in a convex domain G where $f(z)$ is analytic, and γ goes around z_0 once in the foregoing sense,

$$f(z_0) = \frac{1}{2\pi i} \int_\gamma \frac{f(z)}{z - z_0}\,dz.$$

From Cauchy's formula we can show that all analytic functions have local power series expansions. We saw earlier that every power series represents an analytic function, and so functions with a complex derivative can be characterized as those functions that are locally represented by power series.

PROPOSITION 3: *If $f(z)$ is analytic in a domain G, and $a \in G$, then $f(z)$ has a power series $\sum_{n=0}^{\infty} a_n(z - a)^n$ that converges in the biggest disc $D(a, r) \subset G$.*

Proof: For any fixed $a \in G$, the union of all open discs $D(a, r)$ that are contained in G is again an open disc contained in G, and so there is a biggest disc

$D(a, r)$ contained in G. If $|z - a| = r_0 < r$, so that $z \in D(a, r)$, then

$$f(z) = \frac{1}{2\pi i} \int_\gamma \frac{f(\zeta)}{\zeta - z}\, d\zeta,$$

where $\gamma = \{\zeta : |\zeta - a| = s\}$, with $r_0 < s < r$, is a circle in $D(a, r)$ that has z in its inside. We expand $1/(\zeta - z)$ in a series:

$$\frac{1}{\zeta - z} = \frac{1}{\zeta - a - (z - a)}$$

$$= \frac{1}{\zeta - a}\frac{1}{1 - \frac{z-a}{\zeta-a}}$$

$$= \frac{1}{\zeta - a}\sum_{n=0}^{\infty}\left(\frac{z - a}{\zeta - a}\right)^n$$

$$= \sum_{n=0}^{\infty}\frac{1}{(\zeta - a)^{n+1}}(z - a)^n.$$

The preceding geometric series converges uniformly for $\zeta \in \gamma$, since $|z - a| = r_0 < |\zeta - a| = s$ for all $\zeta \in \gamma$. Now we have

$$f(z) = \frac{1}{2\pi i} \int_\gamma \left(\sum_{n=0}^{\infty}\frac{f(\zeta)}{(\zeta - a)^{n+1}}(z - a)^n\right) d\zeta.$$

The series still converges uniformly in ζ, since $f(\zeta)$ is bounded on the compact set γ. Therefore, we can integrate term by term to get

$$f(z) = \sum_{n=0}^{\infty}\left(\frac{1}{2\pi i}\int_\gamma \frac{f(\zeta)}{(\zeta - a)^{n+1}}\, d\zeta\right)(z - a)^n.$$

Thus we have $f(z)$ represented as the power series

$$f(z) = \sum_{n=0}^{\infty} a_n(z - a)^n, \tag{5}$$

where

$$a_n = \frac{1}{2\pi i} \int_\gamma \frac{f(\zeta)}{(\zeta - a)^{n+1}}\, d\zeta. \tag{6}$$

The series (4) converges at z, and z was any point of $D(a, r)$, the largest disc at a contained in G. $\blacksquare$

There are some immediate consequences of (5). Recall (Proposition 6, Chapter 28) that if $f(z)$ is given by a power series, then $f'(z)$ is represented by the term by term differentiation of the series. That is, if, for $|z - a| < r$,

$$f(z) = \sum_{n=0}^{\infty} a_n(z - a)^n,$$ (7)

then

$$f'(z) = \sum_{n=1}^{\infty} na_n(z - a)^{n-1},$$

$$f''(z) = \sum_{n=2}^{\infty} n(n - 1)a_n(z - a)^{n-2},$$ (8)

$$f'''(z) = \sum_{n=3}^{\infty} n(n - 1)(n - 2)a_n(z - a)^{n-3},$$

and so on. It follows from (8) that

$$f'(a) = a_1, \qquad f''(a) = 2a_2, \qquad f'''(a) = 3!a_3, \qquad \text{and so forth.}$$

In general, if $f(z)$ is given by the series (5), then

$$a_n = \frac{1}{n!} f^{(n)}(a).$$

We saw in Chapter 26 that no function has two power series; that is, if $\sum a_n(z - a)^n = \sum b_n(z - a)^n$ in some disc, then $a_n = b_n$ for all n. It follows from (6) that

$$f^{(n)}(a) = \frac{n!}{2\pi i} \int_{\gamma} \frac{f(\zeta)}{(\zeta - a)^{n+1}} \, d\zeta.$$ (9)

Now let us summarize. If $f(z)$ has a power series, $f(z) = \sum a_n(z - a)^n$, that converges for $|z - a| < r$, then $g(z) = \sum na_n(z - a)^{n-1}$ converges in $D(a, r)$, and so $\int g(w) \, dw$ is independent of path, and

$$f(z) = f(a) + \int_a^z g(w) \, dw.$$ (10)

From (10) it follows that $f'(z) = g(z)$ in $D(a, r)$, and so f is analytic in $D(a, r)$. Thus any function that has a local power series at every point of G is analytic in G.

Now suppose $f(z)$ is analytic in G. For any convex domain contained in G, and in particular for any disc $D(a, r) \subset G$, $\int f(z) \, dz$ is independent of path in $D(a, r)$. The same is true for the analytic-except-at-a function $g(z)$, where

$$g(z) = \begin{cases} \dfrac{f(z) - f(a)}{z - a} & \text{if } z \neq a, \\ f'(a) & \text{if } z = a, \end{cases}$$ (11)

and so $\int g(z)\,dz = 0$ around any closed curve in $D(a, r)$. This gives Cauchy's formula for $f(z)$:

$$f(z) = \frac{1}{2\pi i}\int \frac{f(z)}{\zeta - z}\,d\zeta + \frac{1}{2\pi i}\int \frac{f(\zeta) - f(z)}{\zeta - z}\,d\zeta$$

$$= \frac{1}{2\pi i}f(z)\int \frac{1}{\zeta - z}\,d\zeta + 0.$$

From Cauchy's formula for $f(z)$, we get the series $\sum a_n(z - a)^n$ for $f(z)$. Notice that if $f(z)$ is analytic in $D(a, r)$, then $f(z)$ has a power series, so that $f(z)$ has derivatives of all orders. Thus the function $g(z) = (f(z) - f(a))/(z - a)$ of (11) is not just continuous at a, but differentiable at a:

$$g(z) = \sum_{n=1}^{\infty} a_n(z - a)^{n-1}.$$

The hypothesis of Proposition $1'$, that $g(z)$ be analytic in G except at one point where $g(z)$ is only continuous, is never satisfied: if $g(z)$ is continuous on a domain G and analytic except possibly at one point, then $g(z)$ is analytic on all of G.

Another variant of the foregoing arguments is called Morera's Theorem: if $\int_\gamma f(z)\,dz = 0$ for every closed curve in G (or equivalently, if $\int f(z)\,dz$ is independent of path in G), then $f(z)$ is analytic in G.

PROBLEM 4: (i) Let G_1 and G_2 be convex domains with $G_1 \cap G_2 \neq \varnothing$, and let $G = G_1 \cup G_2$, so that G is a domain (connected and open). Show that if f is analytic in G_1 and f is analytic in G_2, then f has an antiderivative F in G, so that $\int_\gamma f(z)\,dz = 0$ for all closed curves γ in G. Hint: Let $\alpha \in G_1 \cap G_2$, and define $F_i(z)$ for $z \in G_i$ by $F_i(z) = \int_{[\alpha,z]} f(w)\,dw$. Notice that $F_1 = F_2$ in $G_1 \cap G_2$.

(ii) Let G_1, G_2, G_3 be three convex domains such that $G_1 \cap G_2 \neq \varnothing$, $G_2 \cap G_3 \neq \varnothing$, but $G_1 \cap G_3 = \varnothing$. Assume that f is analytic on each G_i, and show that f has an antiderivative in $G = G_1 \cup G_2 \cup G_3$.

(iii) Notice that $f(z) = 1/z$ is analytic on any domain not containing 0. Write the domain $G = \mathbb{C} - \{x : x \leq 0\}$ as the union of three half-planes as in (ii) to show that $1/z$ has an antiderivative in G.

(iv) For r not an integer and z not on the negative real axis, we define $z^r = e^{r\log z}$. Show, for example, that $z^{1/2} \cdot z^{1/2} = z$ and that $z^{1/2}$ is analytic on $\mathbb{C} - \{x : x \leq 0\}$.   •

PROBLEM 5 (Identity Theorem for Analytic Functions): Assume f is analytic in a domain G, and $f(z_n) = 0$ for some sequence $\{z_n\}$ that converges to a point $z_0 \in G$. Show that $f \equiv 0$ in G. Hint: By the identity theorem for series (Chapter 26, Proposition 5), we know that $f(z) \equiv 0$ in the largest disc $D(z_0, r)$ that is contained in G. To show that $f(z_1) = 0$ for any $z_1 \in G - D(z_0, r)$, assume that $f(z_1) \neq 0$ and let γ be a curve in G from z_0 to z_1. Let z_2 be the last point on γ such that $f(z) = 0$ for all z on γ between z_0 and z_2. Use the identity theorem at z_2 to reach a contradiction.   •

PROBLEM 6: Use the identity theorem for analytic functions (Problem 5) to prove that $e^{z+w} = e^z e^w$ for all complex z and w. Hint: First fix a real number w and let $f(z) = e^{z+w}$, $g(z) = e^z e^w$. Both f and g are clearly analytic on $\mathbb{C}$, and $f(z) = g(z)$ for all real z (since $e^{x+y} = e^x e^y$ and w is real here). Hence $f(z) = g(z)$ for all $z \in \mathbb{C}$; in other words, $e^{z+w} = e^z e^w$ for all z and all real w. Continue.   •

PROBLEM 7: Let $g(z)$ be continuous on a not necessarily closed curve γ. Define $f(z)$ for $z \notin \gamma$ by

$$f(z) = \int_\gamma \frac{g(\zeta)}{\zeta - z} \, d\zeta.$$

Show that $f(z)$ is analytic on $\mathbb{C} - \gamma$ and

$$f'(z) = \int_\gamma \frac{g(\zeta)}{(\zeta - z)^2} \, d\zeta.$$ •

We know that the uniform limit of a sequence of continuous real or complex functions is continuous. The uniform limit of a sequence of differentiable *real* functions need not be differentiable, since smooth curves can approximate a curve with corners arbitrarily closely. However, the uniform limit of analytic functions is necessarily analytic. These statements are proved in the next problem.

PROBLEM 8: (i) Let $g_n(x)$ be zero for $x \le 0$, $g_n(x) = nx$ for $0 \le x \le 1/n$, and $g_n(x) = 1$ for $x \ge 1/n$. Thus the graph of g_n lies on the x axis for $x \le 0$, on the line with slope n on $[0, 1/n]$, and on the line $y = 1$ for $x \ge 1/n$. Each g_n is continuous. Let $f_n(x) = \int_0^x g_n(t)\,dt$, so that each f_n is differentiable on $\mathbb{R}$, with $f_n'(x) = g_n(x)$ for all x. Show that $\{f_n(x)\}$ converges uniformly on $\mathbb{R}$ to the continuous function $f(x)$ that is 0 for $x \le 0$ and x for $x \ge 0$, so that $f'(0)$ does not exist.

(ii) Let $\{f_n(z)\}$ be a sequence of analytic functions on any domain G, such that $f_n(z)$ converges uniformly to $f(z)$ on G. Show that f is analytic on G. Hint: Analyticity is a local property. Let z_0 be any point of G, and let $D(z_0, r)$ be any disc contained in G. Show that the integral of f is independent of path in $D(z_0, r)$, so that f is a derivative and hence analytic in $D(z_0, r)$.   •

Let $\gamma = \{\zeta(t) = x(t) + iy(t) : a \le t \le b\}$ be a smooth curve in $\mathbb{C}$. We define the slope of the tangent line to γ at $z_0 = \zeta(t_0)$ to be $y'(t_0)/x'(t_0)$, provided $\zeta'(t_0) \ne 0$. Hence the vector (complex number) $\zeta'(t_0)$ is a nonzero vector in the direction of the tangent line at z_0. Recall (Proposition 6, Chapter 26) that if $f(z)$ is analytic in a disc around z_0, and $f'(z_0) \ne 0$, then f is one-to-one in some disc at z_0. We let $[a_0, b_0]$ be an interval such that ζ maps $[a_0, b_0]$ into a disc where f is one-to-one, and we let $\gamma_1 = \{f(\zeta(t)) : a_0 \le t \le b_0\}$. Then γ_1 is the f-image of an arc of γ through z_0, and a tangent vector to γ_1 at $f(z_0)$ is $f'(\zeta(t_0))\,\zeta'(t_0)$. Multiplication of $\zeta'(t_0)$ by the number $f'(z_0)$ rotates $\zeta'(t_0)$ through an angle $\arg f'(z_0)$. Thus the tangent to *every* curve through z_0 is rotated through the same angle $\arg f'(z_0)$ by the mapping $w = f(z)$. If γ_1 and γ_2 are two curves through z_0, with angle θ between their tangent lines, then the angle between $f[\gamma_1]$ and $f[\gamma_2]$ will also be θ. Analytic mappings preserve angles at points z_0 where $f'(z_0) \neq 0$. Such angle-preserving mappings are called **conformal**. An analytic mapping $w = f(z)$ will send the lines $x = $ constant into some family of curves, and send the lines $y = $ constant into another family of curves, and the two image families will be orthogonal at every point of intersection. For example, the function $f(z) = e^z = e^x e^{iy}$ sends all lines $x = $ constant onto circles, and all lines $y = $ constant onto rays from the origin.

The following proposition has as nearly immediate consequences (see Problems 11 and 12) the important **maximum principle** for analytic functions, and the **Fundamental Theorem of Algebra**.

PROPOSITION 4: *If f is analytic on a disc around z_0, and not constant, then there is some $R > 0$ such that f maps each circle $|z - z_0| = r$, with $0 < r \leq R$, onto a set γ_r that intersects every ray from $f(z_0)$.*

Proof: We may assume that $z_0 = 0$ and $f(z_0) = 0$ to simplify the notation. Then we must show that f maps each sufficiently small circle $|z| = r$ onto a set γ_r that intersects every ray from 0.

Assume first that $f'(0) \neq 0$. (Recall from Proposition 6 of Chapter 26 that in this case f is one-to-one on some disc around 0, and so γ_r is actually a curve. We will not need this fact here.) We can also assume without loss of generality (see Problem 9) that $f'(0) = 1$. Then there is some $R > 0$ such that if $0 < |z| \leq R$, then

$$\left| \frac{f(z)}{z} - 1 \right| < \frac{1}{10},$$

and hence

$$|f(z) - z| < \frac{1}{10}|z|,$$
$$\frac{9}{10}|z| < |f(z)| < \frac{11}{10}|z|, \tag{12}$$

for all z with $0 < |z| \leq R$. For $|z| = r$, $f(z)$ lies in the ring $\frac{9}{10}r < |z| < \frac{11}{10}r$. Let $r \leq R$ and let $z_1 = re^{-3\pi i/4}$, $z_2 = re^{3\pi i/4}$. Then $f(z_1)$ lies in the fourth quadrant because of (12), and $f(z_2)$ lies in the second quadrant. For $z = re^{i\theta}$, with $-\frac{3}{4}\pi \leq \theta \leq \frac{3}{4}\pi$, $f(z)$ is well away from the negative real axis, and so $\arg f(z)$ is continuous. Hence the function $A(\theta) = \arg f(re^{i\theta})$ is continuous for $-\frac{3}{4}\pi \leq \theta \leq \frac{3}{4}\pi$, and maps $\left[-\frac{3}{4}\pi, \frac{3}{4}\pi\right]$ onto an interval that surely contains $\left[-\frac{\pi}{2}, \frac{\pi}{2}\right]$, since $\arg f(z_1) < -\frac{\pi}{2}$ and $\arg f(z_1) > \frac{\pi}{2}$. Thus γ_r meets every ray into the right half-plane. The same argument (see Problem 10) can be applied to rays into the left half-plane by considering a new argument function $\arg^* z$ that takes values in $[0, 2\pi)$, and hence is continuous off the positive real axis.

We have proved the proposition in the case $f'(0) \neq 0$, and we now assume $f(0) = 0$, $f'(0) = 0$. Then

$$f(z) = a_k z^k + a_{k+1} z^{k+1} + \cdots$$

where $f^{(k)}(0) = k! a_k$ is the first nonzero derivative of f at 0, and $k \geq 2$. Hence

$$f(z) = z^{k-1}(a_k z + a_{k+1} z^2 + \cdots)$$
$$= z^{k-1} g(z).$$

Since $g(0) = 0$, $g'(0) = a_k \neq 0$, g maps each sufficiently small circle $|z| = r$ onto a curve that intersects every ray from 0. If $z = re^{i\theta}$, with $-\frac{3}{4}\pi \leq \theta \leq \frac{3}{4}\pi$, then

$$f(z) = r^{k-1} e^{i(k-1)\theta} |g(z)| e^{i \arg g(z)}$$
$$= r^{k-1} |g(z)| e^{iA(\theta)},$$

where $A(\theta) = (k-1)\theta + \arg g(z)$. The function $A(\theta)$ is continuous on $\left[-\frac{3}{4}\pi, \frac{3}{4}\pi\right]$, and, since $k - 1 \geq 1$,

$$A\left(-\frac{3}{4}\pi\right) \leq -\frac{3}{4}\pi - \frac{\pi}{2} < -\pi,$$

$$A\left(\frac{3}{4}\pi\right) \geq \frac{3}{4}\pi + \frac{\pi}{2} > \pi.$$

Hence the f-image of the part of $|z| = r$ for $-\frac{3}{4}\pi \leq \theta \leq \frac{3}{4}\pi$ intersects every ray from 0. ∎

PROBLEM 9: Show that if Proposition 4 holds for all f such that $f(0) = 0$ and $f'(0) = 1$, then it holds for all f such that $f(0) = 0$ and $f'(0) \neq 0$. Hint: What does multiplying $f(z)$ by a constant do to γ_r?   •

PROBLEM 10: Show that γ_r intersects every ray from 0 into the left half-plane if $f(0) = 0$, $f'(0) = 1$.   •

PROBLEM 11 (The Maximum Principle): If f is analytic and nonconstant on an open set G, then $|f(z)|$ assumes no maximum value in G. If $f(z) \neq 0$ in G, then $|f(z)|$ assumes no minimum value in G.   •

PROBLEM 12 (Fundamental Theorem of Algebra): If $P(z) = a_n z^n + \cdots + a_1 z + a_0$, with $n \geq 1$ and $a_n \neq 0$, then the equation $P(z) = 0$ has a root. Hint: Show that $\lim_{|z| \to \infty} |P(z)| = \infty$ by considering

$$|P(z)| = |z^n| \left| a_n + \frac{a_{n-1}}{z} + \cdots + \frac{a_0}{z^n} \right|.$$

214 CHAPTER XXIX CAUCHY'S INTEGRAL THEOREMS

It follows that $K = \{z : |P(z)| \leq |P(0)| + 1\}$ is compact, and so $|P(z)|$ has a minimum value on K, and consequently a minimum on the open set $\{z : |P(z)| < |P(0)| + 1\}$. •

PROBLEM 13: What are the directed set and net for the limit $\lim_{|z| \to \infty} |P(z)|$ of the preceding problem? •

XXX

LEBESGUE MEASURE IN (0, 1)

O ur aim in this chapter is to extend the concept of length from intervals to more general subsets. Ideally, we would like to have a measure, $\mu(E)$, for every set $E \subset \mathbb{R}$, with the property that $\mu(I)$ is the length of I if I is an interval, and $\mu\left(\bigcup E_i\right) = \sum \mu(E_i)$ for any countable disjoint family $\{E_i\}$. This turns out to be impossible, but we can define such a function μ, generalizing length for intervals, so that μ is countably additive for lots of sets, including all open sets and all closed sets.

We define $\mu(E)$ initially for subsets E of $(0, 1)$. We will later extend the definition to cover arbitrary subsets of $\mathbb{R}$, but for this chapter all sets are subsets of $(0, 1)$. If I is an interval — any kind — then $\ell(I)$ is the length of I. In the past we have included singletons in the category of intervals. For instance, "A continuous function maps an interval onto an interval." This would be false if singletons were not intervals, since constant functions are continuous. Here, however, we are interested in length, and points (singletons) have negligible length. We will therefore adopt the convention for the remainder of the text that "interval" means "interval with more than one point." Therefore, if I is an interval, $\ell(I) > 0$. If $\{I_n\}$ is a finite or countable family of intervals, we will call $\sum \ell(I_n)$ the **total length** of $\{I_n\}$. For any set $E \subset (0, 1)$, define

$$\mu(E) = \inf\left\{\sum \ell(I_n) : E \subset \bigcup I_n\right\},$$

where the inf is over all finite or countable families $\{I_n\}$ of *open* intervals that cover E.

The function μ is usually called **Lebesgue outer measure**, but we will call μ simply **Lebesgue measure** on subsets of $(0, 1)$. Most texts use μ^* for our function μ, and use the unadorned μ for the function μ^* restricted to a class of sets called **measurable sets**, on which μ^* is countably additive. We will use μ for the (outer) measure of all sets, measurable or not. In practice we will always hypothesize that the sets in question are measurable, since countable additivity is what makes μ worthwhile. The measurable sets will be defined by and by.

The following properties of μ are immediate from the definition.

PROBLEM 1: (i) $\mu(\varnothing) = 0$.

(ii) $\mu(\{x\}) = 0$ for all x.

(iii) $0 \leq \mu(E) \leq 1$ for all $E \subset (0, 1)$.

(iv) If $E \subset F \subset (0, 1)$, $\mu(E) \leq \mu(F)$. (μ is monotone.)   •

We cannot attain the utopian goal that μ is countably additive for all sets, but μ is **countably subadditive** for any collection of sets, which we prove next.

PROPOSITION 1: *The measure μ is countably subadditive; that is, if $\{E_i\}$ is any finite or countable family of subsets of* $(0, 1)$*, then*

$$\mu\left(\bigcup E_i\right) \leq \sum \mu(E_i).$$

Proof: Let $\{E_n\}$ be any finite or countable collection of subsets of $(0, 1)$. For each E_n, let $\{I_{nj}\}$ be a covering of E_n by open intervals such that

$$\sum_{j=1}^{\infty} \ell(I_{nj}) < \mu(E_n) + \frac{\varepsilon}{2^n}.$$

Then the family of all I_{nj} is a covering of $\bigcup E_n$, and

$$\sum_{n, j} \ell(I_{nj}) = \sum_{n=1}^{\infty} \sum_{j=1}^{\infty} \ell(I_{nj})$$

$$< \sum_{n=1}^{\infty} \left(\mu(E_n) + \frac{\varepsilon}{2^n}\right)$$

$$= \sum_{n=1}^{\infty} \mu(E_n) + \varepsilon.$$

Therefore $\mu\left(\bigcup E_n\right) < \sum \mu(E_n) + \varepsilon$ for every $\varepsilon > 0$, and we have the subadditivity inequality.   ∎

In the definition of $\mu(E)$, we specified coverings $\{I_n\}$ of E by open intervals. In fact it does not matter whether we use open intervals or closed intervals or some of each. Recall, however, that henceforth all intervals have strictly positive length.

PROPOSITION 2: *If $E \subset (0, 1)$, then $\mu(E)$ is the inf of total lengths of coverings of E by intervals of any kind.*

Proof: First, we show that the definition of $\mu(E)$ does not change if we use all closed intervals. If $\{I_n\}$ is any covering of E by open intervals, then $\{\bar{I}_n\}$ is a covering of E by closed intervals, and the total length is the same. Therefore, the inf of total lengths of coverings by closed intervals is the same or smaller than $\mu(E)$. Let $\{J_n\} = \{[a_n, b_n]\}$ be any covering of E by closed intervals. For each n, let $I_n = (a_n - \varepsilon/2^n, b_n + \varepsilon/2^n)$, so that $\{I_n\}$ is a covering of E by open intervals and $\sum \ell(I_n) = \sum \ell(J_n) + 2\varepsilon$. Therefore, each covering by closed intervals has a total length that can be approximated arbitrarily closely by the total length of a covering by open intervals. The inf over all open coverings is the same as the inf over all closed coverings. The case of coverings of E by intervals that are neither all open nor all closed is the following problem. ∎

PROBLEM 2: Let $\{I_n\}$ be any countable collection of intervals — any kind — such that $E \subset \bigcup I_n$. Show that $\mu(E) \leq \sum \ell(I_n)$. •

Now we have $\mu(E)$ defined for all subsets of $(0, 1)$, and we have to check that $\mu(I) = \ell(I)$ if I is an interval.

PROPOSITION 3: *If J is an interval in $(0, 1)$, then $\mu(J) = \ell(J)$.*

Proof: First suppose J is a closed interval. Since $\{J\}$ is a one-interval covering of itself, $\mu(J) \leq \ell(J)$. To prove the reverse inequality, let $\{I_j\}$ be any open covering of J. Since J is compact, a finite number of the I_j will cover J, and we can discard the rest since we are in pursuit of the inf of total lengths. We show by induction that if $J \subset I_1 \cup \cdots \cup I_n$ for any finite number of open intervals $I_1, \ldots, I_n$, then $\ell(J) \leq \ell(I_1) + \cdots + \ell(I_n)$, which proves that $\ell(J) \leq \mu(J)$.

Assume as our inductive assumption that $\mu(J) \leq \ell(I_1) + \cdots + \ell(I_n)$ for every closed interval J and any covering of J by n or fewer open intervals. Let $J \subset I_1 \cup \cdots \cup I_{n+1}$ for open intervals $I_1, \ldots, I_{n+1}$. If any n of these cover J, we are done, so that we can assume each I_j intersects J. To be specific, let $J = [a, b]$, and let $I_{n+1} = (c, d)$. We consider the case $a < c < d < b$ and leave the other cases as a problem. Let $J_1 = [a, c]$ and $J_2 = [d, b]$, so that $J_1 \cup J_2 = J - I_{n+1}$. No I_j intersects both J_1 and J_2, for such an interval would cover I_{n+1}, and I_{n+1} could be discarded. Therefore, some of the I_j cover J_1 and the rest cover J_2. By the inductive assumption

$$\ell(J_1) + \ell(J_2) \leq \sum_{j=1}^{n} \ell(I_n).$$

Therefore,

$$\ell(J) = \ell(J_1) + \ell(I_{n+1}) + \ell(J_2) \leq \sum_{j=1}^{n+1} \ell(I_n),$$

and this completes the inductive proof for closed intervals J. If J is an open interval, $J = (a, b)$, let $J_0 = [a + \varepsilon, b - \varepsilon]$ where ε is small enough so $J_0 \subset J$. Then by monotonicity, $\mu(J_0) \le \mu(J)$. Since $\mu(J_0) = \ell(J) - 2\varepsilon \le \mu(J)$ for all ε, $\ell(J) \le \mu(J)$ for open intervals J. If J is half-open, say $J = (a, b]$, then $(a, b) \subset (a, b] \subset [a, b]$, and $\mu(a, b) = \mu[a, b] = b - a$, and so $\mu(a, b] = b - a$. ∎

PROBLEM 3: Let $J = [a, b] \subset I_1 \cup \cdots \cup I_{n+1}$ where the I_j are open intervals. Prove that $\ell(J) \le \sum_{j=1}^{n+1} \ell(I_j)$ in the case that $I_{n+1} = (c, d)$ with $a < c < b < d$, and in the case $a < c < b = d$.   •

PROBLEM 4: Show that the set of all rational numbers in $(0, 1)$ has measure zero.   •

PROBLEM 5: Let E be any subset of $(0, 1)$ and let $\mu(E_0) = 0$. Show that $\mu(E \cup E_0) = \mu(E - E_0) = \mu(E)$.   •

PROBLEM 6: (i) If $E_1 \subset I_1$ and $E_2 \subset I_2$ where I_1 and I_2 are disjoint intervals in $(0, 1)$, then $\mu(E_1 \cup E_2) = \mu(E_1) + \mu(E_2)$.

(ii) If $I_1, \ldots, I_n$ are disjoint intervals in $(0, 1)$, then $\mu(I_1 \cup \cdots \cup I_n) = \ell(I_1) + \cdots + \ell(I_n)$.   •

Now we proceed to construct a set E with the following property. There is a countable disjoint family $\{E_n\}$ of sets, all with the same measure as E, such that $\bigcup E_n = (0, 1)$. These sets E_n cannot have zero measure, since that would imply $\mu(0, 1) \le \sum \mu(E_n) = 0$, and we know that $\mu(0, 1) = 1$. Therefore, $\mu(E_n) = p > 0$ for all n and

$$\mu\left(\bigcup E_n\right) = \mu(0, 1) = 1; \qquad \sum \mu(E_n) = \infty.$$

Our function μ is, alas, not countably additive over *all* disjoint families.

For a set $E \subset (0, 1)$ and a number $r \in (0, 1)$, we define the set $E \oplus r$ to be all numbers $e + r < 1$, together with all the numbers $e + r - 1$ for $e \in E$ and $e + r > 1$. Thus $E \oplus r$ is the set gotten by shoving E to the right a distance r, cutting off the part that sticks out beyond 1, and sticking that part back on the left end of $(0, 1)$. The set $E \oplus r$ is a sort of translate of E, and we will now show that $\mu(E \oplus r) = \mu(E)$ for all r. We will need the following problem as a lemma for Proposition 4.

PROBLEM 7: If $E \subset (0, 1)$ and $E + r \subset (0, 1)$, where $E + r = \{e + r : e \in E\}$, then $\mu(E) = \mu(E + r)$. In other words, μ is translation invariant insofar as that makes sense for the sets on which μ is defined.   •

PROPOSITION 4: *For all $E \subset (0, 1)$ and all $r \in (0, 1)$, $\mu(E \oplus r) = \mu(E)$.*

Proof: Let $E_1 = \{e \in E : e + r < 1\}$ and $E_2 = \{e \in E : e + r > 1\}$, so that $E_1 = E \cap (0, 1 - r)$ and $E_2 = E \cap (1 - r, 1)$. We can assume that $1 - r \notin E$ because adding or subtracting one point from a set does not change its measure, by Problem 5. From Problem 6, $\mu(E) = \mu(E_1) + \mu(E_2)$, and $\mu(E \oplus r) =$

$\mu(E_1 \oplus r) + \mu(E_2 \oplus r)$. From Problem 7, $\mu(E_1) = \mu(E_1 \oplus r)$, and so we only need to show that $\mu(E_2) = \mu(E_2 \oplus r)$. Every covering of E_2 can be replaced by a lighter covering consisting of intervals $I_n \subset (1 - r, r)$. For any such covering, let $J_n = I_n - (1 - r)$. Then $\{J_n\}$ is a covering of $E_2 \oplus r$ and $\sum \ell(J_n) = \sum \ell(I_n)$. Conversely, to every covering of $E_2 \oplus r$ there corresponds a covering of E_2 with the same total length. Therefore, $\mu(E_2 \oplus r) = \mu(E_2)$, which completes the proof that $\mu(E \oplus r) = \mu(E)$. ∎

Now let $r_1, r_2, \ldots$ be all the rationals in $(0, 1)$, and let $r_0 = 0$. We construct a set $E = E \oplus r_0$ such that the sets $E \oplus r_i$ are disjoint and $(0, 1) = \bigcup_i E \oplus r_i$. Let x be any point of $(0, 1)$. Pick a point $y \in (0, 1)$ such that $y - x$ is irrational; x and y will be elements of E. Then pick any point $z \in (0, 1)$ such that $z - y$ and $z - x$ are irrational, and throw z into E. Continue this picking, uncountably many times, to get a set E such that $x_1 - x_2$ is irrational for every pair of elements $x_1, x_2 \in E$. Let E be a maximal such set. If r_i is any of the rationals in $(0, 1)$, or 0, then $\mu(E \oplus r_i) = \mu(E)$. Moreover, the sets $E \oplus r_i$ are disjoint, because if $x + r_i = y + r_j$ for some $x, y \in E$ and some $r_i \neq r_j$, then

$$\left.\begin{array}{c} x + r_i \\ \text{or} \\ x + r_i - 1 \end{array}\right\} = \left\{\begin{array}{c} y + r_j \\ \text{or} \\ y + r_j - 1. \end{array}\right.$$

In any of these cases, $x - y$ would be rational, which only happens if $x = y$ and $r_i = r_j$. Now check that $\bigcup_i E \oplus r_i = (0, 1)$. Let $z \in (0, 1)$ and suppose $z \neq x + r_i$ and $z \neq x + r_i - 1$ for all $x \in E$ and all r_i. In this case $z - x$ is irrational for all $x \in E$, and so z must belong to $E = E \oplus r_0$. Now we see that μ is not countably additive on the sets $E \oplus r_i$. Clearly $\mu(E) \neq 0$, for otherwise we would have $\mu(0, 1) = \mu\left(\bigcup E \oplus r_i\right) = 0$. Therefore, $\mu(E) > 0$ and $\sum \mu(E \oplus r_i) = \infty$, which contradicts countable additivity.

XXXI

MEASURABLE SETS

he measure function μ gives the right number for intervals, is countably subadditive, and additive at least for some sets. For example, if E_1 and E_2 are subsets of disjoint intervals, then $\mu(E_1 \cup E_2) = \mu(E_1) + \mu(E_2)$. In this chapter we single out a family of sets on which μ is additive, and indeed countably additive in the sense that $\mu\left(\bigcup E_i\right) = \sum \mu(E_i)$ whenever the E_i are disjoint. We continue to restrict our attention to subsets of $(0, 1)$. The extension of μ from subsets of $(0, 1)$ to general subsets of $\mathbb{R}$ will be simple once we have all the desired properties for μ in $(0, 1)$.

For any set $E \subset (0, 1)$, let E' denote the complement in $(0, 1)$: $E' = (0, 1) \sim E$. A set E is **measurable** provided $\mu(E) + \mu(E') = 1$. This is certainly a minimal additivity assumption, but it turns out to be sufficient to distinguish the useful sets. Notice that the condition for measurability is symmetric in E and E', so that E is measurable if and only if E' is measurable.

PROBLEM 1: Show that intervals — any kind — are measurable. Hint: Use Problem 6 of the last chapter. •

The definition of measurability requires that a measurable set E cut the interval $(0, 1)$ additively. The next proposition shows that this is the same as requiring that E cut *every* interval additively.

PROPOSITION 1: *E is measurable if and only if* $\mu(E \cap J) + \mu(E' \cap J) = \mu(J)$ *for every interval J in $(0, 1)$.*

Proof: The condition is certainly sufficient for E to be measurable, since we can take $J = (0, 1)$. Therefore, the content of the proposition is that E cannot cut $(0, 1)$ additively unless E cuts every interval additively. Notice that μ is finitely subadditive, because μ is countably subadditive:

$$\mu(F_1 \cup F_2 \cup \varnothing \cup \varnothing \cup \cdots) \leq \mu(F_1) + \mu(F_2) + 0 + 0 + \cdots.$$

Therefore, what we must show is that if $\mu(E) + \mu(E') = 1$, then, for all J,

$$\mu(E \cap J) + \mu(E' \cap J) \leq \mu(J).$$

We consider the case where J has endpoints a and b with $0 < a < b < 1$, and leave the other cases as an exercise. Let J_1 and J_3 be the intervals to the left and right of J, and let $J = J_2$, so that $J_1 \cup J_2 \cup J_3 = (0, 1)$ and the J_i are disjoint. By subadditivity,

$$\mu(E \cap J_1) + \mu(E' \cap J_1) \geq \mu(J_1),$$

$$\mu(E \cap J_2) + \mu(E' \cap J_2) \geq \mu(J_2),$$

$$\mu(E \cap J_3) + \mu(E' \cap J_3) \geq \mu(J_3).$$

Adding the columns and using the fact that the J_i are disjoint, we get

$$\mu(E) = \sum_{i=1}^{3} \mu(E \cap J_i); \qquad \mu(E') = \sum_{i=1}^{3} \mu(E' \cap J_i).$$

Hence,

$$\mu(E) + \mu(E') = \sum_{i=1}^{3} \left[\mu(E \cap J_i) + \mu(E' \cap J_i) \right]$$

$$\geq \sum_{i=1}^{3} \mu(J_i) = 1. \tag{1}$$

Equality holds in (1) if and only if $\mu(E \cap J_i) + \mu(E' \cap J_i) = \mu(J_i)$ for each i. If E is measurable, equality holds in (1), and we have the desired result. ∎

PROBLEM 2: Write out the proof of Proposition 1 in case $J = (0, a)$ with $0 < a < 1$.   •

PROBLEM 3: If E_1 and E_2 are disjoint measurable sets, then $\mu(E_1 \cup E_2) = \mu(E_1) + \mu(E_2)$. Hint: Let $E_1 \cup E_2 \subset \bigcup I_n$ with $\sum \ell(I_n) < \mu(E_1 \cup E_2) + \varepsilon$. Use $E_1 \subset \bigcup(E_1 \cap I_n), E_2 \subset \bigcup(E_2 \cap I_n)$, and

$$\mu(E_1 \cap I_n) + \mu(E_2 \cap I_n) \leq \mu(E_1 \cap I_n) + \mu(E_1' \cap I_n) = \mu(I_n).$$ •

Next we show that E is measurable if and only if E splits *every* set T additively; that is, $\mu(E \cap T) + \mu(E' \cap T) = \mu(T)$ for all T. This is the famous Carathéodory criterion for measurability, and is the *definition* of a measurable set in all modern abstract settings.

PROPOSITION 2: *E is a measurable set if and only if*

$$\mu(E \cap T) + \mu(E' \cap T) = \mu(T) \tag{2}$$

for every set $T \subset (0, 1)$.

Proof: Clearly the criterion (2) is sufficient for E to be measurable, since we can take $T = (0, 1)$. Since μ is subadditive, the inequality $\mu(E \cap T) + \mu(E' \cap T) \geq \mu(T)$ is automatic, and so we need only show that if E is measurable, then, for all T,

$$\mu(E \cap T) + \mu(E' \cap T) \leq \mu(T). \tag{3}$$

Assume E is measurable, and let T be any set. Let $T \subset \bigcup I_n$ with $\sum \ell(I_n) < \mu(T) + \varepsilon$. Then $E \cap T \subset \bigcup (E \cap I_n)$ and $E' \cap T \subset \bigcup (E' \cap I_n)$, so that

$$\mu(E \cap T) + \mu(E' \cap T) \leq \sum \mu(E \cap I_n) + \sum \mu(E' \cap I_n)$$

$$= \sum \left[\mu(E \cap I_n) + \mu(E' \cap I_n) \right]$$

$$= \sum \mu(I_n) < \mu(T) + \varepsilon. \qquad \blacksquare$$

An **outer measure** is a nonnegative, monotone, countably subadditive function m defined on all subsets of a given set, with $m(\emptyset) = 0$. Our function μ is an outer measure on subsets of $(0,1)$, and the properties just listed are explicitly these:

$$\mu(E) \geq 0 \qquad \text{for all } E,$$

$$\mu(\emptyset) = 0,$$

$$\mu(E) \leq (F) \qquad \text{if } E \subset F, \tag{4}$$

$$\mu \left(\bigcup_{i=1}^{\infty} E \right) \leq \sum_{i=1}^{\infty} \mu(E_i).$$

Outer measures abound in nature, and Problem 4 exhibits another one.

PROBLEM 4: Let X be the plane $\mathbb{R}^2$. A rectangle is a set $R = I \times J$ where I and J are intervals. Let $A(R) = \ell(I)\ell(J)$ if $R = I \times J$. For any set $E \subset \mathbb{R}^2$, define

$$m(E) = \inf \left\{ \sum A(R_i) : E \subset \bigcup R_i \right\},$$

where the inf is over all finite or countable families $\{R_i\}$ of rectangles that cover E. Show that m is an outer measure and develop some of its properties.   •

We now proceed to show that the measurable sets, as characterized by the Carathéodory criterion (2), are closed under countable unions and intersections, and contain all open and closed sets, and that μ is countably additive on the measurable sets. From here on, we use only the properties (4) of any outer measure, and the measurability criterion (2). Our arguments do not depend on the fact that the sets E are subsets of $(0,1)$, or even that $\mu(E) < \infty$ for all sets. All the following arguments apply equally well, for example, to the plane outer measure m defined in Problem 4. When m is restricted to the measurable sets as characterized by (2), m is Lebesgue planar measure, and has all the properties of μ we develop in this chapter.

PROPOSITION 3: *If E_1 and E_2 are measurable, then $E_1 \cup E_2$ and $E_1 \cap E_2$ are measurable.*

Proof: Assume that E_1 and E_2 are measurable, so that E_1' and E_2' are also measurable. If $E_1' \cup E_2'$ is measurable, then $(E_1' \cup E_2')' = E_1 \cap E_2$ is measurable, and so it is sufficient to prove the union of two measurable sets is measurable.

Let T be any test set, and let

$$T_1 = T \cap (E_1 - E_2),$$
$$T_2 = T \cap (E_1 \cap E_2),$$
$$T_3 = T \cap (E_2 - E_1),$$
$$T_4 = T \cap (E_1 \cup E_2)'.$$

Then $T = T_1 \cup T_2 \cup T_3 \cup T_4$ as indicated in Figure I. We have to show that

$$\mu(T_1 \cup T_2 \cup T_3) + \mu(T_4) = \mu(T). \tag{5}$$

Since E_2 is measurable, E_2 cuts $T_1 \cup T_2$ additively, so that

$$\mu(T_1 \cup T_2) = \mu(T_1) + \mu(T_2). \tag{6}$$

Similarly, E_1 cuts $T_1 \cup T_2 \cup T_3$ additively, so that

$$\mu(T_1 \cup T_2) + \mu(T_3) = \mu(T_1 \cup T_2 \cup T_3), \tag{7}$$

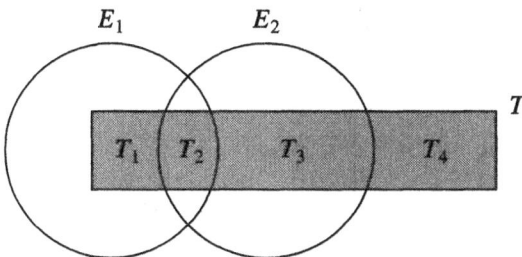

FIGURE I

and hence

$$\mu(T_1 \cup T_2 \cup T_3) = \mu(T_1) + \mu(T_2) + \mu(T_3). \tag{8}$$

Now cut $T_3 \cup T_4$ with E_2, and get

$$\mu(T_3) + \mu(T_4) = \mu(T_3 \cup T_4). \tag{9}$$

Cutting T with E_1 gives

$$\mu(T_1 \cup T_2) + \mu(T_3 \cup T_4) = \mu(T), \tag{10}$$

whence from (6), (9), and (10) we have

$$\mu(T_1) + \mu(T_2) + \mu(T_3) + \mu(T_4) = \mu(T). \tag{11}$$

Combining (8) with (11) gives the desired equality (5). ∎

PROBLEM 5: If E_1 and E_2 are measurable sets, then $E_1 - E_2$ is measurable, and $\mu(E_1 - E_2) = \mu(E_1) - \mu(E_1 \cap E_2)$ and $\mu(E_1 \cup E_2) = \mu(E_1) + \mu(E_2) - \mu(E_1 \cap E_2)$. Hint: Use Problem 3. •

PROPOSITION 4: *If* $\{E_n\}$ *is a finite or countable family of disjoint measurable sets, then* $\mu\left(\bigcup E_i\right) = \sum \mu(E_i)$.

Proof: From Problem 3 we know that μ is finitely additive, and so for each n,

$$\mu\left(\bigcup_{i=1}^{n} E_i\right) = \sum_{i=1}^{n} \mu(E_i).$$

By monotonicity we have

$$\mu\left(\bigcup_{i=1}^{\infty} E_i\right) \geq \mu\left(\bigcup_{i=1}^{n} E_i\right) = \sum_{i=1}^{n} \mu(E_i).$$

Since this holds for all n,

$$\mu\left(\bigcup_{i=1}^{\infty} E_i\right) \geq \sum_{i=1}^{\infty} \mu(E_i).$$

The reverse inequality is the subadditivity of μ. ∎

PROBLEM 6: Let $\{E_i\}$ be a finite or countable family of disjoint measurable sets, and let T be any set. Show that

$$\mu\left(T \cap \bigcup E_i\right) = \sum \mu(T \cap E_i). \quad •$$

PROPOSITION 5: *If* $\{E_i\}$ *is a countable family of measurable sets,* $\bigcup E_i$ *is measurable.*

Proof: If $E = \bigcup E_i$, then E is the union of a countable family of disjoint measurable sets, namely,

$$E = E_1 \cup (E_2 - E_1) \cup \left(E_3 - (E_1 \cup E_2)\right) \cup \cdots.$$

We can assume, therefore, that the E_i are disjoint. Let $F_n = E_1 \cup \cdots \cup E_n$, so that $F_n \subset E$ and $F_n' \supset E'$. Therefore, for any T,

$$\mu(T) = \mu(T \cap F_n) + \mu(T \cap F_n')$$
$$\geq \mu(T \cap F_n) + \mu(T \cap E')$$
$$= \sum_{i=1}^{n} \mu(T \cap E_i) + \mu(T \cap E').$$

Since this holds for all n,

$$\mu(T) \geq \sum_{i=1}^{\infty} \mu(T \cap E_i) + \mu(T \cap E').$$

Since $\sum_{i=1}^{\infty} \mu(T \cap E_i) \geq \mu(T \cap E)$ by subadditivity,

$$\mu(T) \geq \mu(T \cap E) + \mu(T \cap E'),$$

and E is measurable.   ∎

As corollaries of Proposition 5 we immediately have that $\bigcap E_i$ is measurable for any countable family of measurable sets, since $\bigcap E_i = \left(\bigcup E_i'\right)'$. We saw earlier that open intervals are measurable. Every open set is a countable or finite union of open intervals, and so all open sets are measurable. Complements of measurable sets are measurable, and so all closed sets are measurable.

A σ-**algebra** of subsets of a given set X is a family of sets that contains X and $\varnothing$, and is closed under complementation, countable unions, and countable intersections. The measurable subsets of $(0,1)$ form a σ-algebra, and moreover, a σ-algebra containing all open and closed sets.

The foregoing arguments depend only on the properties (4) of any outer measure, and the Carathéodory criterion (2). Therefore, if m is any outer measure, and a set E is defined to be m-measurable if and only if $m(E \cap T) + m(E' \cap T) = m(T)$ for all T, then the m-measurable sets form a σ-algebra, and m is countably additive on the m-measurable sets. Let us use that fact to extend μ to all of $\mathbb{R}$. For any set $E \subset \mathbb{R}$, let $\mu(E) = \inf\{\sum \ell(I_n) : E \subset I_n\}$. Then μ is the same on subsets of $(0,1)$ as before, and it is easy to show that μ is an outer measure on all subsets of $\mathbb{R}$. (See Problem 7.)

PROBLEM 7: Verify that μ as defined above does not change for subsets of $(0,1)$, and satisfies all the conditions (4). Caution: $\mu(E) = \infty$ is now a possibility, and $\sum \ell(I_n) < \mu(E) + \varepsilon$ is clearly not a possibility with strict inequality if $\mu(E) = \infty$.   •

With μ now defined on all subsets of $\mathbb{R}$, we agree that E is measurable provided that $\mu(E \cap T) + \mu(E' \cap T) = \mu(T)$ for every set T. The measurable sets form a σ-algebra, which of course contains all the old measurable subsets of $(0, 1)$.

PROBLEM 8: Show that any open interval $(a, b) \subset \mathbb{R}$ is measurable, and hence the measurable subsets of $\mathbb{R}$ contain all open and closed sets. Hint: It is sufficient to verify the Carathéodory criterion for sets T such that $\mu(T) < \infty$. Why?   ●

The following proposition gives two basic utilitarian limit theorems for measurable sets.

PROPOSITION 6: *Let $\{E_i\}$ be a countable family of measurable sets.*

(i) *If $E_1 \subset E_2 \subset E_3 \subset \cdots$, then $\mu \left(\bigcup E_i \right) = \lim \mu(E_i)$.*
(ii) *If $E_1 \supset E_2 \supset E_3 \supset \cdots$, then $\mu \left(\bigcap E_i \right) = \lim \mu(E_i)$ if $\mu(E_1) < \infty$.*

Proof: (i) If $E_1 \subset E_2 \subset \cdots$, then

$$\bigcup E_i = E_1 \cup (E_2 - E_1) \cup (E_3 - E_2) \cup \cdots,$$

and the right side is a disjoint union. Therefore,

$$\mu \left(\bigcup E_i \right) = \mu(E_1) + \mu(E_2 - E_1) + \mu(E_3 - E_2) + \cdots$$
$$= \lim \left[\mu(E_1) + \cdots + \mu(E_n - E_{n-1}) \right]$$
$$= \lim \mu(E_n).$$

(ii) If $E_1 \supset E_2 \supset \cdots$, let $E = \bigcap E_n$. Then

$$(E_1 - E_2) \cup (E_2 - E_3) \cup \cdots = E_1 - E,$$

and the left side is a disjoint union. Since all E_n have finite measure, and $E_n \supset E_{n+1}$,

$$\mu(E_n - E_{n+1}) = \mu(E_n) - \mu(E_{n+1}), \qquad (12)$$

and

$$\mu(E_1 - E) = \sum_{n=1}^{\infty} \mu(E_n - E_{n+1})$$
$$= \lim_{N \to \infty} \sum_{n=1}^{N} \mu(E_n) - \mu(E_{n+1})$$
$$= \lim_{N \to \infty} \left[\mu(E_n) - \mu(E_{N+1}) \right]$$
$$= \mu(E_1) - \lim_{N \to \infty} \mu(E_{N+1}).$$

Thus

$$\mu(E_1) - \mu(E) = \mu(E_1) - \lim_{N \to \infty} \mu(E_{N+1}),$$

and the result follows, since all terms are finite. ∎

The hypothesis $\mu(E_1) < \infty$ (or $\mu(E_n) < \infty$ for some n) is essential in the last result. For example, if $E_n = [n, \infty)$, then $\bigcap E_n = \varnothing$, but $\lim \mu(E_n) = \infty$. Since now some sets have infinite measure, a little more care must be used in the arithmetic. Addition poses no problem. For example, if E, F are measurable sets and $E \subset F$ (cf. (12)), then

$$\mu(E) + \mu(F - E) = \mu(F),$$

whether or not any of the terms are $+\infty$. However, $\mu(F - E) = \mu(F) - \mu(E)$ is true only if $\mu(E) < \infty$. When subtracting you have to rule out ∞ in the same way you rule out zero when dividing; $\infty - \infty$ makes no sense.

XXXII

THE LEBESGUE INTEGRAL

O ur definition of the Lebesgue integral will closely parallel the Darboux
definition of the Riemann integral in Chapter 14. We partition the
domain of the function, and we form upper and lower sums as before. If
the nets of upper sums and lower sums have a common finite limit, then the function
is integrable. The only difference is that now we have a more sophisticated concept
of length in $\mathbb{R}$. This allows us to integrate over measurable sets, not just intervals,
and to partition these sets more delicately. The result is that more functions
are integrable, and more importantly, the limit theorems are more general and
easier to apply. We will take advantage of the countable additivity of Lebesgue
measure to allow countable partitions. With countable partitions we can treat all
functions the same, including unbounded functions, and functions defined on sets
of infinite measure. There will be no improper integrals, only integrals, and f will
be integrable if and only if $|f|$ is integrable. We consider only functions defined
on measurable sets, but the measurable set can have infinite measure, and indeed
can be all of $\mathbb{R}$.

A **partition** of a measurable set S is now a finite or countable family $P = \{E_i\}$
of disjoint measurable sets whose union is S. Some sets E_i can have zero measure,
but all sets E_i must have finite measure. For any function f on S, we define the

upper and lower sums for f and the partition P as before:

$$M_i = \sup\{f(x) : x \in E_i\},$$

$$m_i = \inf\{f(x) : x \in E_i\},$$

$$U(f, P) = \sum M_i \mu(E_i),$$

$$L(f, P) = \sum m_i \mu(E_i).$$

Now some M_i can equal $+\infty$ and some m_i can equal $-\infty$; some upper and lower sums can be infinite.

We want the integral of f to denote the finite net area (finite positive area minus the finite negative area) between the graph of f and the x axis. We therefore consider only functions f whose graphs lie in some union of measurable rectangles $E_i \times [-B_i, B_i]$ whose aggregate area is finite. Specifically, we will say that f is **admissible** over S if and only if there is a partition $P = \{E_i\}$ of S and a sequence $\{B_i\}$ of nonnegative numbers such that

$$-B_i \leq m_i \leq M_i \leq B_i$$

for all i, and $\sum B_i \mu(E_i) < \infty$. It follows that both series $\sum m_i \mu(E_i)$ and $\sum M_i \mu(E_i)$ converge absolutely (see Problem 1). It is easy to see that f is admissible if and only if $|f|$ has at least one finite upper sum, that is, there is a partition $P = \{E_i\}$ such that $U(|f|, P) < \infty$. If $\overline{M}_i = \sup\{|f(x)| : x \in E_i\}$, then $\sum \overline{M}_i \mu(E_i) < \infty$ and for all i,

$$-\overline{M}_i \leq m_i \leq M_i \leq \overline{M}_i.$$

Any partition P such that $U(|f|, P) < \infty$ will be called an **admissible partition** for f.

Since we now allow countable partitions $\{E_i\}$, upper and lower sums are now generally infinite series. If such a series converged conditionally, then rearranging the terms could give any value for the sum. Changing the order in which the sets E_i are labeled clearly has nothing to do with the geometry, and so conditionally convergent upper or lower sums are simply not acceptable. We will insist, then, on admissible functions and admissible partitions for these functions.

A partition $Q = \{F_j\}$ is a **refinement** of the partition $P = \{E_i\}$ provided every F_j is a subset of some $E_i \in P$. If Q is a refinement of P, denoted $Q > P$ or $P < Q$, then we write $Q = \{F_{ij}\}$ to indicate that $E_i = \bigcup_j F_{ij}$ for each $E_i \in P$.

PROBLEM 1: Let f be an admissible function on the measurable set S, and let P_0 be an admissible partition for f.

    (i) Every refinement of P_0 is an admissible partition for f.
    (ii) If $P = \{E_i\}$ and P is admissible for f, then the series $L(f, P) = \sum m_i \mu(E_i)$ and $U(f, P) = \sum M_i \mu(E_i)$ converge absolutely.

(iii) If P and Q are admissible partitions and $P \prec Q$, then

$$L(f, P) \leq L(f, Q) \leq U(f, Q) \leq U(f, P).$$

(iv) If P and Q are admissible partitions, then $L(f, P) \leq U(f, Q)$. •

If $P = \{E_i\}$ is an admissible partition for f, then the series

$$L(f, P) = \sum m_i \mu(E_i)$$

$$U(f, P) = \sum M_i \mu(E_i)$$

converge absolutely. It follows that all m_i and M_i must be finite, unless $\mu(E_i) = 0$. We will agree that $0 \cdot (\pm\infty) = 0$, so that if $\mu(E_i) = 0$, then $m_i \mu(E_i) = M_i \mu(E_i) = 0$ no matter what m_i and M_i are. In any partition P, we can therefore lump together all zero measure sets E_i without changing the upper or lower sum. Accordingly, we will assume that E_0 is the only zero measure set in any partition, and so $\mu(E_i) > 0$ if $i \neq 0$.

We say f is **integrable** over the measurable set S provided f is admissible and

$$\sup_P L(f, P) = \inf_P U(f, P),$$

where the sup and inf are over all admissible partitions. Both $\sup L(f, P)$ and $\inf U(f, P)$ are finite by Problem 1, but the equality is not automatic. It is easy to see that f is integrable over S if and only if f is admissible and, for every $\varepsilon > 0$, there is an admissible partition P such that $U(f, P) - L(f, P) < \varepsilon$. If f is integrable over S, we denote the common (finite) value of $\sup L(f, P)$ and $\inf U(f, P)$ by $\int_S f$. We may occasionally omit the S from the integral sign when the domain of the function is understood or is irrelevant.

PROBLEM 2: Show that $1/\sqrt{x}$ is admissible on $(0, 1)$, and that $1/x^2$ is admissible on $[1, \infty)$. In both cases find an admissible partition. •

PROBLEM 3: If f is integrable over S, then $|f|$ is integrable over S, and $|\int_S f| \leq \int_S |f|$. •

If f is a bounded function on $[a, b]$ and $P = \{x_0, x_1, \ldots, x_n\}$ is a partition of $[a, b]$ in the Riemann-integral sense, we will now interpret P to be the partition of the measurable set $[a, b]$ into the measurable sets (x_{i-1}, x_i), together with the $(n + 1)$-point zero measure set $\{x_0, \ldots, x_n\}$. With this agreement, $U(f, P)$ and $L(f, P)$ mean exactly what they meant in the treatment of the Riemann integral.

PROBLEM 4: If f is Riemann integrable over $[a, b]$, then f is Lebesgue integrable over $[a, b]$, and the integrals are the same. •

The **characteristic function** of a set A, denoted X_A, is defined to be the function that is one on A and zero off A. A function of the form $\sum a_i X_{A_i}$, where the A_i are disjoint measurable sets of finite measure and $\sum |a_i| \mu(A_i) < \infty$, is called a **simple function**. Most texts define a simple function to be a finite sum $\sum_{i=1}^{n} a_i \chi_{A_i}$, but it will be convenient for us to extend the definition to include countable sums provided $\sum a_i \mu(A_i)$ is absolutely convergent. We will allow a_i to be $+\infty$ or $-\infty$ if $\mu(A_i) = 0$ in the same way we allow M_i and m_i to take the values $\pm\infty$ in upper and lower sums. With this agreement, if $P = \{E_i\}$ is an admissible partition for the admissible function f, then $g = \sum m_i \chi_{E_i}$ and $h = \sum M_i \chi_{E_i}$ are simple functions, and $g \leq f \leq h$. The following problem shows that upper and lower sums for f are the same as the integrals of suitable simple functions above or below f.

PROBLEM 5: (i) Every simple function $g = \sum a_i \chi_{A_i}$ is integrable over $\mathbb{R}$, and $\int g = \sum a_i \mu(A_i)$. Hint: Let P_0 be the partition of $\mathbb{R}$ consisting of all A_i together with all sets $[n, n + 1) \cap (\mathbb{R} - \bigcup A_i)$. (The sets in a partition must have finite measure.) Consider upper and lower sums for $P > P_0$.

(ii) A function f is integrable on $\mathbb{R}$ if and only if for each $\varepsilon > 0$ there are simple functions g and h with $g \leq f \leq h$ and $\int h - \int g < \varepsilon$.

(iii) If f is integrable over $\mathbb{R}$, then

$$\int f = \sup \left\{ \int g : g \leq f, g \text{ simple} \right\}$$

$$= \inf \left\{ \int h : h \geq f, h \text{ simple} \right\}. \qquad \bullet$$

The phrase "**almost everywhere**," abbreviated "**a.e.**," means "except on some set of measure zero." Thus we say "$f = g$ a.e." if $\{x : f(x) \neq g(x)\}$ has measure zero, and "$f_n \longrightarrow f$ a.e." if $\lim f_n(x) = f(x)$ except for x in some set of measure zero.

PROBLEM 6: (i) If f is integrable over S and $f = g$ a.e. on S, then g is integrable over S and $\int_S f = \int_S g$.
(ii) If $f(n) = n$ for $n \in \mathbb{N}$ and $f = 0$ elsewhere on $\mathbb{R}$, then f is integrable on $\mathbb{R}$. What is the integral?
(iii) If f is the characteristic function of the rational numbers, then f is integrable over $\mathbb{R}$. $\qquad \bullet$

PROBLEM 7: Let $f(x) = (-1)^n/n$ on $[n, n + 1)$ for all $n \in \mathbb{N}$. Show that f is improperly Riemann integrable on $[1, \infty)$, but not Lebesgue integrable. $\qquad \bullet$

Our definition of the Lebesgue integral as the common (finite) limit of upper and lower sums is exactly analogous to our earlier Darboux definition of the Riemann integral. The Riemann integral can also be characterized as the limit of the net of Riemann sums, and as we saw earlier this characterization is frequently more convenient to work with. The Lebesgue integral can also be characterized as the limit of Riemann sums, and that is our next goal.

Let f be any admissible function on S — and again f is not assumed to be bounded, and S can have infinite measure. Let $P = \{E_i\}$ be an admissible partition of S, and c a choice function for P, so that $c_i \in E_i$ for all i. The **Riemann sum** for f, P, c is

$$R(f, P, c) = \sum f(c_i)\mu(E_i). \tag{1}$$

As before, the pairs (P, c) are directed by refinement: $(P, c) > (P', c')$ if and only if $P > P'$. The choice function plays no role in the ordering of the pairs (P, c). The Riemann sums $R(f, P, c)$ form a net for any given function f. Notice that we require the function and the partitions to be admissible, so that the series (1) does not depend on the ordering of the terms.

PROPOSITION 1: *If f is an admissible function on the measurable set S, then f is integrable with integral I if and only if $R(f, P, c) \longrightarrow I$.*

Proof: Assume f is integrable, so that there is a partition P_0 with $U\left(|f|, P_0\right) < \infty$, and all upper and lower sums converge absolutely for $P > P_0$. Let $P_1 > P_0$ be a partition with $U(f, P_1) - L(f, P_1) < \varepsilon$. For all $P > P_1$ and all choices c for P,

$$\int f - \varepsilon < L(f, P) \leq R(f, P, c) \leq U(f, P) \leq \int f + \varepsilon.$$

Hence

$$\left| R(f, P, c) - \int f \right| < \varepsilon$$

if $P > P_1$, and $R(f, P, c) \longrightarrow \int f$.

Now assume f is admissible over S and $R(f, P, c) \longrightarrow I$. Let P_0 be a partition of S such that $U\left(|f|, P_0\right) < \infty$. Choose $P_1 > P_0$ such that $|R(f, P, c) - I| < \varepsilon$ for all $P > P_1$. Of course all upper and lower sums will converge absolutely for $P > P_1 > P_0$. Let $P = \{E_i\} > P_1$, and for each i with $\mu(E_i) > 0$ (so m_i and M_i are finite) pick $c_i \in E$ so that

$$f(c_i) + \frac{\varepsilon}{2^i \mu(E_i)} > M_i,$$

$$f(c_i)\mu(E_i) + \frac{\varepsilon}{2^i} > M_i\mu(E_i),$$

$$R(f, P, c) + \varepsilon > U(f, P).$$

Similarly, for each i pick c_i' so that

$$R(f, P, c') - \varepsilon < L(f, P).$$

Therefore,

$$I - \varepsilon < R(f, P, c') \qquad \text{and} \qquad R(f, P, c) < I + \varepsilon,$$

$$I - 2\varepsilon < L(f, P) \leq U(f, P) \leq I + 2\varepsilon.$$

Since ε is arbitrary, f is integrable with integral I. ∎

Notice that all the foregoing arguments that show that

$$I - 2\varepsilon < L(f, P) \leq U(f, P) \leq I + 2\varepsilon$$

work perfectly well for the nonintegrable function of Problem 7, provided $P > P_0$, where $P_0 = \{E_i\}$ with $E_i = [i, i + 1)$. What is missing is the proof that there are absolutely convergent upper and lower sums, and that is why f must be assumed to be admissible in addition to assuming that $\{R(f, P, c)\}$ converges.

PROBLEM 8: (i) If f and g are integrable, then af and $f + g$ are integrable, and $\int af = a \int f$, $\int (f + g) = \int f + \int g$. Hint: Do not forget to show that af and $f + g$ are admissible. The rest is easy.

(ii) If f and g are integrable and $f \leq g$ a.e., then $\int f \leq \int g$. ●

PROPOSITION 2: *If f is integrable over S, and T is a measurable subset of S, then f is integrable over T, $f \chi_T$ is integrable over S, and $\int_T f = \int_S f \chi_T$.*

Proof: If f is admissible over S, then clearly f is admissible over T and $f \chi_T$ is admissible over S. Let P_0 be a partition of S with $U\left(|f|, P_0\right) < \infty$, and we can assume that $P_0 > \{T, S - T\}$. If $P > P_0$ and P_T is the partition of T consisting of the sets of P that are subsets of T, then

$$U(f, P_T) = U(f \chi_T, P),$$
$$L(f, P_T) = L(f \chi_T, P). \tag{2}$$

For any $P > P_0$,

$$U(f \chi_T, P) - L(f \chi_T, P) \leq U(f, P) - L(f, P),$$

and so $f \chi_T$ is integrable over S, and by (2), f is integrable over T, and the integrals are the same. ∎

PROBLEM 9: If S and T are disjoint measurable sets, then f is integrable over $S \cup T$ if and only if f is integrable over S and T, and then $\int_{S \cup T} f = \int_S f + \int_T f$. Hint: You need only show that integrability over S and T implies integrability over $S \cup T$. Then use Proposition 2, linearity, and $f = f \chi_S + f \chi_T$. ●

We have seen that there are functions that are improperly Riemann integrable but not Lebesgue integrable. This is because the improper Riemann integral allows conditionally convergent integrals, and the Lebesgue integral does not. However, if a *nonnegative* function is improperly Riemann integrable, then it is (properly) Lebesgue integrable. Improper Riemann integrals are of two types: either the interval is unbounded $((-\infty, b]$ or $[a, \infty))$, or the function is unbounded at one end of a bounded interval $[a, b]$. We consider the unbounded interval case, and leave the other case as a problem.

PROPOSITION 3: *If f is nonnegative and Riemann integrable on* $[1, \infty)$, *then f is Lebesgue integrable on* $[1, \infty)$, *and the integrals are the same.*

Proof: Our assumptions about f are that f is bounded and Reimann integrable on $[1, b]$ for each $b > 1$, and $\lim_{b \to \infty} \int_1^b f = I$. Let $S = [1, \infty)$. We want to show $\int_S f = I$. Since f is Reimann integrable on $[1, \infty)$, f is integrable — both senses — on $[n, n + 1]$ for each n. Moreover

$$I = \int_1^\infty f = \lim_{N \to \infty} \int_1^{N+1} f$$

$$= \lim_{N \to \infty} \sum_{n=1}^{N} \int_n^{n+1} f$$

$$= \sum_{n=1}^{\infty} \int_n^{n+1} f. \tag{3}$$

The Riemann and Lebesgue integrals of f over $[n, n + 1]$ are equal to the Lebesgue integral of f over $[n, n + 1)$. For each n, let $\{E_{ni}\}$ be a partition of $[n, n + 1)$ such that

$$\int_n^{n+1} f - \frac{\varepsilon}{2^n} < \sum_{i=1}^{\infty} m_{ni} \mu(E_{ni})$$

$$\leq \sum_{i=1}^{\infty} M_{ni} \mu(E_{ni})$$

$$< \int_n^{n+1} f + \frac{\varepsilon}{2^n}. \tag{4}$$

Let P be the partition of $S = [1, \infty)$ consisting of all the sets E_{ni}. Adding both sides of the inequalities (4), and using (3), we get

$$I - \varepsilon < L(f, P) \leq U(f, P) < I + \varepsilon.$$

All upper and lower sums converge absolutely, since all m_{ni} and M_{ni} are nonnegative. Therefore, f is Lebesgue integrable with integral $\int_1^\infty f$. ∎

PROBLEM 10: Let f be nonnegative on $[0, 1]$ and Riemann integrable on $[\varepsilon, 1]$ for each $\varepsilon \in (0, 1)$. Let $\lim_{\varepsilon \to 0+} f(x) = \infty$. Show that if $\lim_{\varepsilon \to 0+} \int_\varepsilon^1 f = I$, then f is Lebesgue integrable over $[0, 1]$ with integral I. •

XXXIII

MEASURABLE FUNCTIONS

R ecall the proof that a continuous function f is Riemann integrable on an interval $[a, b]$. If f is continuous on $[a, b]$, then f is uniformly continuous on $[a, b]$, and so given $\varepsilon > 0$ there is $\delta > 0$ such that $|f(x) - f(x')| < \varepsilon$ whenever $|x - x'| < \delta$. It follows that if $P = \{x_0, \dots, x_n\}$ is a partition of $[a, b]$ with $x_i - x_{i-1} < \delta$ for all i, then $M_i - m_i \leq \varepsilon$ for all i, and

$$U(f, P) - L(f, P) = \sum(M_i - m_i)\Delta x_i \leq \varepsilon(b - a).$$

The same argument works for the Lebesgue integral of a *bounded* function f on a set S of *finite measure*. Let $-M < f(x) < M$ for all $x \in S$, with $\mu(S) < \infty$. Let $P = \{E_i\}$ be the partition of S defined as follows:

$$E_i = \{x \in S : -M + (i - 1)\varepsilon \leq f(x) < -M + i\varepsilon\}. \tag{1}$$

If M_i and m_i are the sups and infs of f on the E_i, then $M_i - m_i \leq \varepsilon$ for all i, all upper and lower sums are bounded by $M\mu(S)$, and

$$U(f, P) - L(f, P) = \sum(M_i - m_i)\mu(E_i) \leq \varepsilon\mu(S).$$

Therefore, f is integrable over S provided the sets E_i defined in (1) are measurable for all i and all ε. This suggests the following definition. A function f defined on $\mathbb{R}$ is **measurable** provided $\{x : a \leq f(x) < b\}$ is a measurable set for all a and b. A function f defined on a measurable set S is **measurable on** S provided $\{x \in S : a \leq f(x) < b\}$ is measurable for all a and b.

235

With the preceding nomenclature we know that any bounded measurable function on a set of finite measure is integrable. More is true, and we will show in this chapter that an admissible function is integrable if and only if it is measurable. That is, as long as the graph of f lies in some finite area, then measurability is necessary and sufficient for integrability.

PROPOSITION 1: *Every measurable admissible function is integrable.*

Proof: Let f be an admissible function on a measurable set S. The set S can have infinite measure, and f can be unbounded on S. Let $P = \{E_i\}$ be an admissible partition of S, so that $\sum \overline{M}_i \mu(E_i) < \infty$, where

$$\overline{M}_i = \sup \left\{ |f(x)| : x \in E_i \right\}.$$

Let E_0 be the single zero measure set in P, so that $\overline{M}_i < \infty$ for all $i \neq 0$. Let $\varepsilon > 0$, and pick N so $\sum_{i=N+1}^{\infty} \overline{M}_i \mu(E_i) < \varepsilon$. If $M = \max \{\overline{M}_1, \ldots, \overline{M}_N\}$, then $-M \leq f(x) \leq M < \infty$ on $T = E_1 \cup \cdots \cup E_N$. Thus f is bounded on T, and T has finite measure since all E_i have finite measure. As we saw, f is integrable over T, so that there is a partition P_T of T such that $U(f, P_T) - L(f, P_T) < \varepsilon$. Let Q be the partition of S consisting of all the sets of P_T together with the sets E_0 (with $\mu(E_0) = 0$) and $E_{N+1}, E_{N+2}, \ldots$. Since

$$\sum_{i=N+1}^{\infty} (M_i - m_i)\mu(E_i) \leq \sum_{i=N+1}^{\infty} 2\overline{M}_i \mu(E_i) < 2\varepsilon,$$

we have

$$U(f, Q) - L(f, Q) = U(f, P_T) - L(f, P_T) + \sum_{i=N+1}^{\infty} (M_i - m_i)\mu(E_i)$$

$$\leq \varepsilon + 2\varepsilon = 3\varepsilon,$$

and f is integrable over S. ∎

PROBLEM 1: (i) If f is a measurable function and $f = g$ a.e., then g is a measurable function.

(ii) If f is measurable and admissible over a measurable set S, and $f = g$ a.e. on S, then g is integrable on S.

(iii) If $0 \leq f \leq g$ and f is measurable and g is integrable, then f is integrable. •

PROPOSITION 2: *The following conditions are equivalent to f being measurable.*

(i) $\{x : a \leq f(x)\}$ is measurable for all a.
(ii) $\{x : f(x) < a\}$ is measurable for all a.
(iii) $\{x : f(x) \leq a\}$ is measurable for all a.
(iv) $\{x : f(x) > a\}$ is measurable for all a.
(v) $\{x : a < f(x) < b\}$ is measurable for all a, b.

Proof: We will show that measurability implies (i), which implies (ii), which implies (iii), which implies (iv), which implies measurability. Hence, (i), (ii), (iii), and (iv) are equivalent to measurability, and we leave the equivalence of (v) as a problem. Assume f is measurable, and write

$$\{x : a \le f(x)\} = \bigcup_n \{x : a \le f(x) < a + n\},$$

so that the left side is a countable union of measurable sets. If each set $\{x : a \le f(x)\}$ is measurable, then the complements $\{x : f(x) < a\}$ are measurable, and (i) implies (ii). If each set $\{x : f(x) < a\}$ is measurable, then

$$\{x : f(x) \le a\} = \bigcap_n \left\{x : f(x) < a + \frac{1}{n}\right\},$$

and the left side is measurable. If each set $\{x : f(x) \le a\}$ is measurable, then the complements $\{x : f(x) > a\}$ are measurable for all a, so (iii) implies (iv). If $\{x : f(x) > a\}$ is measurable for all a, then

$$\{x : f(x) \ge a\} = \bigcap_n \left\{x : f(x) > a - \frac{1}{n}\right\}$$

is measurable for all a, and the complementary sets $\{x : f(x) < b\}$ are measurable for all b, so that any intersection $\{x : a \le f(x) < b\} = \{x : a \le f(x)\} \cap \{x : f(x) < b\}$ is measurable. Thus (iv) implies that f is measurable. ∎

PROBLEM 2: Show that f is measurable if and only if $\{x : a < f(x) < b\}$ is measurable for all a and b.   •

PROBLEM 3: (i) If f is measurable on the measurable set S, then $f \vee 0 = \max\{f, 0\}$ and $f \wedge 0 = \min\{f, 0\}$ are measurable.
(ii) If f is measurable and integrable over a measurable set S, then $f_1 = f \vee 0$ and $f_2 = f \wedge 0$ are integrable over S, and $\int_S f = \int_S f_1 + \int_S f_2$.   •

PROBLEM 4: (i) If f is continuous on $\mathbb{R}$, then f is measurable on $\mathbb{R}$. Hint: Show that $\{x : f(x) > a\}$ is open for each a and therefore measurable.
(ii) If f is continuous on an open set U, then f is measurable on U.
(iii) If f is continuous on a closed set F, then f is measurable on F. Hint: $\{x \in F : f(x) \ge a\}$ is closed.   •

PROPOSITION 3: *If f and g are measurable, then $f + g$ is measurable, and kf is measurable for every constant k.*

Proof: We show that $\{x : f(x) + g(x) > a\}$ is measurable for any number a. Clearly, $f(x) + g(x) > a$ if and only if $f(x) > a - g(x)$, and hence if and only if there is some rational number r such that

$$f(x) > r > a - g(x).$$

That is,

$$\{x : f(x) + g(x) > a\} = \bigcup_r \{x : f(x) > r\} \cap \{x : g(x) > a - r\},$$

where the union is over all rationals r. Each of the sets on the right is measurable, and so their countable union is measurable. The measurability of kf is the following problem. ∎

PROBLEM 5: (i) Show that kf is measurable if f is measurable.

(ii) If $f = \sum a_i \chi_{A_i}$ where $\{A_i\}$ is a finite or countable family of disjoint measurable sets, then f is measurable. In particular, all simple functions are measurable. (Recall that f is not simple unless $\sum |a_i| \mu(A_i) < \infty$.) •

PROPOSITION 4: *Let $\{f_n(x)\}$ be a sequence of measurable functions with a common domain S, where S is a measurable set. Assume* $\sup f_n(x) < \infty$ *and* $\inf f_n(x) > -\infty$ *for all x. Then the following are true:*

(i) $\sup_n f_n(x)$ *and* $\inf_n f_n(x)$ *are measurable functions.*

(ii) $\limsup_{n \to \infty} f_n(x)$ *and* $\liminf_{n \to \infty} f_n(x)$ *are measurable functions.*

(iii) *If* $\lim_n f_n(x)$ *exists for all x, then* $\lim_n f_n(x)$ *is a measurable function.*

Proof: Let $f(x) = \sup f_n(x)$ and assume $f(x) < \infty$ for all x. Then for any a,

$$\{x : f(x) > a\} = \bigcup_n \{x : f_n(x) > a\},$$

and so f is measurable. The same sort of argument shows that $\inf f_n(x)$ is measurable. If $g_n(x) = \sup_{k \geq n} f_k(x)$, then g_n is measurable for each n, and $\limsup_{n \to \infty} f_n(x) = \inf_n g_n(x)$ is measurable. Similarly, $\liminf f_n(x)$ is measurable, and hence $\lim f_n(x)$ is measurable if the limit exists everywhere. ∎

It can happen that a sequence $\{f_n\}$ of measurable functions converges a.e., but diverges to $+\infty$ or $-\infty$ on some set of measure zero. The limit function may still be integrable (i.e., equal a.e. to an integrable function f), with $\int f_n \to \int f$. We will therefore agree that f is measurable on S in case there is a zero measure set $E \subset S$ such that f is measurable on $S - E$, and f is $+\infty$ or $-\infty$ or just not defined on E. With this agreement we can restate Proposition 4 as follows:

PROPOSITION 4′: *If $\{f_n\}$ is a sequence of measurable functions on a measurable set S, and $\sup f_n < \infty$ a.e. and $\inf f_n > -\infty$ a.e., then $\sup f_n$, $\inf f_n$, $\limsup f_n$, and $\liminf f_n$ are measurable, and $\lim f_n$ is measurable if $\{f_n\}$ converges a.e.*

Now we show that every integrable function is the a.e. limit of simple functions. Since simple functions are measurable, every integrable function is measurable. For admissible functions, integrability and measurability are equivalent. We will need the following preliminary result.

PROPOSITION 5: *If h is a nonnegative function that is integrable and measurable over a set S, and $\int_S h = 0$, then $h = 0$ a.e. on S.*

Proof: Assume that h is as advertised. Let $T = \{x \in S : f(x) > 0\}$, and assume that $\mu(T) = p > 0$. Since $T = \bigcup T_n$, where

$$T_n = \left\{ x \in S : f(x) \geq \frac{1}{n} \right\},$$

and $T_1 \subset T_2 \subset \cdots$,

$$\mu(T) = \mu\left(\bigcup T_n\right) = \lim \mu(T_n) = p > 0.$$

Therefore, there is N with $\mu(T_N) > \dfrac{p}{2}$. If $f = \dfrac{1}{N}\chi_{T_N}$, then $\int f \geq \dfrac{1}{N} \cdot \dfrac{p}{2}$. Since $h \geq \dfrac{1}{N}$ on T_N, $h \geq f$ on S and

$$\int h \geq \int f \geq \frac{p}{2N} > 0.$$

Thus $\int h > 0$ unless $h = 0$ a.e.   ∎

PROPOSITION 6: *If f is integrable over S, then there is an increasing sequence $\{g_n\}$ of simple functions and a decreasing sequence of simple functions $\{h_n\}$, such that $g_n \leq f \leq h_n$ for all n, and*

$$g_n \longrightarrow f \text{ a.e.,} \qquad h_n \longrightarrow f \text{ a.e.,}$$

and

$$\int g_n \longrightarrow \int f, \qquad \int h_n \longrightarrow \int f.$$

It follows that every integrable function f is measurable.

Proof: Assume that f is integrable over S. Then for each n there is a partition P_n of S such that $U(f, P_n)$ and $L(f, P_n)$ converge absolutely, and $U(f, P_n) - L(f, P_n) < 1/n$. We can assume that $P_1 < P_2 < \cdots$ by replacing each P_n by the common refinement of $P_1, P_2, \ldots, P_n$. Let $P_n = \{E_{ni}\}$ and let M_{ni}, m_{ni} be the sups and infs of f on the E_{ni}, so that

$$L(f, P_n) = \sum_i m_{ni} \mu(E_{ni}),$$

$$U(f, P_n) = \sum_i M_{ni} \mu(E_{ni}).$$

Let E_{10} be the one set in P_1 of measure zero, so that f is bounded on each of the remaining sets E_{1i}. There is no need to refine E_{10} in successive partitions P_n, and so assume $E_{n0} = E_{10}$ is the single zero measure set in each P_n. Then all M_{ni} and

m_{mi} for $i \neq 0$ are finite. Let

$$g_n = \sum_{i=1}^{\infty} m_{ni}\chi_{E_{ni}},$$

$$h_n = \sum_{i=1}^{\infty} M_{ni}\chi_{E_{ni}}.$$

Both g_n and h_n are measurable for each n, and $g_n \leq f \leq h_n$ a.e. (i.e., off E_{10}), and $g_n = h_n = 0$ on E_{n0}. Since $P_1 < P_2 < \cdots$, $\{g_n\}$ is increasing and $\{h_n\}$ is decreasing. Let $g = \lim g_n$, $h = \lim h_n$, so that g and h are measurable functions, and $g \leq f \leq h$ a.e. All the g_n and h_n are integrable simple functions, with

$$\int g_n = \sum_i m_{ni}\mu(E_{ni}) = L(f, P_n),$$

$$\int h_n = \sum_i M_{ni}\mu(E_{ni}) = U(f, P_n).$$

Moreover, g and h are integrable (see Problem 6), and since

$$g_n \leq g \leq f \leq h \leq h_n$$

for all n,

$$\int (h - g) = \int h - \int g$$

$$\leq \int h_n - \int g_n$$

$$= U(f, P_n) - L(f, P_n) < \frac{1}{n}.$$

Thus $h - g$ is a nonnegative integrable function with $\int (h-g) = 0$, and so $h-g = 0$ a.e., and consequently $f = g = h$ a.e., and f is measurable. ∎

PROBLEM 6: Show that the functions g and h of the preceding proof are integrable. Hint: Show

$$L(f, P_n) \leq L(h, P_n) \leq U(h, P_n) \leq U(f, P_n)$$

for all n, and similarly for g.   •

XXXIV

CONVERGENCE THEOREMS

W e know that for functions with graphs in some finite area — admissible functions — integrability and measurability are equivalent. Moreover, the pointwise limit of measurable functions is measurable, and so the limit of integrable functions will be integrable if the area remains finite. What we want is a theorem that says if $f_n \longrightarrow f$, then $\int f_n \longrightarrow \int f$. In this chapter we will show that this always holds for the Lebesgue integral, provided all the f_n and f have their graphs in some fixed finite area.

For the Riemann integral, the basic tool in showing convergence of integrals is uniform convergence of the functions. That is, if $f_n \longrightarrow f$ uniformly on $[a, b]$, and all f_n are integrable, then f is integrable and $\int f_n \longrightarrow \int f$. The key to convergence theorems for the Lebesgue integral is the fact that pointwise convergence is nearly uniform on sets of finite measure. Specifically, if $f_n \longrightarrow f$ pointwise on a set of finite measure, then $f_n \longrightarrow f$ uniformly off sets of arbitrarily small measure. This is enough to show that for uniformly bounded sequences $\{f_n\}$ of measurable functions on a finite measure set, $f_n \longrightarrow f$ implies $\int f_n \longrightarrow \int f$. The following problem illustrates the idea.

PROBLEM 1: (i) Let $I = [0, 1]$ and $f_n(x) = x^n$ on I. Show that for each $\delta > 0$ there is a set $E \subset I$ such that $\mu(E) < \delta$ and $f_n(x) \longrightarrow 0$ uniformly on $I - E$. Use this and

$$\int_I f_n = \int_{I-E} f_n + \int_E f_n,$$

and the fact that $|f_n(x)| \leq 1$ for all x, to show that $\int_I f_n \longrightarrow 0$.

(ii) Let $g_n(x) = nx^n$ on $I = [0, 1]$. Show that $\{g_n\}$ also converges uniformly to zero off sets of arbitrarily small measure, but $\int_I g_n \longrightarrow 1$. (The uniform boundedness of the sequence in (i) is essential.) Hint: $\log(nx^n) = n[(\log n)/n + \log x]$.   •

PROPOSITION 1 (Egoroff's Theorem): *If $\{f_n\}$ is a sequence of measurable functions on a measurable set S of finite measure, and $f_n \longrightarrow f$ a.e. on S, then for every $\delta > 0$, there is a measurable set $E \subset S$ with $\mu(E) < \delta$ such that $f_n \longrightarrow f$ uniformly on $S - E$.*

Proof: We can assume that $f_n(x) \longrightarrow f(x)$ for all $x \in S$ since a zero measure exceptional set could just be thrown in with E. The first step is to show the following: for each $\alpha > 0$ and $\beta > 0$ (small numbers), there is $N \in \mathbb{N}$ and a measurable set F with $\mu(F) < \beta$ such that $|f_k(x) - f(x)| < \alpha$ for all $k \geq N$ and all $x \notin F$. To this end, let

$$F_n = \left\{ x \in S : |f_k(x) - f(x)| \geq \alpha \text{ for some } k \geq n \right\}.$$

Since each function $f_k - f$ is measurable, F_n is a countable union of measurable sets, so F_n is measurable. For any fixed x, $|f_k(x) - f(x)| < \alpha$ for all sufficiently large k, and so $x \notin F_n$ if n is sufficiently large. That is, $\bigcap F_n = \emptyset$. Since $\mu(F_1) < \infty$ and $F_1 \supset F_2 \supset \cdots$,

$$\mu \left(\bigcap F_n \right) = \mu(\emptyset) = 0 = \lim \mu(F_n).$$

There is, therefore, N such that $\mu(F_N) < \beta$. We have shown that for any given $\alpha > 0$ and $\beta > 0$, there is a set F_N with $\mu(F_N) < \beta$ and a number N such that $|f_k(x) - f(x)| < \alpha$ if $k \geq N$ and $x \notin F_N$.

Now let $\delta > 0$. We want to find a set E with $\mu(E) < \delta$ such that $f_n \longrightarrow f$ uniformly off E. For each $\alpha = 1/n$ and $\beta = \delta/2^n$ we find F_n such that $\mu(F_n) < \delta/2^n$ and $|f_k(x) - f(x)| < 1/n$ if $k \geq n$ and $x \notin F_n$. Let $E = \bigcup F_n$, so that $\mu(E) < \delta$. To see that $f_n \longrightarrow f$ uniformly off E, let $\varepsilon > 0$. If $1/n < \varepsilon$, and $x \notin F_n$, then $x \notin E$ and $|f_k(x) - f(x)| < 1/n < \varepsilon$ for all $k > n$.   ∎

PROBLEM 2: Give an example of a sequence $\{f_n\}$ on a measurable set S (of infinite measure) such that $f_n \longrightarrow f$ on S, but there is no set E of finite measure such that $f_n \longrightarrow f$ uniformly off E.   •

The following proposition follows readily from Egoroff's Theorem, and is a powerful example of how nicely pointwise convergence works with the Lebesgue integral. The proof is exactly that used in Problem 1.

PROPOSITION 2 (The Bounded Convergence Theorem): *If $\{f_n\}$ is a uniformly bounded sequence of measurable functions on a measurable set S of finite measure, and $f_n \longrightarrow f$ a.e. on S, then f is integrable over S, and*

$$\int_S f_n \longrightarrow \int_S f.$$

Proof: Assume that $|f_n(x)| \leq M$ for all $x \in S$ and all n. Then all f_n are integrable, since they are measurable, and $U\left(|f_n|, P\right) < M\mu(S)$ for any partition P. Clearly f is also integrable for the same reason, since $|f(x)| \leq M$ for all x, and f is measurable. Let $\varepsilon > 0$. Let E be a set of measure less than ε such that $f_n \longrightarrow f$ uniformly on $S - E$. Pick N so that $|f_n(x) - f(x)| < \varepsilon$ for all $x \in S - E$ and all $n \geq N$. Then if $n \geq N$,

$$\left| \int_S f_n - \int_S f \right| \leq \int_S |f_n - f|$$

$$= \int_{S-E} |f_n - f| + \int_E |f_n - f|$$

$$\leq \varepsilon\mu(S - E) + 2M\mu(E)$$

$$< \varepsilon\left(\mu(S) + 2M\right).$$

Since ε is arbitrary, and $\mu(S)$ and M are fixed, $\int f_n \longrightarrow \int f$. ∎

In the bounded convergence theorem, all the action takes place in the rectangle $S \times [-M, M]$, which has finite area $2M\mu(S)$. To see what can go wrong if the f_n are allowed to wander around in an infinite area, consider the following examples: If $f_n = \chi_{[n,n+1]}$, then $f_n \longrightarrow 0$ for all $x \in \mathbb{R}$; $\int f_n = 1$ for all n, so that $\int f_n \longrightarrow 1 \neq \int 0$. A similar example works if the f_n are allowed too much latitude in the up-down direction. If $f_n = n\chi_{(0,1/n)}$, then $f_n \longrightarrow 0$ for all x, $\int f_n = 1$ for all n, and again $\lim \int f_n > \int \lim f_n$.

PROBLEM 3: Modify the foregoing examples to find nonnegative functions f_n such that $f_n(x) \longrightarrow 0$ for all x but $\int f_n = n \longrightarrow \infty$. •

Let $f^+ = f \vee 0$ and $f^- = (-f) \vee 0$, so that f^+ and f^- are nonnegative functions, and $f = f^+ - f^-$. The following problem shows how this decomposition allows us to consider just nonnegative functions for questions of integrability and convergence.

PROBLEM 4: (i) f is measurable if and only if both f^+ and f^- are measurable.

(ii) f is integrable if and only if both f^+ and f^- are integrable, and then $\int f = \int f^+ - \int f^-$.

(iii) $f_n \longrightarrow f$ on S if and only if $f_n^+ \longrightarrow f^+$ and $f_n^- \longrightarrow f^-$ on S. •

As Problem 4 indicates, we can now restrict our convergence proofs to nonnegative functions. To make the link between the bounded convergence theorem (uniformly bounded functions on a set of finite measure) and general convergence theorems, we introduce the following terminology. We say that g is a **primary function** if g is nonnegative, bounded, measurable, and zero off some set of finite measure. We saw in Proposition 6 of the last chapter that if f is integrable, then

$$\int f = \sup\left\{\int g : g \leq f, g \text{ simple}\right\}. \tag{1}$$

If f is nonnegative and integrable, then the simple functions g in (1) can all be taken to be nonnegative. In the next proposition we beef (1) up from simple functions to the even simpler primary functions.

PROPOSITION 3: *If f is a nonnegative measurable function, then f is integrable if and only if*

$$\sup\left\{\int g : 0 \le g \le f, g\ primary\right\} < \infty. \tag{2}$$

If the sup *above is finite, it equals $\int f$.*

Proof: Assume f is integrable, so that $\int f$ is the sup of $\int g$ for g simple, $0 \le g \le f$. If $g = \sum a_i \chi_{A_i}$, with $a_i \ge 0$, then

$$\int g = \sum a_i \mu(A_i).$$

If $g_N = \sum_{i=1}^{N} a_i \chi_{A_i}$, then g_N is primary, $0 \le g_N \le f$, and $\int g = \sup_N \int g_N$. It follows that

$$\int f = \sup\left\{\int g : 0 \le g \le f, g\ primary\right\} < \infty. \tag{3}$$

Now we must show that if the sup in (2) is finite, then f is integrable. Equivalently, we show that if f is not integrable (i.e., not admissible) then the sup in (2) is infinite. Suppose f is nonnegative measurable on S, but not admissible, so that every upper sum is infinite. Let $\{E_{ni}\}$ be the partition of $S \cap [n, n + 1)$, $n = 0, \pm 1, \pm 2, \ldots$, defined by

$$E_{ni} = \left\{x \in S \cap [n, n + 1) : \frac{i}{2^{|n|}} \le f(x) < \frac{i + 1}{2^{|n|}}\right\}.$$

Let M_{ni}, m_{ni} be the sup and inf of f on E_{ni}. Then $M_{ni} \ge m_{ni} \ge 0$, $\sum_{i=0}^{\infty} \mu(E_{ni}) \le 1$ for all n, and

$$\sum (M_{ni} - m_{ni})\mu(E_{ni}) = \sum_{n=-\infty}^{\infty} \sum_i (M_{ni} - m_{ni})\mu(E_{ni})$$

$$\le \sum_{i=-\infty}^{\infty} \frac{1}{2^{|n|}} \sum_{i=0}^{\infty} \mu(E_{ni})$$

$$\le \sum_{i=-\infty}^{\infty} \frac{1}{2^{|n|}} = 3.$$

For any finite set F of pairs (n, i),

$$\sum_{(n,i)\in F} (M_{ni} - m_{ni})\mu(E_{ni}) \le 3.$$

However,

$$\sup_{F} \sum_{(n,i)\in F} M_{ni}\mu(E_{ni}) = \infty$$

by our assumption that all upper sums for f are infinite. Therefore, there is for each $N \in \mathbb{N}$ a finite set F such that

$$\sum_{(n,i)\in F} m_{ni}\mu(E_{ni}) > N.$$

If $g = \sum_{F} m_{ni}\chi_{E_{ni}}$, then g is a primary function under f, and $\int g > N$. Thus if f is not integrable, the sup in (2) is infinite. ∎

If $\{f_n\}$ is a sequence of nonnegative integrable functions, and $f_n \longrightarrow f$, then the areas under the f_n get at least as big as the area under f. It can happen that the integrals get too big, as we have seen. If $f_n = f + \chi_{[n,n+1]}$, then $f_n \longrightarrow f$ but $\int f_n = \int f + 1$ for all n. The following is the basis for the general Lebesgue convergence theorems.

PROPOSITION 4 (Fatou's Lemma): *Let $\{f_n\}$ be a sequence of nonnegative integrable functions on S such that $f_n \longrightarrow f$ a.e. on S. If $\liminf \int f_n < \infty$, then f is integrable and*

$$\liminf \int f_n \geq \int f.$$

Proof: Assume $\liminf \int f_n < \infty$. We can assume that $f_n(x) \longrightarrow f(x)$ for all $x \in S$. The function f is certainly measurable. Let g be any primary function under f, so that g is bounded, and zero off some finite measure set $T \subset S$. The functions $f_n \wedge g$ are uniformly bounded since g is bounded, and $f_n \wedge g$ is zero off T. Moreover, $f_n \wedge g \longrightarrow g$ since $f_n \longrightarrow f \geq g$. By the bounded convergence theorem,

$$\int f_n \wedge g \longrightarrow \int g.$$

Since $\int f_n \geq \int f_n \wedge g$,

$$\liminf \int f_n \geq \int g.$$

By Proposition 3, f is integrable and $\int f$ is the sup of $\int g$ for primary $g \leq f$. Therefore, $\int f \leq \liminf \int f_n$. ∎

PROPOSITION 5 (The Monotone Convergence Theorem): *If $\{f_n\}$ is a sequence of nonnegative integrable functions, $f_n \longrightarrow f$ a.e., and $f_n \leq f$ for all n, then f is integrable if and only if $\lim \int f_n < \infty$, and if f is integrable, $\int f = \lim f_n$.*

Proof: The usual application is to a sequence that increases to f, hence the name. If $\liminf \int f_n < \infty$, then f is integrable, and $\liminf \int f_n \geq \int f$ by Fatou's Lemma. However $f_n \leq f$ for all n, so that $\limsup \int f_n \leq \int f$, and so $\lim \int f_n = \int f$ if the $\liminf$ is finite. If $\lim \int f_n = \liminf \int f_n = \infty$, then by Proposition 3 there is for each n a primary function $g_n \leq f_n$ with

$$\int g_n > \int f_n - 1.$$

If $\lim \int f_n = \infty$, then $\lim \int g_n = \infty$, and by Proposition 3 f is not integrable. ∎

The final result is the general workhorse theorem on convergence of Lebesgue integrals: pointwise convergence in some fixed finite area implies convergence of the integrals.

PROPOSITION 6 (The Lebesgue Dominated Convergence Theorem): *If $\{f_n\}$ is a sequence of measurable functions on a measurable set S, and there is a nonnegative integrable function g on S such that $|f_n| \leq g$ for all n, and $f_n \longrightarrow f$ on S, then all f_n and f are integrable and*

$$\int f_n \longrightarrow \int f.$$

Proof: Since g is integrable and $|f_n| \leq g$, $f_n^+ \leq g$ and $f_n^- \leq g$ for all n, so that f_n^+ and f_n^- are integrable, and hence $f_n = f_n^+ - f_n^-$ is integrable for each n. Similarly, f is integrable because f is measurable and $|f| \leq g$. Since $f_n^+ \longrightarrow f^+$, and

$$\limsup \int f_n^+ \leq \int g < \infty,$$

f^+ is integrable and $\int f^+ = \lim \int f_n^+$ by Fatou's Lemma. Similarly, f^- is integrable and $\int f^- = \lim \int f_n^-$. Therefore, $f = f^+ - f^-$ is integrable, and

$$\int f = \int f^+ - \int f^-$$

$$= \lim \int f_n^+ - \lim \int f_n^-$$

$$= \lim \left(\int f_n^+ - \int f_n^- \right)$$

$$= \lim \int f_n. \qquad ∎$$

PROBLEM 5: Show that if g is integrable, f is measurable, and $0 \leq f \leq g$, then f is integrable, $g - f$ is integrable, and $\int g = \int f + \int (g - f)$. •

Index